NATHANIEL SOUTHGATE SHALER
AND THE
CULTURE OF AMERICAN SCIENCE

History of American Science and Technology Series

General Editor, LESTER D. STEPHENS

The Eagle's Nest: Natural History and American Ideas, 1812–1842, by Charlotte M. Porter

Nathaniel Southgate Shaler and the Culture of American Science, by David N. Livingstone

NATHANIEL SOUTHGATE SHALER

and the Culture of American Science

DAVID N. LIVINGSTONE

The University of Alabama Press
Tuscaloosa and London

Library of Congress Cataloging-in-Publication Data
Livingstone, David N., 1953–
  Nathaniel Southgate Shaler and the culture of
American science.

  (History of American science and technology series)
  Bibliography: p.
  Includes index.
  1. Shaler, Nathaniel Southgate, 1841–1906.
2. Geographers—United States—Biography.   I. Title.
II. Series.
G69.S49L58   1987   910'.924   85-28982
ISBN 0-8173-0305-7 (alk. paper)

British Library Cataloguing-in-Publication Data is available.

To
my father
and
mother

I have just read Shaler's autobiography, and it has fairly haunted me with the overflowing impression of his myriad-minded character. Full of excesses as he was, due to his intense vivacity, impulsiveness, and imaginativeness, his centre of gravity was absolutely steady, and I knew no man whose sense of the larger relation of things was always so true and right. Of all the minds I have known, his leaves the largest impression, and I miss him more than I have missed anyone before.

—William James

Contents

Illustrations

Preface

Like all organisms, the historian of ideas thrives only in a conducive environment. In my case, numerous friends and colleagues have helped to provide just such a setting. My interest in the history of geographical thought in its broader intellectual and social context was first stimulated by William Kirk, who encouraged me to pursue the conceptual relations between nature, humanity, and God in history. More specifically, this study of Shaler has benefited from the advice and judgments of Peter Bowler, Ronald Buchanan, Gordon Herries Davies, Neal Gillespie, Clarence Glacken, Reijer Hooykaas, William Koelsch, David Stoddart, Yi-Fu Tuan, and Brian Whalley. To all of these my thanks are due. I am grateful to Gill Alexander for drawing the two charts in the Appendix, and to Jenitha Orr for careful reading of the text.

To the staff of various libraries I also owe a debt of gratitude: to Clark Elliott and the staff of the Harvard University Archives, to Ann Blum of Harvard's Museum of Comparative Zoology, to Jaunette Eaglesfield, formerly at Harvard's Geology Department Library, to Otto Solbrig of the Gray Herbarium, to the staff of the Yale University Archives, to Diana Yount of Andover-Newton Theological School, and to Claire McConaghy at the Queen's University of Belfast Library. In particular I must record my deep appreciation to Susan Ekin, assistant librarian at the Queen's University Science Library, without whose indefatigable efforts to secure many obscure items, this work would have greatly suffered.

My greatest intellectual debt, however, is to John Campbell, who willingly spent many hours discussing the content and context of Shaler's work. Without his constant guidance, encouragement, and sustained interest, this book might never have been completed.

Parts of chapters 5, 6, and 7 are derived from material that has already appeared as articles and is used here with the permission of the journal and book publishers concerned: "Environment and Inheritance: Nathaniel Southgate Shaler and the American Frontier," in *The Origins of Academic Geography in the United States*, edited by Brian W. Blouet (Hamden, Conn.: Archon Books, 1981), pp. 123–38; "Nature and Man in America: Nathaniel Southgate Shaler and the Conservation of Natural Resources," *Transactions of the Institute of British Geographers*, n.s., 5 (1980): 369–82; and "Science and Society: Nathaniel S. Shaler and Racial Ideology," *Transactions of the Institute of British Geographers*, n.s., 9 (1984): 181–210. Quotations from unpublished manuscripts have been made by permission of the Harvard University Archives and by permission of the Houghton Library, the Museum of Comparative Zoology, and the Gray Herbarium Library, all of Harvard University.

Finally, I am grateful to my parents for their very practical interest and encouragement and, most of all, to Frances, who, with our daughter Emma, has had to share our home for too long with the ghost of Nathaniel Southgate Shaler.

PART ONE

Man and Milieu

Introduction: Nature, Humanity, and God

1

"Geographie is better than divinity."
—Peter Heylyn, *Cosmographie*

"A text without a context is only a pretext," an old Scottish divine once told his young ministerial assistant. Historians of geography, no less than students of divinity, would do well to heed this maxim. Unlike their colleagues in the more general fields of intellectual history and the history of science, historians of geography have been rather tardy in putting this principle into practice. In these other branches of history, there have been repeated calls for a "contextualist" approach to the understanding of the "texts" as repositories of intellectual endeavor. Thus their practitioners have been concerned, for example, to liberate the history of ideas from the slavish search for precursors, from that tireless ransacking of the documents of history in the eternal hunt for anticipations of grand theories and great ideas. They have also shunned the all-too-common temptation of seeking disciplinary self-justification in the pronouncements of past pioneers. Hagiography has given way to critical biography. And they have rejected as too stultifying those purely internalist accounts of conceptual development that cut knowledge off from knowers, theory from practice, ideas from society.

To achieve this vision of a history unblemished by Whiggish manipulation, contextualists have displayed a range of styles in their

3

remaking of history. Thus some have highlighted the role of the zeitgeist in intellectual development and therefore the way in which the thought of epoch and individual interpenetrate. Others have recast the nature of the scientific enterprise in terms of the socioeconomic conditions of the time and its ideological underpinnings. Others have focused on the institutionalization of the academic experiment and on its role in the culture of professionalism. Still others look to the psychological and sociological structuration of human knowledge. But whatever the diversity of their emphases, these approaches unite in their repudiation of cumulative views of knowledge, internalist disciplinary history, and presentist reconstructions of scientific progress.[1]

Historians of geography have, by and large, worked in isolation from these historiographical currents. The result has often been a history of geography prehistorical in its methodology, insensitive in its execution, and manipulative in its motivation. Yet, mirroring the concerns of the broader field of intellectual history, there have been several recent calls for greater sophistication in the reconstruction and interpretation of the history of geographical thought and practice.[2] There have been those, for instance, who have concentrated on the possibility of applying the Kuhnian paradigmatic perspective to the structure of geography or on the institutional links between its internal domain and external relations. Others have begun to stress the importance of informal socioscientific circles or colleges, the biographical milieus of the discipline's crusading predecessors, or the anxiety over careers and credentials characteristic of disciplinary professionalization. The way in which geographical praxis crystallized the ideological commitments of its practitioners has attracted the attention of some others.[3] Still, if learning these lessons is one of the first signs that, as Glick puts it, "the history of geography has come of age as an independent subdiscipline,"[4] they are in no sense to be taken as representing what might be called "the state of the art." The strands are as yet too diffuse and too solely programmatic to constitute a coherent alternative; they are merely symptoms of a growing sensitivity to historiographical subtleties.

In addition, simple though it may seem, it is surely worth remembering that theory is one thing, practice quite another. Expressing a theoretical commitment to contextualism does not of itself purge a work from presentism in practice. Confession is easily mistaken for absolution. So it is important to recognize from the outset that aspects of presentism are inherently inescapable, if only because historical reconstruction is value-laden communication. It is value-laden in the sense that historians are always involved in the business of selecting and interpreting those "facts" deemed significant in the light of the (necessarily current) questions they are asking. There is no history on a mortuary table.[5] And it is also presentist in communication for, as David Hull reminds us: "Histories are written not only *by* people and *about* people but also *for* people. The people about whom a history is written lived in the past, but the historian and his readers live in the present. No purpose is served by pretending otherwise. The very fact that the historian shares knowledge of the present with his readers is what allows him to communicate successfully with them. . . . Communication is a relation."[6]

That elements of presentism are inevitable, however, does not necessitate capitulation to the worst excesses of "writing history backward." Care can and should be taken to reduce the dangers of transferring contemporary cultural values to the past; of tracing "contributions" to an as yet unformulated, idealized doctrine; of harmonizing tensions and contradictions in the light of a later paradigm; or of imposing a unified conceptual structure on the fragmentary anticipations of present-day orthodoxy. By contrast, the contextual perspective forces us to ask what an individual, in writing at a particular time and for a particular audience, meant to communicate, how that meaning was to be taken, and what, if only in principle, were the problems with which he or she could have been wrestling. It encourages us to recognize the multicontextual nature of knowledge and, if possible, to delineate personal, ideological, religious, and other elements. It motivates us to make at least some effort to extricate from the interaction of "text" and "context" in the nexus of the historical complex the relative significance of the various component elements.

These methodological considerations intersect with my reexamination of the life and work of Nathaniel S. Shaler in two ways. First, I contend that the revitalization of geography during the nineteenth century, as the study of human interaction with environment, was part of an ongoing debate on "man's place in the natural order." Second, a contextual interpretation of the discipline's modern reemergence requires the historian to transcend not only the now conventional disciplinary boundaries but also putative distinctions between science and pseudoscience.[7] Indeed, J. K. Wright, suspicious perhaps that falsity may be truth in disguise, prophetically asked a half century ago: "Is not the history of error, folly, and emotion often as enlightening as the history of wisdom?"[8]

Even more specifically, I shall argue that the work of Shaler as a founding father of American academic geography mirrors the nineteenth century currents of thought in geology, theology, biology, and sociology—disciplines hitherto conducted within a framework derived from natural theology—and that it, like them, exhibits the same fundamental reorientation in the conception of the relations between nature, humanity, and God that was necessitated by the Darwinian revolution. In a series of substantive papers on the ideological context of the great Victorian debate on "man's place in nature," Robert Young developed the thesis that this reorientation in natural theology "became more closely defined as a change from mechanistic analogies employed within an explicitly theistic natural theology to the use of organic analogies based on a secularized, implicit natural theology."[9] In this transitional context, I believe, Shaler's contributions can best be understood.

Old metaphors of course die hard. The image of a protracted death struggle between science and religion—encapsulated in the "warfare thesis" of Draper and White—has often obscured the subtle accommodations to evolutionary natural history that thoughtful and literate religious figures could, and often did, negotiate. That Darwinism did play a key role in the ripening of Victorian naturalism is certainly not to be denied. But as Young has again written, "The idea of opposing theology could not have been further from the minds of the main evolutionists. Their aim was to reconcile nature, God, and man."[10] The fact that confidence in material, social, and spiritual progress often provided a surrogate for the new remoteness of God from nature's laws or for his identification with them, merely attests to the changing composition of a broadly theistic cosmology.

The fragmentation of this common context plainly elicited different responses from those engaged in the nineteenth-century debates.[11]

Some rejected outright the evolutionary edifice and did what they could to perpetuate the older natural theology. Others rejected natural theology and, with it, all forms of teleology. But just as typical were those who endeavored to retain philosophical continuity with the past by reinterpreting evolution in terms congenial to purpose and de-sign—in effect by constructing an evolutionary natural theology. Shaler's own contribution to understanding the reciprocal relations between society, environment, and a purposeful material world stands as one such attempt to integrate nature, humanity, and God. Thereby he hoped to rediscover man's place in nature and to rationalize the existing social order in the wake of the naturalistic onslaught. In his case, as in many others, the easiest way of transforming Darwinism into the immanent divine power that could preserve a teleological interpretation of nature and society was by suffusing evolution with the orthogenetic glow of the neo-Lamarckian version. The providen-tial metaphysics of post-Kantian theology could readily be grafted onto this evolutionary trunk.

Beyond doubt, Shaler was one of the great men among the intel-lectual forebears of American geography. Carl Sauer called him "one of the greatest geographers of America"; Clarence Glacken described him as "an outstanding figure in the last quarter of the nineteenth century"; and Lewis Mumford regarded him as "one of the seminal minds of his generation."[12] Yet despite these conspicuous tributes and the availability since 1909 of an autobiography generously embellished with anecdote, reminiscence, and mythology, no full-length treatment of Shaler's contributions has appeared, with the exception of Walter Berg's 1957 doctoral dissertation—a study largely biographical rather than conceptual in emphasis.[13] These considerations would of them-selves seem sufficient justification for embarking on a reevaluation of Shaler's life and work.

In other, more specific ways, however, Shaler seems to me to provide a particularly appropriate lens for focusing on the context of geography's modern renascence in the United States. While being a major contributor to such themes as the conservation of natural re-sources, educational reform, and the geographical background to so-cial history, Shaler has the advantage, for the purposes of this study, of occupying a secondary role in terms of developments in evolution

Nathaniel Southgate Shaler as a student. Courtesy of the Harvard University Archives.

theory and theology. His synthesis of the prevailing scientific-religious ethos as the context for his geographical writings thus provides a more general perspective on American attitudes to humanity and nature in the late nineteenth century.[14] Naturally, I do not mean that the nature–humanity–God trinity appears in every nook and cranny of Shaler's output; bread-and-butter work in geology, administration, and teaching manifestly forms the external fabric of his biography. But some awareness of this assumed context helps give an overall coherence to his wide-ranging contribution to postbellum scientific culture. As I have said, Shaler emerges as a figure in transition between two intellectual eras. On the one hand, he linked the older nature philosophy of Agassiz with the modern geography of Davis; on the other, he stands between the theodicean interpretations of Guyot and Maury and the secular naturalism of later generations. Finally, Shaler's sustained efforts to relate contemporary scholarship to the social problems of his day provide a particularly sharp image of the ways in which the understanding of the relations between science, religion, and society helped crystallize scientific theory in ideological praxis.

This book, therefore, is structured rather differently from conventional biography. In it I have attempted to extricate the fundamental inspirations and underpinnings of Shaler's numerous pronouncements and to demonstrate the ways in which they were rehearsed in a wide variety of contexts. As a thematic evaluation of Shaler's work and thought, it is submitted as a contribution to the history of geography in its socioscientific context. I am aware, however, of its failure to meet Charles T. Copeland's exacting appeal in 1906 for a sensitive biography of Professor Shaler:

The true biographer—he must not be too old nor too young—will qualify his narrative in just proportions with Shaler's racy wit (not watering it down to placate the squeamish), with his unforced humor, his homely shrewdness, his persuasive wisdom, and the poetic feeling with which so much of what he said and wrote was tinctured. This true biographer will know men and be a master of language, for his task will be to transmit a personality, one of the most brilliant, winning, conquering personalities of our time.[15]

2 Shaler and His Contemporaries

> Of course a sad deal is lost. Mr. Shaler's presence was magnetic
> and heartening; his speech was wine; his laugh a cordial. These
> may be suggested in writing; they cannot be recaptured. A man
> is always better than a book.
> —Charles T. Copeland, "Biography of Shaler"

During the Harvard Summer School of 1906, Charles T. Copeland
reminded his hearers of the words of a recently published book, *The
Individual*. Its author had condemned costly funerals and monuments
and had suggested instead—as the best memorials—scholarships,
charities, and the like in the name of those who had gone before.
"Good general counsel this," Copeland remarked, "and for the common man, the best memorial is some beneficent thing or function
that shall bear his name. But in the case of Professor Shaler [author
of the volume] we shall be content with no remembrance short of so
much of himself as can be put into a book."[1]

So highly regarded was this genial southerner, forty years professor
at Harvard, that, on the afternoon of his funeral, flags on city buildings
and student fraternities were hung at half-mast, and shops in "old
Cambridge" were closed. Such tributes had not been paid to another
teacher in the university within the preceding generation. As one
memorial expressed it, "Of the lives of but few men is it so true that

the cold recital of facts utterly fails to portray what the man was or what his presence meant to the community of which he was a part. Perhaps no teacher has in recent years so indelibly impressed himself upon the lives of college men as did Dean Shaler."[2] Indeed the image of this stately figure, stick under arm, flowing white hair beneath a soft slouch hat, walking briskly in the early morning through the Harvard Yard, became part of the very mythology of the university at the turn of the century. Around his magnetic personality was spun a whole web of anecdote and reminiscence, perhaps best epitomized in the first exhibit of a collection of caricatures and verse compiled by undergraduates and entitled *Harvard Celebrities*:

> This is Shaler,
> Fairy-taler,
> Scientific mountain-scaler,
> Penetrator
> Of each crater
> From the poles to the equator,
> Tamer of the hurricane,
> Prophet of the wind and rain,
> Hypnotizer
> Of the geyser,
> Wizard of the frozen plain.
> Hark! What is that deep and distant subterranean roar,
> Arising near Memorial and reaching out to Gore?
> 'Tis the rumble of applause
> When the speaker makes a pause
> In relating an adventure from his fund of earthquake lore.

As one writer in the *New York Evening Post* remarked, "These lines are a not inadequate characterization of that geologist, philosopher and poet, that man to whom all the various aspects of life were as fascinating as a fairy-tale, Nathaniel Southgate Shaler."[3]

Ancestors: An Anglo-Saxon Heritage

Shaler's status as a Harvard professor was of itself sufficient to allow him to penetrate the tight social and intellectual coteries of the Boston intelligentsia. Still, he was pleased, like many other patrician intellectuals of his era, that he could trace his forebears back to English roots—Anglo-Saxon stock that furnished an ancestral pedigree admirably suited to a prominent Bostonian. His father's people, who had come from Warwickshire via Jamaica to Connecticut, where they were farmers and quarrymen, showed the same restlessness throughout later generations. All three sons of Nathaniel's great grandfather— Captain Timothy Shaler, a privateer during the American Revolution—had careers that were nothing if not colorful. The eldest, William, distinguished himself as a merchant seaman, later as a government official (first in lower Mississippi and then as United States consul in Havana and Algiers), and also as an author.[4] The career of the second son, though less noble, was equally dashing and is best encapsulated in Nathaniel Shaler's own words when he remarked about this "dare-devil ne'er-do-well" that he "wound up an adventurous career as a prisoner of war at Dartmoor in England, where he was shot leading a revolt against the jailers."[5] The youngest of this family, Nathaniel's paternal grandfather, also took to the sea as a boy, soon commanded his own vessel, and later became a prosperous merchant in New York. After the War of 1812, however, impecuniosity forced him to turn privateer. This change led to a revival in his fortunes, but he and his ship were subsequently lost at sea. He left behind a wife, three sons, and a daughter.

Nathaniel Burger Shaler, born in 1803, was still a youth when his mother died from grief over the disappearance of her husband. But she had left sufficient means to guarantee her children a good education, and in 1827, Nathaniel Burger went up to Harvard College. At the commencement of his final year, however, he withdrew to enter the Medical School and eventually followed his uncle to Havana, where he hoped to make his fortune. Still restless, he left after two years; only the challenge of battling with the Asiatic cholera that plagued the people of Newport, Kentucky, encouraged him to settle there. His success in treating the disease soon established him in the community, and in 1832 he married Ann Hinde Southgate. Five children were born of this marriage, the first of whom died in infancy. Three of the other children survived to maturity, the eldest being Nathaniel Southgate, whose birth on 20 February 1841 was duly recorded in the family Bible.

The Southgates, from whom the young Shaler got his middle name, had originally come from London. On the whole they were rather less distinguished than the Shaler line. One or two became clergymen; one, Henry Southgate, a minor poet; and Shaler's great grandfather, who settled in Virginia during the early eighteenth century, a successful planter and merchant. To this pigtailed gentleman the young Nathaniel was closely drawn, perhaps because of his sensitive nature, his love of literature, and his broad interest in humanity. His son Richard, to his great chagrin, had forsaken the family business to study law. But this was Richard Southgate's only departure from family tradition, for both he and his seven children perpetuated the social conventions of their forefathers and remained solidly English in both taste and style.

These ancestral credentials were meticulously rehearsed in Shaler's *Autobiography*. Intoxicated with the hereditarian concerns of the era, he insisted that it was "a fair presumption that when a half a dozen stages of a life are fairly well known, they afford a tolerable basis for reckoning as to what comes to the last stage in the way of birthright." Certainly he rejected any cramping hereditary determinism, insisting that personal qualities had as great an influence on individuals as "that shadowy domination" that comes from the past.[6] Nevertheless, in an age when Americans exhibited a fascination with ancestors and immigrants, Shaler was patently conscious of his personal pedigree. That typically Victorian intellectual pastime, autobiographical art, is itself witness to the genealogical curiosity that captivated the minds of so many around the turn of the century.

The Early Years

Shaler spent his early years in Newport, Kentucky, a village with a population of about a thousand; on the southern bank of the Ohio River, it was little more than an outskirt of Cincinnati. Although physically separated by less than half a mile, the two settlements were, according to Shaler, socially more different from each other than New York and New Orleans. One was a prosperous modern town, the other a relic of the sixteenth century; in one community all had equal legal rights, in the other slavery still prevailed. In Campbell County, where Shaler was born, the Southgate family was a long-established slaveholder, and it is no surprise that the image of his black nursemaid remained one of his most vivid memories.

Delicate health and the fact that his birth closely followed the death of his mother's firstborn meant that Shaler was brought up within the sheltered confines of the home without formal education. Youthful rambles along the riverbank, however, helped foster a country boy's interest in wild and domestic animals and an early taste for geology and natural history. Since his father dabbled in amateur mineralogy and possessed the remnants of a rather good mineral collection originally assembled by Charles T. Jackson, Shaler's interest in the world of nature rapidly developed to the point where he sought out such geological classics as Roderick Murchison's two-volume *Silurian System* in Cincinnati's Mercantile Library.

To boyhood days can also be traced another of Shaler's adult intellectual concerns—religion. His father had little interest in religious matters, but his mother, although Episcopalian, often took her children to the Methodist chapel. By the age of twelve, however, Shaler had come to "utter disbelief" and was never to return to traditional Christian orthodoxy. But he did possess what he described as "a curiously intense belief in immortality, altogether instinctive, for I had no teachings concerning it except in sermons which did not affect me. I was not at any time interested in the matter of a life after death, but accepted it as an absolute reality based on feeling."[7] Such deep convictions later became the grounds for his exploration of humanity's place in nature and the role of evolution in explaining the meaning of individual existence.

No doubt to compensate for his son's rather desultory education, Shaler's father engaged a personal tutor for the now fifteen-year-old boy. A German-Swiss clergyman and a competent philosophical theologian, Johannes Escher exerted a profound influence over his young pupil. Under his tutelage Shaler assiduously explored Latin, Greek, and German literature and was soon initiated into the thinking of such *Naturphilosophen* as Goethe and Schelling. This exposure to post-Kantian idealism was to have a lasting influence, for while his "youthful love of philosophy, though for two or three years very great, bore no immediate fruit," Shaler judged that "the ground on which the seedling fell was essentially strong, and it had certain secondary effects which have been of permanent value." Not the least of these effects was

Shaler's later neo-Kantian approach to the relationship between science and religion and his propensity for the idealistic pragmatism of Josiah Royce.[8]

Besides Escher's formal instruction, the Zurich scholar also influenced his pupil in other ways. He introduced him, for example, to the German society of Cincinnati and thereby encouraged a life-long admiration for Germanic ideals and practices. Personally, too, Escher greatly impressed him. So when Shaler had the opportunity a few years later of visiting Germany, he sought out his old teacher, now a pastor of a small Protestant church in the hamlet of Sitzbirg. As he reflected on the close ties between pastor and people, Shaler grew warm in his praise. "A man of profound learning and extensive experience in the world," Shaler reported, "he had already seen that the time when the sermon could be trusted as the main agent of religious guidance had passed, and sought to replace it by the influence of example and that continual incitation which the rural clergyman can still bring to bear on his flock." Certainly some might think Escher's congregation doctrinally undernourished. But, Shaler went on, "If we may judge the work by its fruits, the pure lives, leading through contentment and cheerfulness to a hopeful end, surely warranted the omission."[9]

In 1858, in spite of, and certainly because of, a training that amounted to little more than a smattering of superficial knowledge, plans were laid to complete the seventeen-year-old Shaler's education at some institution of higher learning. The proposal that he should be sent to West Point was quickly abandoned, as was Escher's suggestion of Heidelberg. Eventually it was decided that he should follow in his father's footsteps and go to the nation's center of learning. And so it was that Shaler enrolled as a Harvard sophomore in the class of 1859, a year symbolically straddled between two eras in the history of Western culture; the demise of the old seemed reflected in the passing of Humboldt, while the publication of Darwin's *Origin of Species* heralded the advent of an intellectual revolution.

Midcentury Currents of Thought at Harvard

The cultural elite of nineteenth-century Boston was as intensely religious as it was intellectually vigorous. Among members of its cultivated circles, as Kuklick has reminded us, "Religion was at the heart of their way of life and theological issues were part of their daily concerns."[10] By the close of the first third of the century, the religious ideals of Unitarianism had triumphed over those of traditional Congregationalist Calvinism. As a product of a more commercial and secular order, Unitarianism deprecated the Reformation's somber confession of man's total depravity and applauded the innate goodness of human beings. More optimistic in philosophy than the Puritans, though less disposed to miracles, Unitarians soon found divinity in the universal reign of natural law. So by leading men and women back to nature—the place par excellence of divine revelation—they reconstructed a form of natural theology founded on the classical Newtonian model of the universe.

Unitarian victory in securing control of the religious institutions of eastern Massachusetts seemed complete when the newly founded Harvard Divinity School emerged as the training ground for their ministry. But the victory was bittersweet. No sooner had they established themselves than, as by spontaneous generation, they spawned a movement that cast their own "neoorthodoxy" in the shade. The rise of transcendentalism constitutes a familiar chapter in American intellectual history. And if its influence on American culture has been overpublicized, its impact on the Harvard intelligentsia should not be underestimated.

The Transcendentalist Interlude

In the mid-1830s a group of young New England intellectuals— among them Theodore Parker, William Ellery Channing, and Henry David Thoreau—met together in Concord, Massachusetts, for the explicit purpose of discussing recent developments in philosophy, theology, and literature. The sheltered study of lapsed Unitarian clergyman Ralph Waldo Emerson was the birthplace of this "Transcendental Club," as the neighbors playfully dubbed it. Dissatisfied with Enlightenment rationalism and with the religion of their day, these "transcendentalists" sought a new faith. The result was a high romantic idealism grounded in the practical expressions of a sensitive social conscience. Finding the locus of religious truth neither in authority nor in tradition but in the intuitive witness of the human soul, the transcendentalists sought to redefine the American ideal of individualism in such a way as to encompass humanitarianism and a humanitarian religion—the religion of humanity.[11]

While for intellectual inspiration they drew upon Platonism, Oriental mysticism, the Kantian critique, and Coleridge's *Aids to Reflection*, their more immediate motivation lay in their disillusionment with the sterile complacency and cold credal formality of Unitarianism. Galvanized by the colonial sense of God-given mission, the church had traditionally attended to the spiritual needs of its community. Eighteenth-century rationalism, however, had undermined Puritan faith, minimized spiritual devotion, and cooled evangelical zeal. So while the Unitarians remained tolerant, broad-minded, and literate, they increasingly lacked burning conviction. The transcendentalists sought to rekindle the flame.

Even the tempered liberal doctrines of a Unitarianism fearful of degenerating into a cultured paganism were too formal for the transcendentalists. So Emerson seceded from his Concord pastorate in 1832 to write *Nature*. It became the transcendentalist Bible. Metaphysically pantheistic and religiously syncretistic, Emerson pursued the higher spiritual reality behind nature in his celebration of spontaneity, poetic imagination, and intuitive insight. Still with a vision of finding cosmic comfort in the wake of the Puritan demise, Emerson reinvigorated the individualistic cast of America's spiritual heritage. To find solitude, to retire from society, to gaze at the heavens in isolation, to encounter nature face to face—this, the new beatific vision, would nurture a religion of personal revelation liberated from the bondage of historical tradition. Why, mused Emerson in *Nature*, "should we grope among the dry bones of the past, or put the living generation into masquerade out of its faded wardrobe? The sun shines to-day also."[12] How apposite, then, from his own standpoint, was Emerson's aphorism that Calvinism rushes to Unitarianism, as Unitarianism rushes to naturalism.

For some, Emerson seemed too exaggerated in his condemnation of the "pale negations" of "corpse-cold" Unitarianism. Not all with transcendentalist sympathies broke from established denominations. They too developed Channing's stress on human dignity and its ramifications for social action. In this way the transcendentalist critique, with its disclosure of divine likeness in the human soul, was synthesized with the scattered vestiges of the philosophical pietism that had its roots in Edwardsian Calvinism. James Marsh, for example, a leader of the Presbyterian New Light movement and president of the University of Vermont, used Coleridge's reflections to bridge the gap between philosophy and theology, while Theodore Parker, who remained a Unitarian minister, notwithstanding almost total opposition from his fellow clergymen, won an acknowledged place for transcendentalist doctrines within that denomination.

Transcendentalism was, of course, merely one, if in Massachusetts the major, expression of contemporary romanticism. It was both a cast of mind and a complex of sentiments. Genially disposed to emotion and intuition, the romantic temperament turned to subjective experience in the face of rationalistic empiricism. Thus for the romantics and a fortiori the transcendentalists, the direct encounter with physical nature functioned as a source both of inspiration and of substantive truth. Under nature's tutelage humanity could achieve a more heroic destiny. Not that the romantics had rediscovered nature; rather, through nature they had rediscovered themselves. Still with dreams of creating a theocracy on earth, they found in the ideology of democracy a particularly amenable social philosophy. If their humanistic optimism was well founded, then it followed that democracy was the most appropriate vehicle for the expression of individuals' aspirations and dispositions. Displaying none of the exaggerated pessimism often associated with romanticism, the transcendentalists were incorrigibly optimistic, remained preoccupied with the present and the future rather than the past, and shared the romantic revulsion against the intellect as the sole means of arriving at truth.

The metaphysical certainties of transcendentalism, however, were soon to be, if not shattered, at least badly shaken. The coming of Darwinism, with its naturalistic implications, presented an unprecedented challenge not just to the New England transcendentalists but to the entire fabric of idealist conceptions of the natural order.

The Impact of Darwinian Evolution

Charles Darwin did not invent evolution. As a metaphysical doctrine of progressive development, the idea of evolution had already been applied to embryology, to astronomy, and to the study of society.

These themes have been explored by numerous scholars and need not detain us here.[13] Suffice it to note that many of these early evolutionists were opposed to Darwin's theory when it first appeared, for in many ways his version ran counter to the supposition of a necessary upward progression in organic history. Moreover, the furor generated by the pre-Darwinian evolutionary speculations of Robert Chambers conditioned very substantially the initial North American response to the *Origin of Species*.

Although Chambers, in his anonymously published *Vestiges of the Natural History of Creation,* entertained no doubts about divine creation itself, he resisted any suggestion of direct divine intervention in natural processes. The world, as he pictured it, was the product of universal natural law.[14] With the prevailing common context for both theological and scientific discourse provided by natural theology in the Paley mold, his work came under sustained, bitter attack from those who, like Francis Bowen, saw organic adaptations as proof of divine intervention in the world. But overall, as Pfeifer writes, Chambers's book "had a naivete that kept specialists from taking it seriously."[15]

Darwin's *Origin of Species,* by contrast, could not be so summarily dismissed. And yet, while persuasively argued and supported with an impressive array of empirical evidence, it was both grand and simple. He showed how the myriad features of living things that fit them to cope with their environments could be explained merely as the product of ordinary cause and effect relations and not as specifically designed. Apparent purpose was explicable in a purely naturalistic fashion. Darwin's book, then, put the theory of evolution, based on the mechanism of natural selection, on a new and solid foundation. Its scientific elegance, moreover, was enhanced by the fact that he presented his case deductively in the form of a coherent necessary argument. Given the self-evident facts of inheritance and variation and the Malthusian parameter of population increase, it was inevitable that some struggle for existence must take place. Plainly those who survived in the battle for life were better fitted to their environment than their competitors. As a result, relatively superior adaptations increase, and relatively inferior ones decrease. So the environment, viewed historically, "selects" those organisms well adapted; they survive and reproduce their kind; the others are steadily eliminated.[16]

It is well known that, in its day, Darwin's theory was a source of raging controversy. The precise nature of the disagreements, however, is less clearly understood. Certainly it cannot be denied that it engendered some sort of confrontation between science and religion. But the unsophisticated military metaphor proposing a warfare between science and theology is inadequate and has been exposed on several fronts.[17] First, as Robert Young has shown, those very scientists whose views most troubled Victorian orthodoxy were far from wishing to demean God and humanity. Lyell, Herschel, Babbage, Wallace, and Baden Powell, for example, all argued for a grander view of God.[18] Second, the colorful accounts of the Huxley-Wilberforce encounter at the Oxford meeting of the British Association have been shown to be more a product of later historical predisposition than a description of what really happened. Indeed on the theological front, the publication of the controversial *Essays and Reviews,* which set the style for much theological liberalism, probably did a good deal more to shake the conservative Victorian mind.[19] Third, as I have tried to show elsewhere, many of those with the most conservative evangelical opinions were happy to align themselves with the evolutionary cause, not least in the United States.[20] These considerations notwithstanding, Darwinism did play a significant part in the advance of nineteenth-century naturalism. Clearly the impression of an embittered warfare became part of the fabric of at least some Victorian minds. My claim is rather that the military model does little to advance our understanding of the period in general and, as we shall see, of the role played by a figure like Shaler in particular. Whatever confrontation did take place seems likely to have focused more on the growing concern for a professional science unhampered by the constraints of the ecclesiastical hierarchy.[21]

The situation at Harvard during the early 1860s was rather different from the setting of the supposed Wilberforce-Huxley fracas at Oxford. In Boston the leading participants in what turned out to be a rather lengthy debate over Darwinism were Louis Agassiz and Asa Gray, both respected Harvard professors enjoying international scientific reputations. Here I want to review briefly the background to, and the salient points of, this particular drama and the major confrontation between Agassiz and William Barton Rogers (later, the first president of the Massachusetts Institute of Technology) for two reasons. First, because Shaler's presence in Cambridge during the 1860s permitted him to witness firsthand the reception of Darwin's theory at the intellectual center of New England's scholarly life. And second, because he was a favorite student and eventually the successor of Agassiz and was a close friend of Gray.

Louis Agassiz, one of the foremost savants of his generation, had grown up in a dazzling intellectual tradition boasting names like Zwingli, von Haller, Rousseau, and Pestalozzi. At the University of Heidelberg he was first initiated into the botanical elements of Goethe's nature philosophy and through it acquired a lasting admiration for the philosophical and scientific vitality of German high culture. So during the summer of 1827, he steeped himself in the work of Lorenz Oken. Arguing from metaphysical assumptions for a common structural and functional plan for all living things, Oken saw the successive stages of human embryological development reflected in the morphology of lower animals. Animals were thus the persistent fetal stages of human beings. At the same time, Oken believed that all creatures aspired to the fulfillment of an ideal type in their own development and, like other Naturphilosophen, sensed an archetypal plan behind the flux of organic forms that gave coherence to the seeming chaos of prolific nature.[22]

Oken's outlook was to have a lasting effect on Agassiz. So in turn, Agassiz insisted that species were the permanent manifestations of divine ideas. There was no genetic link between them. Certainly he identified progressive levels of organization in the natural world, which he took as reflections of the plan God had followed in the order of his creation—a predetermined sequence culminating in the human form. But he never viewed his paleontological work as furnishing any framework for an evolutionary conception of life history. As the son of a Protestant divine, moreover, Agassiz easily infused German idealism with religious sentiments. To him the Absolute Being was present in all nature, and the task of the natural historian was to discover the way an individual life-form approximated the pattern stamped on the universe at the initial act of creation. His religion was thus a synthetic blend of theism and romantic idealism.[23]

There were, too, other influences on the young Agassiz. Perhaps the most significant was his lasting friendship with Georges Cuvier.

From him Agassiz learned the need for rigorous empirical precision, the value of teleological explanation, and the doctrine of the permanence of species. Like his master, he repudiated Geoffroy Saint-Hilaire's "de-theologizing" of nature in 1840 because it "shut out all argument from design and all notion of a Creative Providence."[24] And while the painstaking empiricism of Cuvier made him a little impatient with the vague speculative metaphysics of *Naturphilosophie,* he still found he could marry these diverse traditions—idealist and empiricist—in the idea of structural plans in the natural world. Because they existed in nature, the scientist who uncovered them was doing nothing less than expounding the very mind of the Creator by elucidating his plan for the natural order. It was a heady enterprise, and Agassiz threw himself into it with all his might.

With this particular synthesis of science and metaphysics, it was almost inevitable that Agassiz should emerge as Darwin's most implacable scholarly opponent. And if, as Mayr maintains, he stands closer to eighteenth-century speculative philosophy than to modern biology, that did nothing to diminish his prestige when he first came to Harvard's Lawrence Scientific School in 1847.[25] Indeed the ethos of natural history research in the United States for the next half century owed much to the charisma of this sparkling Swiss naturalist. Yet by 1859 his scientific authority, never previously questioned, began to be doubted. One reason was the suspicion that he seemed unprepared even to give Darwin a fair hearing. His whole intellectual upbringing militated against Darwinism, and his outlook, as Asa Gray later recalled, was "singularly entire and homogeneous—if not uninfluenced yet quite unchanged by the transitions which marked the period."[26] But the Darwin question only served to bring matters to a head. Agassiz's intransigence on the immutability of species had already earned him some opponents. Later, indeed, his extreme creationist interpretation of the origin of the human races forced Lyell to confess that "Agassiz. . . drove me far into the Darwinian camp. . . for when he attributed the origins of every race of man to an independent starting point, or act of creation. . . I could not help thinking that Lamarck must be right."[27] Still, in January 1859, some months before the appearance of the *Origin,* the growing professional dissatisfaction with Agassiz's views was first publicly aired. The setting was the Boston meeting of the American Academy of Arts and Sciences, and the spokesman was Asa Gray, Harvard's Fisher Professor of Natural History. His presentation was deliberately designed to "knock out the underpinning of Agassiz' theories about species and their origin,"[28] and, drawing on his unparalleled knowledge of North American and Japanese botany, he presented the case for the diffusion of species from an original single center. This view directly challenged Agassiz's multiple origin thesis, and so over the next fourteen months, Agassiz and Gray engaged in a series of debates in Boston over the Darwinian theory.

The formative intellectual influences on Gray were singularly different from those on Agassiz.[29] He had begun as a medical practitioner and from those early days remained firmly committed to the British tradition of natural theology and its empirical counterpart. Thus the empiricist tenor of Sir William Lawrence's *Lectures on Physiology, Zoology, and the Natural History of Man,* first delivered at the Royal College of Surgeons in 1828, greatly attracted him. So at an impressionable age Gray was confronted with Lawrence's assertion that the human race was a single species, subject to the same physical laws that governed other organisms. During the winter of 1831–32, Gray was forced to abandon his declining Bridgewater practice, and over the next five years he held a number of part-time teaching and library positions. But his increasing passion for botany brought him into close contact with John Torrey, who exerted a great influence over the young scientist—botanically in that Gray soon became a full collaborator on the *Flora of North America* and spiritually in that he came to accept orthodox Christianity in its Congregationalist mold. Indeed when appointed to the chair of natural history at Harvard in 1842, Gray did not hesitate to display his Congregationalist sympathies by deserting the college's Unitarian chapel for Boston's Park Street Church.

By the 1850s, Gray was the leading American botanist, enjoying a worldwide reputation, especially in the field of botanical taxonomy. Eschewing mysticism and German idealism alike, he had a great deal more in common with British empiricism than with those romantic currents sweeping many into transcendentalism in New England. This tie nourished his long and lasting link with Darwin. Darwin in fact had outlined the major tenets of his controversial theory of natural selection to Gray in 1857, and so when the *Origin* first appeared, Gray entered the controversy primarily to ensure that Darwin should receive a fair hearing.

Gray did not accept the arguments of the *Origin* either immediately or without reservation. With his longstanding approval of Paleyan natural theology as expressed in the *Bridgewater Treatises,* Gray attempted to reconcile natural selection and the argument from design. Such a synthesis, he hoped, would purge the theory of any unfortunate materialistic connotations. His basic strategy therefore was to claim that, despite its title, Darwin's book really gave no account of the *origin* of life but only of "its diversification into the forms and kinds which we now behold."[30] So when Gray issued his pamphlet entitled "A Free Examination of Darwin's Treatise on the Origin of Species, and of American Reviewers," he had printed across the top in Gothic script: "Natural Selection Not Incompatible With Natural Theology."[31] As part of his defense, too, Gray joined forces with the Harvard philosopher Chauncey Wright. Wright, of course, had no concern to defend the argument from design, much less Christianity, but his empiricist challenge to a priori systems was attractive to Gray, as were his assaults on those with a penchant for inflating evolutionary biology into a cosmic weltanschauung after the fashion of Herbert Spencer.[32]

As we have noted, the main confrontation between Agassiz and Gray occurred prior to the publication of the *Origin.* But Gray continued to defend Darwin against both Agassiz and Francis Bowen, and soon the alignments of opposing intellectual factions began to be determined.[33] By now the scene was set for the more specific (and for Shaler the more significant) clash between Agassiz and William Barton Rogers at the Boston Society of Natural History. In contrast to the Huxley-Wilberforce encounter at Oxford, these four debates were extended, carefully stage-managed, and between opponents of comparable scientific standing. The spectacle was well worth seeing: Agassiz—romantic, eloquent, passionate; Rogers—cool, calculating, rational. Still, as the meetings wore on, Rogers more and more stamped his authority on the debate. Turning Agassiz's progressivist paleontology against him and pressing him to explain precisely his terms, Rogers outflanked every move Agassiz made and finally clinched the argument in a detailed interpretation of the New York fossil-bearing strata. The specific issue concerned the Paleozoic series and the conditions under which they had been formed. Agassiz maintained that the facts supported upheaval, but Rogers was convinced they had been deposited on a subsiding seafloor. Quite astoundingly, during the final session Agassiz began his concluding remarks with an inadvertent admission that the fossils had been laid down during a local upheaval of the shore when the bottom was sinking because of thermal contraction of the earth's crust. Rogers was not slow to press the point home. Agassiz, he recalled, had denied the view that fossils would be substantially destroyed during a slow upheaval of the seabed. The singularly complete New York Paleozoic series, he had maintained, was a classic case. But now, Rogers went on, he had conceded that they had been deposited during a subsidence of the ocean bed. Agassiz's "insuperable objection" to Darwin's theory of fossil formation was undermined.[34]

The outcome of this battle of the greats suggests that the bellicose Agassiz was now beginning to be isolated in his struggle against the underrated menace of Darwin, for even Agassiz's protégé and devotee Jules Marcou was forced to concede victory to Rogers. Shaler too did not miss the significance of Agassiz's admission. After all, he was unique among Agassiz's students in specializing in geology and later did substantial work himself along the New England–Acadian coastline.

Hitherto, ignorance had tended to play its part in the drama as much as propaganda. But in Rogers and Gray, Agassiz encountered formidable opponents. Of course they did not establish the validity of Darwinism. But they did demonstrate that the *Origin* could not be lightly dismissed and substantiated its right to be judged by scientific scrutiny rather than by philosophical speculation, by reason and not by passion.

Of Fish and Fossils

Into this atmosphere, fired with the metaphysics of transcendentalism and the evolution of Darwin, Shaler came in 1859. Financially secure with an annual allowance of $1,500, he soon settled into a routine that was both comfortable and stimulating. At first, however, he was uncertain about which course of study to follow. His father had made arrangements for a period of preparatory tuition in Cambridge, but the prescribed training—traditional and classical—was far from palatable to Shaler, so tutor and pupil duly parted company. Only when he encountered Agassiz's philosophical-scientific essay on classification did he ultimately turn to a more serious study of natural history. Agassiz's captivating personality was just as impressive as his scholarship, and so Shaler decided to enroll as a student in the Lawrence Scientific School. From that moment until Agassiz's death in 1873, he retained a profound admiration for the charismatic Swiss naturalist, perhaps best summed up in his note to Mrs. Elizabeth Cary Agassiz on the death of her husband: "Through him it is that every citizen of this nation holds science in respect. . . . He never was a greater teacher than now. He never was more truly at his chosen work. I know how great your burden must be from my own grief. It is the greatest blow I have ever received. . . . While he lived I always felt myself a boy beside him. I can only hope that you will look at his great work greatly done and find some consolation there."[35]

The Lawrence Scientific School, established in 1847 to promote education in applied science, had introduced to Harvard the new degree of Bachelor of Science. Now the scientific school, together with Harvard College (founded in 1636), the Medical School (1782), the Law School (1817), and the Divinity School (1819) composed Harvard University. In the scientific school the methods of teaching were mainly tutorial—"intimate contacts of teacher and pupil through many hours each day spent in the laboratories, with a maximum of independent work on the part of the student."[36] This arrangement, of course, admirably suited Shaler, whose education to date had been conducted on these very principles. Many of the teachers, moreover, were highly distinguished. Numbered among them were Asa Gray, the mathematician and astronomer Benjamin Peirce, and the anatomist and physiologist Jeffries Wyman. With such a faculty the school rapidly gained a high reputation.

Shaler's first and now-famous lesson quickly evaporated any youthful conceit. Agassiz presented him with a small fish, with the stern injunction that he neither discuss it with anyone nor read anything about it. His cryptic command was "Find out what you can without damaging the specimen; when I think you have done the work I will question you." After an hour's analysis of the evil-smelling object, Shaler felt he had completed the assigned task. But Agassiz, although always close at hand, remained silent for the remainder of the day, and for the next, and . . . for a whole week. Frustrated, Shaler returned to the job, examining and reexamining the specimen, each time noting new subtleties of form and details of function previously unnoticed. At the end of the week came the master's terse question, "Well?" Shaler's hour-long recital, however, brought the devastatingly taciturn reply, "That is not right." The task had to be recommenced, and Shaler set to work with renewed vigor. It was finished only when Agassiz eventually placed a pile of bones before him on which he was required to work unassisted for a further two months.[37] This method of instruction by discovery had a profound impact on Shaler, and in many ways his own educational philosophy and practice grew out of these early experiences.[38]

It was nevertheless a rather unconventional introduction to university study, and Shaler was glad to find that, after the first few months, Agassiz became quite loquacious. A close friendship soon developed between teacher and student, and when Agassiz learned that Shaler wanted to specialize in geology, he set him to work on fossil brachiopods—research that became the basis for Shaler's first paper, read to the Boston Society of Natural History in 1861. The Boston Society, where the debates over Darwin's theory were conducted, was the center of a vigorous intellectual circle; here Shaler became part of a group that, in addition to the Lawrence Scientific School staff, included Charles T. Jackson, Jules Marcou, Henry and William Barton Rogers, Alpheus Hyatt, and Alexander Agassiz.

Student life for Shaler largely revolved around the Zoological Club. This group of about a dozen of Agassiz's students occupied the old building that had previously been used to house specimens prior to their removal to the new museum. It had a number of bedrooms, a dining room, and a central lounge—the meeting place of the club. Here the halcyon days of student life passed by as Shaler's friendship with Agassiz deepened in conversations ranging from paleontology to speculative metaphysics. The only shadow was caused by Darwin's

Origin, for Agassiz's implacable hostility to natural selection and to Asa Gray extended to the students of both professors. They were members of warring camps, and, as Shaler recalled, "It was dangerous for a student to be seen in parley with the enemy." To Shaler, this was the only issue on which Agassiz was suspect. And indeed only later, when he came to a full acceptance of evolutionary theory (albeit a Lamarckian version) and when he was appointed to the teaching staff at Harvard, did Shaler become closely acquainted with Gray—an association that lasted until the latter's death in 1888. During student days, of course, he had read Chambers's *Vestiges,* Lamarck's *Philosophie Zöologique,* the first of the Darwin-Wallace papers, and finally *The Origin* itself. But these had to be perused in private, and so William Stimpson and Shaler kept their discussions of Darwinism a closely guarded secret. To be caught at this pursuit, Shaler later reflected, "was as it is for the faithful to be detected in a careful study of a heresy." When he pressed Agassiz on the topic, moreover, he received the well-known reply that a species was the "thought of God," which, Shaler felt, revealed "the curious mysticism which lay at the foundation of his nature."[39]

Among Shaler's many student friends, undoubtedly one of the most influential was Alpheus Hyatt. He was subsequently to emerge as leader of the American neo-Lamarckian school of biology and, although Shaler's senior by about three years, played "a large part" in Shaler's life.[40] Together they participated in a number of field trips. With Addison Verrill they spent part of the summer of 1860 dredging, exploring, and collecting specimens on Mount Desert Island, Maine. During the next winter they laid plans for a far more ambitious expedition to the Gulf of St. Lawrence. But this proposal became practicable only when the needed financial backing became available. The source was a twenty-thousand-dollar grant provided by the Massachusetts legislature for the work of the Harvard Museum. The gulf indeed was a rich source of potential museum specimens, but beyond that it held another fascination for Agassiz's students. The geological structure of the locality around Anticosti Island was frequently mentioned in the protracted disputes between Rogers and Marcou over Darwinism. Here, then, the anti-Darwinian theories of the Agassiz fraternity could be tested by examining, as Hyatt recorded in his diary, "whether the same species passed from one geological horizon into another or whether each formation was specifically distinct from all others in every member of its fauna."[41]

With university permission, financial support, and a scholarly problem, the party duly left Eastport on 14 June 1861 and, despite inclement weather, reached its destination within ten days. For nearly two months they explored Anticosti and undertook subsidiary excursions to the coast of Labrador and the Gaspé peninsula.[42] This early expedition did not result in any extensive publications—Shaler published a list of brachiopoda and Hyatt wrote up his analysis of Beatrica (a fossil hydrozoan)—but it did have a lasting influence on the young naturalists.[43] Shaler, for example, described it as "the most profitable journey" he ever made, and while in the end the team made no pronouncements about their aim of putting Agassiz's anti-Darwinism to the test, "this field experience," as Dexter observes, "certainly instilled the belief of organic evolution into the thinking of these three young men."[44]

Shaler's final year as a student at Harvard coincided with the first year of the Civil War. Torn between the desire to enlist and the duty of completing his education, he departed for Kentucky and there, following the counsel of his grandfather Southgate, sided with the Union cause. At the same time, he was advised to postpone his commitment to the Northern army in order to study for his final examinations. So he duly returned to Cambridge to face the rigorous questioning of Agassiz, Wyman, Gray, Peirce, Cooke, Lovering, and Horsford on subjects ranging from mineralogy to structural geology. Shaler had prepared himself well, and as noted in the faculty records for 8 July 1862, he took the degree of Bachelor of Science in geology summa cum laude.

After a short vacation in the White Mountains, Shaler returned to Cambridge, only to learn of the most recent events in Kentucky. He immediately departed for Frankfort and there obtained a commission as a captain of the Fifth Kentucky Battery—afterward known as Shaler's Battery. War experience had a deep effect on him, some aspects of which are recorded in his posthumously published collection of poems, *From Old Fields: Poems of the Civil War*. Although these verses have not stood up to critical scrutiny, their sense of immediacy prompted one reviewer to describe them as "some of the most genuine poetry that this generation in America has produced."[45] Shaler's period of active service, however, was brief, and the onset of bronchitis forced him to resign his command. All in all it had been an eventful year, for also during that autumn he married Sophia Penn Page, whom he had known since childhood. It was to be a lifetime marriage. Mrs. Shaler, as well as rearing two daughters, Gabriella and Ann Penn, displayed her own literary talents in her book *The Masters of Fate: The Power of the Will* and in editing her husband's autobiography.

Having tasted the intellectual delights of Bostonian high culture, Shaler soon grew dissatisfied with life in Kentucky and determined to return to Cambridge. So, early in the summer of 1864, he departed for the well-known surroundings of the Harvard Yard to begin what was to become an almost legendary career as the university's professor of geology. Cambridge was to be his adopted home, and here, with only temporary absences, he spent the rest of his life.

Building a Career

After the "malarial influence" of the Kentucky climate and the social upheavals of the Civil War, Harvard was for Shaler a haven of health and happiness. "The weather is delightfully cool and bracing," he wrote to Sophia on his arrival in June, "and old Boston looks magnificent. It is a great gratification to see a clean town once more after having lived in Western mud and dust for two years."[46] He felt as if he was returning to his true home—Old Boston, the last bastion of Teutonic grace and sophistication. More immediately, however, it was a place of employment. No sooner had he arrived than Agassiz made him his assistant in paleontology at the Museum of Comparative Zoology. Although currently at the height of his public fame, Agassiz's health and scientific credibility were beginning to wane, and Shaler's sensitivity to his old teacher,[47] along with his competence to undertake the classification of the museum's store of fossils,[48] meant that he was tailor-made for the job. Then Agassiz's convalescent expedition to Brazil the following year opened the door to Shaler's teaching career, and from then until 1872, he took charge of regular instruction in zoology and geology at the scientific school.[49] These early rewards, however, brought their own problems. The pressures of research, general administration, and teaching were both arduous and fatiguing, and so, convinced that he needed a complete change of scene, he and his wife departed in the autumn of 1866 for Europe.

With a letter of introduction from Agassiz describing him as "the one of my American students whom I love the best,"[50] Shaler set sail for England both for reasons of health and for the opportunity of studying the important geological sites of the Old World. After a few days in Britain, the Shalers crossed the Channel to stay with Nathaniel's sister-in-law in Montreux on Lake Geneva. Soon he was busy investigating mountain structures and Alpine glaciers (in particular the Aletsch glacier, the Grindelwald, and the Mer de Glace) as he had been encouraged to do by Agassiz. Long invigorating walks, stimulating discussions with European naturalists, and the stunning beauty of the landscape soon brought renewed vigor and vitality. In Italy he explored the narrow streets of Old Rome, admired the architectural work of Palladio in Vicenza, and peered into the crater of Mount Vesuvius; in France he visited the art galleries, the museums, and the School of Mines; but in Germany he felt most at home, whether speculating on the similarities between Anglo-Saxon and Teutonic temperaments, conversing with Bernhard von Cotta in Freiberg, or being entertained at the curious faith-healing hotel run by a Father Blumhardt.[51] It was indeed a very successful trip, and when he returned to Boston in 1868, he had visited about fifty of the main European museums, met many Continental savants, and collected some thirty thousand fossil specimens for the Museum of Comparative Zoology.[52]

Shortly after resuming work in Cambridge, Shaler received his first formal appointment from Harvard, as "Lecturer on Paleontology and Animal Life Considered in Its Geological Relations." The following year, 1869, at the age of twenty-eight, he became professor of paleontology (changed in 1888 to professor of geology).[53] Although this promotion brought further teaching duties, including extramural courses on practical geology to the School of Mining, he still found time to spend part of 1868 on a paleontological excavation at Big Bone Lick in Kentucky. This expedition provided the raw material for a paper presented the next year to the Boston Society of Natural History on the geographical distribution of the American buffalo. His observations on the topic were subsequently adopted by Joel A. Allen and later by Frank Gilbert Roe, who maintained that they had "by no means been wholly superseded by the scientific progress of half a century."[54] Also during 1869, Shaler briefly returned to his home state, this time with Joseph Winlock and Charles Peirce, as an extra employee of the United States Coast Survey. Their purpose was to record astronomical observations of a solar eclipse, and Peirce later recalled that, when the time for the eclipse came, Shaler "generously relinquished his opportunity of witnessing the sublime phenomenon" so that he could see it.[55]

All this increased work, however, brought little in the way of financial reward. The coast survey paid one thousand dollars per year, but there was no salary attached to the professorship. So in a letter to Harvard's new president Charles Eliot in 1870, he pointed out that, with a wife and now two children to support, he was rapidly depleting his small capital by spending two or three times his salary each year.[56] Still, Shaler vigorously pursued his vocation as teacher both during and outside the term, and his summer field outings with undergraduates in the late 1860s and early 1870s became the mainspring of the Harvard, and indeed of the American, summer school. From the earliest days these voluntary excursions attracted as many as eighty students straggling behind him like a school crocodile. Indeed Theodore Roosevelt confessed to him, "You taught me how to walk"; in later years one student is reported as whispering to a friend, "If he hears you call him old man, he'll walk your d——d legs off."[57]

These years, however, were a time of change not only for Shaler; they were a period of transition for Harvard itself. In March 1869, Eliot had been nominated for the presidency of the university. Strongly committed to university reform, Eliot was opposed both by the classicists who feared the erosion of Harvard's long-established tradition of liberal education and by the pure scientists concerned about his ideas on technology and applied science.[58] Nevertheless, when faced with an even less desirable candidate, the opposition receded, and the appointment was confirmed by the corporation and by the board of overseers. Perhaps because of their rather similar educational aspirations (despite their later disagreement over the Lawrence Scientific School) and their closely parallel appointments to Harvard, Eliot and Shaler remained firm friends throughout their lives.

By the end of 1872, Shaler's health had again begun to fail. But he had by now displayed his talents as a scientist, a writer, and an administrator. A number of technical papers on geology, several popular accounts of landscape features, and one or two more general pieces drawn from various excursions had appeared in print. And he had also been heavily involved in the planning of what came to be called the Anderson School of Natural History—a precursor of the American summer school.[59] The only hindrance to the realization of this vision was the lack of financial backing, and he was forced to leave that problem unresolved when he sailed for England toward the end of the year. The visit was important to Shaler in at least two respects. First, it afforded him the opportunity of spending some time in the English Fens; out of this experience grew his concern to reclaim some of the United States' Atlantic coastal marshes. Second, he encountered such leading British savants as Darwin, Huxley, Tyndall, Lyell, and Galton at the many scientific meetings and social functions he attended. But, as the following extract from a letter to William James makes plain, he was kept fully informed about events at home:

I have just learned that the scheme of a summer school of Nat. Hist. has had a certain success in getting a large sum of money from somebody. This is good news indeed. I do not expect much from it the first year but hope great things in the future. It may serve as a basis for communication between Europe and America. I hope we may be able to bring each year some of the workers of Europe to give us the good of their experience and take back the best of our own. For their expenses paid we can get any one we may be pleased to invite. Huxley half promised to come.[60]

While Shaler was in England, according to his wife's *Memoir,* some of his friends suggested to the governor of Kentucky that he should be invited to take charge of the Kentucky Geological Survey, which was on the verge of being reactivated after several years' interruption. With strong support from Louis Agassiz and Benjamin Peirce, Shaler was offered the directorship while still in Europe, and he gratefully accepted the position.[61] It seems, however, that the invitation did not exactly surprise him. In November 1868 he had written to John W. Stevenson, then governor of Kentucky, urging that the geological survey be revived and apologizing for taking the liberty of enclosing a synopsis of his own qualifications![62] Again in January 1869 he had suggested to Stevenson that the state survey should be linked to the work of the coast survey—and a week later forwarded personal letters of recommendation from Agassiz and Peirce. The bill to recreate the Kentucky Geological Survey was not passed until 1873, by which time Preston H. Leslie was the new governor. But with renewed support from friends, Shaler predictably received the directorship.

On his return from England Shaler immediately began the business of reorganizing and restaffing the survey. A broad legislative brief gave him all the freedom he needed to guide the structure and orientation of the whole project. So with an appropriation of ten thousand dollars for the first year, he planned a general reconnaissance of the entire state, the preparation of topographical and geological maps, a study of animal and plant life, and the publication of a full report. With additional federal assistance from the coast survey for a triangulation survey, field operations commenced in July 1875 under the direction of Professor William Byrd Page in Cumberland Gap, where Shaler held a summer school that same year.[63] The further stimulus of the Centennial Exposition in 1876 prompted Shaler to add yet another project to his already busy schedule—an inventory of Kentucky's natural resources. The outcome was the publication in 1876 of a monograph entitled *A General Account of the Commonwealth of Kentucky*. It represents the germ of another recurrent theme in his later work, that of resource conservation. Shaler's experience with the survey also sensitized him to problems in the organization of higher education in the United States. The unsuitability of many young applicants suggested the need for more technical and scientific schools attached to universities for the training of professional geologists, engineers, and so on.[64] This concern for college reform was also to become a driving force in his later years.

While Shaler did enjoy many aspects of the work, his time as director of the Kentucky Geological Survey was, overall, not a happy one. Opposition from various quarters to the whole venture, persistent doubts as to its financial viability, administrative frustrations usually over assistants' salaries, and the constant traveling to and from Cambridge all eroded his enthusiasm for the job. His often daily letters to his wife suggest a prevailing mood of tiredness and depression. "This endless running around makes one's life a thing of shreds and patches," he wrote to her on 15 May 1875; "I feel sick of it."[65] So when the legislature passed a residential requirement for the office of director, Shaler was only too glad to resign the post. As he observed in a letter to the editor of the *Frankfort Yeoman*, "I have been anxious to lay down the labor as soon as the good work would permit. The passage of this amendment offers me the long desired opportunity of escaping from the duties which have proved an over heavy burden under the peculiarly difficult conditions which I have had to face."[66] His original program of work, of course, was far from complete. But Shaler's period of office was still in many ways a success. The first triangulation survey of the state had been instigated, and various reports ranging from geology to forestry, an inventory of natural resources, and cross sections of the western Kentucky coalfield had all been produced. The wide-ranging interests of its director were thus indelibly imprinted on the character of the revitalized Kentucky Geological Survey.[67]

The Kentucky Geological Survey was merely one of the post–Civil War geological surveys to flourish during the 1870s. Others included Newberry's survey of Ohio, Chamberlin's in Wisconsin, and Pumpelly's in Missouri.[68] The Civil War had highlighted the critical nature of coal, lead, and iron; the realization of their value, together with plans to connect the two coasts by rail, had encouraged a major focus on practical geology in the postbellum era. Then, too, financial cutbacks had begun to threaten the wide variety of surveys existing from before the war, and the need for rationalization became apparent. As a result there eventually emerged in 1879, after a decade of scientific and political infighting, the United States Geological Survey. In essence it amounted to a fusion of the four major western territorial expeditions: Clarence King's Geological Survey of the Fortieth Parallel, F. V. Hayden's Geological Survey of the Territories, John Wesley Powell's United States Geological and Geographical Survey of the Rocky Mountain Range, and George M. Wheeler's Geological Surveys West of the One Hundredth Meridian. The whole matter had been handed over to the scrutiny of the National Academy of Science, and its support for civic science over against the military, the Public Land Office, and by implication, both Wheeler and Hayden, meant that the proposals of Powell carried the day. By now King was solidly supporting Powell's master plan.

The details of the establishment of the United States Geological Survey and of King's nomination as its first director constitute a well-known chapter in the history of American science and need not detain us here.[69] The important element for Shaler's career was the survey's fundamental commitment to social planning, eloquently articulated by Congressman Abraham Hewitt in the House of Representatives in February 1879. Not that the bill passed without comment. It encountered the determined opposition of western senators disenchanted with the image of the yeoman farmer and with Powell's aim of excluding settlers from the West. This mostly Republican opposition stood for the pioneer fringe, for the frontier with its incipient capitalism, and forced the tabling of a compromise bill. Intrigue and political machination followed, but Hewitt managed eventually to gain what in the last analysis was "only a partial victory for the forces of reform."[70] Nevertheless, the systematic exploitation of resources, the value of applied science, and the concern to prevent the dissipation of the nation's material base were established as major priorities. Utility, research, and application all went hand in hand. And this foundation, laid by King, was ably built upon by the self-educated visionary John Wesley Powell, who succeeded to the directorship in 1881. But he went even further, liberalizing the legal language and thereby transforming his bureau into the truly national enterprise he had always envisioned.[71]

From its inception Shaler remained a loyal supporter of the survey, of its utilitarian philosophy, and of Powell, its chief architect. His first association with it, however, was while King was still in charge. In August 1880 he had been appointed expert special agent for the New England subdivision to accumulate statistical information on the mining industry for the Tenth United States Census—an economic review assigned to the geological survey.[72] But the post was short-lived, for King interpreted the New England coastlands as beyond the ambit of the survey, and the contract was duly terminated. By the time Powell had redefined its scope, Shaler's health had again begun to fail, and he set sail once more for Europe in the summer of 1881.

As before, Shaler spent a short time in England prior to his Continental tour. The exceptionally fine weather, especially during his winter stay in southern Italy, permitted long walks and geological field work, and his enjoyment of the break was only interrupted by the news of his father's death. In the early spring he left Florence to return to Vesuvius. It brought back many memories of his last visit a decade and a half earlier. But while the geology had not changed, Shaler certainly had. "I wonder where my geological eyes were fifteen years ago," he wrote. "It must have been some other fellow who was here."[73] Once again Shaler spent his time in Europe visiting scientific institutions, scholars, universities, and schools of mines, as well as making his now customary pilgrimage to Malvern for a course of water-cure treatment. Both refreshed and renewed, he returned to the United States in the spring of 1882.

After a brief visit to Kentucky to console his widowed mother, Shaler resorted to Campobello Island off the coast of Maine for the summer months. The locality was very familiar, of course, for he had published a preliminary report of its topography the previous year.[74] But it was not just a place of study; he was "charmed by its beauty" and would have made a holiday retreat there had it not been for its inconvenient distance from Cambridge.[75] These summer weeks all too soon passed by, and with the new academic term, the scholarly round of teaching, research, and writing had to be resumed. Encouraged by the favorable reception of his 1881 volume on glaciers (for which W. M. Davis selected and annotated the photographs),[76] he was now working on an introductory geology textbook and on a major historical study of Kentucky for the New Commonwealth Series. In addition, he was involved in a number of public service activities, including consultancy work for mining enterprises in Kentucky, Virginia, Colorado, Montana, and Ontario, and also for the Kentucky Union, the Cincinnati, and the Southwestern Railway companies.[77] These commitments notwithstanding, Shaler also sought a position with the United States Geological Survey and, in the autumn of 1882, offered to make a study of the Narragansett coal basin in Rhode Island and Massachusetts if financial assistance could be made available. Powell, to whom Shaler sent his proposal, however, was away on field work, and his secretary James Pilling replied saying that funds for the current year had already been allocated.[78]

Various structural changes were by now taking place within the survey under Powell's new regime—changes that would ultimately lead to the appointment of Shaler as director of the Atlantic Coast Division. For one thing, Powell was always far more concerned for agriculture and irrigation than for mining and technology, and the character of the survey soon began to mirror his predilections. Nor was he fastidious about legal minutiae, and the result was that the survey's staff expanded dramatically. He managed, for example, to get Congress in 1882–83 to authorize the preparation of a topographical map of the United States, which necessitated the extension of its activities. This whole enterprise, of course, had political ramifications, for Powell was committed to the reform of the land system, and a series of topographical maps by quadrant represented one move in that direction. Thus, early in March 1884, Powell wrote to Shaler outlining the proposed geographical and cartographic survey of New England and suggesting that Francis A. Walker and he should supervise the project.[79] Shaler readily agreed, and as soon as he had joined the survey in the summer, he began negotiating with Powell an ambitious research schedule involving, for the first year alone, a report of the Narragansett coalfield, an inventory of mineral resources, and a study of coastal and marine geology. Questioned as to the viability of such an extended undertaking, he replied that he had periodically worked on the Narragansett basin for some fifteen years, that he had already accumulated substantial data on the geology of the coastline, and that his earlier work for the Kentucky Geological Survey and the coast survey had involved resource assessment.[80] Despite this defense, however, Powell granted permission only for research on coastal geology. In fact, it took the rest of Shaler's career with the survey to bring his original plan to a successful conclusion.

Once settled into the job, Shaler devoted his energies to organizing the material already available on the coastal region around the Cobscock Bay area in Maine and to initiating what turned out to be a ten-year project on the Narragansett coalfield and the New England saltwater marshlands. He was able to maintain a consistently high output, which reflects the extensive assistance of his students in providing bibliographies and in undertaking basic geological research. R. A. F. Penrose, for instance, completed a self-financed research project on lime phosphate deposits, which contributed to his doctoral degree under Shaler.[81] Moreover, Shaler's teaching load at Harvard was now fairly light, and he could easily integrate field teaching with various projects for the survey.

Although 1884 brought Shaler's appointment as director of the Atlantic Coast Division, it was a difficult year for the survey as a whole. Political opponents had continued to create obstacles, and not content with engineering congressional investigation into the various government science agencies, they instigated a commission of inquiry, before which Powell himself was called. Beginning in December 1884, the feud dragged on for over a year, Powell being questioned on sixteen occasions and the economic efficiency of the survey being scrutinized through a searching audit by the Treasury Department, before suggestions of malpractice were finally dispelled.[82] Throughout the proceedings Powell had gone on the offensive. He startled the Allison Commission with remarkable statistics about aridity and settlement, backed them up with supporting topographic maps, and challenged the Land Office Survey on one point after another.

During the whole scandal Shaler uncompromisingly aligned himself with Powell and his cause. He was, for example, just as committed as Powell to land reform. Then too he repudiated the baseless rumor printed in the *Boston Advertiser* that he might succeed Powell as director.[83] And later, as the following extract from one of Powell's letters indicates, he offered help in the current controversy:

There has been no charge brought against the Survey which is not curiously erroneous. I have no fear that the Survey will be overthrown, no fear that an attempt will be made to remove me, and no fear of any kind except that a general sentiment may be worked up adverse to the operations of the Survey in such a manner as to injure it when appropriations are asked.

Still, matters may be more serious than I am willing to believe; but I shall watch matters pretty closely, and if I think that an adverse sentiment is growing up I shall after a while appeal to the press and answer the many foolish things that have been said. I may also be glad to avail myself of your help in this matter.[84]

The following year too, when this particular fracas had died down, Shaler continued to support Powell in the face of the rather more academic criticisms made by Alexander Agassiz. He felt that the survey had no business dealing with economic geology, that it published wastefully large reports, and that paleontological research should be left to individuals prepared to "do such work merely for the prestige of associating with a university or museum."[85] In response Shaler penned the following to Agassiz on 27 June 1886:

Powell showed me your letter; I advised him to keep out of a fight with you. He evidently felt your criticism to be a fatal blow to his chance of a hearing before the committee unless he has some answer to it. I have known Powell for many years and admire many of his qualities. . . . I believe that his future work will profit by the searching criticism to which he is now exposed. I think we have seen an end of certain indiscriminate publications which you have well criticized.[86]

Despite such political and academic setbacks, Shaler's work for the survey along the Atlantic coast and in Florida continued unabated. During the second half of the decade he delivered reports on the seacoast swamps; on the geology of the Cobscock Bay district, Martha's Vineyard, Nantucket, the island of Mount Desert, and Cape Ann; and on the freshwater morasses of the Dismal Swamp region of Virginia and North Carolina. These coastal studies enabled him not only to test his theories on changes of shoreline level but also to increase the scope of his conservation thinking. His proposals for reclaiming inundated lands were thus, in many ways, an extension of Powell's earlier call for a planned approach to the lands of the arid region. Such a utilitarian philosophy was particularly important, Shaler told the readers of *Science,* because of the imminent closure of the frontier and mounting immigration. Before 1890, he predicted, all the best land would be occupied: "There will be no more rich frontier lands ready to welcome the immigrant; therefore the tide of immigration will be turned upon the areas which have been passed in the swift westward movement of our population."[87]

As these comments suggest, Shaler's attention was increasingly directed to matters of social relevance throughout the 1880s. The eugenic quality of the American population in the wake of an unprecedented influx of immigrants was one such concern; the future of university education another. Thus his name soon began to feature regularly in the columns of such literary journals as the *Atlantic Monthly* and the *North American Review,* and his periodic popularizations of geology in *Scribner's Magazine* increased his reputation as purveyor of science to the nation. In 1889 several of these latter pieces were put together in a volume entitled *Aspects of the Earth*. It drew from the reviewer in the *Nation* the observation that "there is probably no American geologist at the present time who is more successful in presenting the facts of geology in a popularly comprehensible and attractive form, and at the same time keeping himself within the bounds of scientific accuracy, than Prof. Shaler."[88]

Of course there were detractors, dismissive of mere "popularizers." Not untypical was a reviewer of Shaler's *Sea and Land* who rather caustically remarked that it was "very possible that babes in science will find [such] well-sweetened pap more digestible than strong meat."[89] In dissipating their energies on bringing science to the public, Harvard professors met their most formidable opponent in Alexander Agassiz. Convinced that the university should restrict itself to matters of pure science and research, he also emerged as a vitriolic opponent of Shaler's later educational reforms at the Lawrence Scientific School. According to Penrose, Shaler's student devotee, however, such criticisms were consistently disregarded by him. As he later reflected:

I was a student under Shaler over a third of a century ago and I can well remember how many smug, self-satisfied workers in science sneered at him for daring to popularize the study of geology and take it out of the cloister of the recluse. Professor Shaler often spoke smilingly to me about this criticism, and after lighting his famous long-stemmed pipe, which reached to his knees, would tell me that the time was coming when such work as he was doing in popularizing geology would become essential in all sciences.[90]

By now Shaler had acquired a reputable niche as a scientific humanitarian among the cultured and lettered men of his generation. It was a place of both respect and influence. And his soaring popularity as a university teacher further added to his image and prestige, particularly among the college fraternity.

In 1888 Shaler purchased a tract of land along the northern shore of Martha's Vineyard. Here, for many years, he spent the summer months away from the bustle of Cambridge life. At "Seven Gates" he found "a 'civil wilderness'—that is, spacious possessions tamed to comfort, but not made artificial, nor yet closely packed with humanity." Whether experimenting with solutions to agricultural problems, exploring the geological formations of the island, entertaining numerous friends, or planting thousands of daffodil bulbs, Shaler felt most at home here—close to nature. As his wife recalled, he "sought by the active use of his intelligence to interpret [nature] truly and lovingly; but he never yielded to the mawkish sentimentality that exalts grass and stones and trees at the expense of the human interest."[91]

As well as giving opportunity for philosophical reflection on "man's place in nature," these summer breaks allowed Shaler time to indulge the poetic side of his temperament. These efforts culminated in a five-volume historical drama entitled *Elizabeth of England,* published in 1903, and in *From Old Fields* (1906), a collection of poems from the Civil War. While neither work has stood the literary test of time, these excursions into creative writing do stand as a significant attempt to straddle the two cultures. In the preface to his eight-hundred-page *Elizabeth,* Shaler testified that his aim was to expose as a misconception the idea that the sciences and humanities were irreconcilable or that reports of Darwin's affective decline could be interpreted as evidence that scientific training necessarily undermined artistic appreciation. This misconception, he insisted, sprang from the similarity between "the poetic and the scientific imagination"; while in Darwin's case, he continued, "a youthful interest in the great literary products of the imagination might be so overlaid by other interests that they would cease to be attractive . . . the essential capacity for such picturing should not be destroyed but augmented by scientific work."[92] But Shaler had another reason for venturing into the realms of literary fiction. He had long thought that knowledge of literature, however scholarly or profound, was no substitute for imaginative writing— the creative act itself. So when, in the wake of the flurry of excitement generated by the appearance of this handsome five-volume boxed set in the bookstore windows of Harvard Square, some literary colleagues expressed their reservations about its fidelity to the Shakespearean mood, Shaler was quick to reply, "It's better than they have done—to date."[93]

Shaler continued as director of the Atlantic Coast Division until 1900 and, during the last decade of the century, regularly presented research findings in the survey's annual reports, on topics ranging from the origin and nature of soils to the geological history of harbors, from the geology of the common roads to the distribution and commercial value of peat. Despite this sustained productivity, however, both the survey and Shaler himself came under critical attack from those outside its ranks. The first skirmish in this renewed campaign prompted a letter of support to Shaler from G. K. Gilbert. "I rejoiced at your interview on the management of the Geological Survey," he wrote. "The whole controversy is bad for science, because with many people its quarrels will make a deeper impression than its achievements." The high scientific standing of Powell, Gilbert went on, had been obscured even by his friends who had concentrated on defending his administrative gifts. "As his scientific writings are almost unknown to the general public," Gilbert concluded, "his scientific reputation has necessarily suffered."[94] But the major setback was yet to come. In 1892 financial support from Congress was cut from $719,000 to $430,000. This drastic reduction reflected both the weakening of the survey's political clout—paradoxically in the face of its growing professionalism and also personal opposition to Powell by western senators disgruntled at his proposed reform of the Homestead Act and at his right to allocate funds without congressional scrutiny. Again Shaler stood squarely behind Powell. He told O. C. Marsh of Yale, for instance, that he was ready at a moment's notice to travel to Washington to assist Powell and, in a short piece printed in the *Engineering Magazine* for 1892, defended the "costly geological survey" by reviewing its accomplishments.[95] And even when the cuts came, he remained optimistic that Congress would soon revoke its decision, for, as he wrote to Eliot, Powell was "a natural born combatant" and "the finest human grizzly bear" he had ever known.[96]

In the current row Shaler himself did not escape criticism—especially from his former teacher Jules Marcou, who had earlier shown his contempt for American geologists in an altercation with Dana.[97] In a privately printed pamphlet entitled *A Little More Light on the United States Geological Survey,* Marcou attacked both Shaler's professional ability as a geologist ("I have given up all hopes to see him an exact observer and a reliable practical geologist") and his activities for the USGS as a "politico-scientist" interested only in the "division of spoils."[98] Powell too came under the whiplash of Marcou's tongue, but in all, Berg's contention that "Marcou's charges represented an attitude of a disgruntled outsider" seems justified.[99]

Powell's friends quickly rallied around him. But their efforts proved

ineffectual in the face of sustained opposition from Senators Edward
O. Wolcott of Colorado and Joseph M. Carey of Wyoming. Clouded
by disappointment over this lack of confidence, Powell began laying
plans to ensure the appointment of Charles D. Walcott as his successor.
On 14 May 1894 he tendered his resignation.[100] Under Walcott's
administration the survey became increasingly concerned with conservation policy, the fostering of independent scientific agencies, and
meeting the more general economic and educational needs of the
nation. All these developments accorded well with Shaler's outlook,
and as with Powell, he developed a firm friendship with the new
director, inviting him to spend a vacation at "Seven Gates" on Martha's
Vineyard.[101]

As in the 1880s, Shaler's survey work during the final decade of the
century provided material for the further popularization of geology.
Sea and Land (1894) outlined aspects of the physical and human geography of coastlands and was generally well received.[102] *American
Highways* (1896), which partially grew out of his service for the Massachusetts Highway Commission, provided a brief review of historical
and contemporary methods of road construction and of the relationship between government and transport. It too was warmly received,
not least because nontechnical accounts of the subject were conspicuous by their absence.[103] Rather less successful was *Outlines of the
Earth's History* (1898), not because of its purpose or orientation, which
was to convey the ideas of continuity and change in nature, but rather
because of its idiosyncratic treatment of theories of the earth's origin
and because of its poor organization. It was bitingly censured by I. C.
Russell, who dismissively relegated it to the "yet small library of
nature-novels," while H. R. Mill, writing in *Nature*, questioned "the
wisdom of piecing together portions of discussions unequal in degree
of detail with the object of showing the uniformity and continuity of
natural processes. There is a want of some more definite coordinating
idea," he went on, "such as would be supplied by considering the
progressive evolution of the world and its processes from the condition
of a raw planet up to its completion as the home of man."[104]

During the late 1880s and 1890s, it should be noted, Shaler was
more preoccupied than ever with questions of a more cultural character: the problems of race and immigration restriction, the organi-

zation of education, the interpretation of American history in its geographical context, and the integration of science and religion. While he did present papers on geomorphological themes to the Geological Society of America, he was simply unable to keep abreast of contemporary research in mainstream geology. Moreover, from 1891 he had been dean of the Lawrence Scientific School—a date that may be conveniently isolated as inaugurating the final phase of his life.

Dean of the Scientific School: Prophet and Patriarch

The Lawrence Scientific School was established in Harvard University in 1847 by a gift of fifty thousand dollars from Abbott Lawrence, an industrialist and philanthropist deeply concerned to promote applied science in eastern Massachusetts.[105] The school that bore his name made the first provision for postgraduate science teaching in Harvard College. Under the charge of a separate faculty (which included some members of the Harvard College faculty), courses were offered in chemistry by Eben Horsford, in zoology and geology by Louis Agassiz, and later in civil engineering by Henry Eustis. All newcomers to Harvard, these men attracted many students who later earned distinction, among them Shaler, Hyatt, Davis, Le Conte, A. Agassiz, and Putnam. The prestige of the school, moreover, was enhanced by the presence of Asa Gray, Benjamin Peirce, and Jeffries Wyman as faculty members. There certainly could be no doubt as to the eminence of such a teaching staff; but some raised questions about the academic quality of the students. This criticism came largely because the initial policy of keeping the bachelor of science degree a post-graduate qualification was soon abandoned as students without an arts degree began to be admitted to the school. To such critics the Lawrence Scientific School seemed to provide a back door to a Harvard education.

By the time Eliot assumed the presidency of Harvard in 1869, the Lawrence Scientific School had already witnessed the coming and going of some five deans; Everett, Sparks, Walker, Fenton, and Hill had each served for periods of between two and seven years. With Eliot's appointment its status was even further eroded. Certainly he believed in the value of applied science, but he had no real concern for the school and made several abortive attempts to merge it with the

Massachusetts Institute of Technology. He felt it simply could not compete either with MIT or with the Sheffield Scientific School at Yale. So by 1886, when Winfield S. Chaplin, strongly supported by Shaler, succeeded Henry L. Eustis as dean, the number of registered students had dropped to 14. Under the new regime, however, things quickly began to change, and by the time he left five years later to take up the chancellorship of Washington University at St. Louis, the number of students had risen to 118.

Shaler was fifty years old when he was appointed Chaplin's successor in 1891. He took up the post with characteristic enthusiasm, fired by his own vision of transforming the school into a great scientific institution. The first step toward the crystallization of this heady dream was realized when he managed to infect the industrial magnate Gordon McKay with a zeal for scientific patronage. Shaler had known McKay since 1865, when their friendship grew out of a mutual interest in mining technology. Ever since those early days McKay had constantly sought Shaler's advice on the best uses of money for public service and now at last in 1891 had consented to make the scientific school his beneficiary.[106]

Certainly conditions at the school were more healthy than they had ever been when the mantle of leadership fell on Shaler's willing shoulders. Under his direction, the school went from strength to strength. By 1902, enrollment had risen to 584 students—a fivefold increase in just over a decade. Shaler himself believed that the school had flourished because it afforded students the rare opportunity of supplementing technical training with literary culture.[107] A more likely reason was the reorganization of Harvard during the years 1889–91, which permitted college students to take courses at the school. In practice, therefore, the only difference between students at the two institutions was that entrants to the scientific school did not require Latin. Shaler's own outstanding gifts as a teacher doubtless also contributed to its meteoric rise in popularity. He had that rare talent of making students feel that his love for nature was really theirs. According to W. M. Davis, Shaler's lectures were like the canvas of an impressionist, painted in bright colors, drawn from wide experience, sprinkled with anecdote, garnished with fascinating asides, and all presented with the desire to make the study of geology comprehensible. "It was never Shaler's idea to make his subject difficult," Davis reflected, "but always and by intent to make it as easy and as interesting as possible; nor did he address his classes as if they were all to become geologists; he knew well that most of them would be in all sorts of other occupations, and he therefore showed them how geology would follow wherever they went."[108]

The welfare of the school was to remain Shaler's passion for the rest of his life. All in all he gave it some twenty-three years' service without a single sabbatical vacation. He implemented, for example, several schemes to attract new students, such as compiling mailing lists of potential students, widely distributing copies of the prospectus, holding an annual dinner, and publishing a small quarterly journal on the work of the school. He instituted many other changes, too. He revised the entire record-keeping system, took measures to ensure sanitary conditions, and drew up a plan for providing student financial assistance. Fastidious in administration, he neglected neither the pastoral side (he regularly visited the Stillman Infirmary) nor the social dimensions of hospitality.[109] Of course, he was not one to suffer the foolish or lazy gladly; indeed he is said to have advised one student to leave the school on the grounds of "miscellaneous worthlessness."[110] But he did love his students and was generally prepared to be tolerant to a fault. More often than not he championed their cause against the whims of the Harvard faculty. This dedication even extended to his choice of residence. For many years the Shaler family occupied 25 Quincy Street in the confines of the Harvard Yard, for here, he believed, he could best get alongside his students. His success in dealing with them and in building up the departments under his control, he told Eliot, "has been due to my residence in the College Yard, and of the opportunity this afforded me of immediate contact with my charges."[111]

Shaler's tenure of the deanship did nothing to diminish his almost feverish vitality. He was writing extensively now, and books and articles flowed freely from his fertile pen. His celebrated *Nature and Man in America* appeared in 1891, just as he took up his new appointment. The outgrowth of his 1888–89 Lowell Institute lectures "Geographical Conditions and Life," this book was destined to become one of his most popular works, reappearing in new editions until well into

the twentieth century. The transatlantic transfer of Old World institutions, the eastward range of the bison, prairie homesteading, the
"Great American Desert," folk tillage traditions, the uses of anthropometric and actuarial data to classify population groups, and aesthetic
responses to landscape were just some of the topics Shaler chose to
illustrate the interplay of the social and natural worlds.[112] They were
to remain dominant motifs in the American cultural geography
tradition.

That same year, 1891, Shaler was also invited to deliver the Winkley
Foundation Lectures at Andover Theological Seminary. They had been
established to keep theology abreast of more general intellectual currents, and Shaler's course was announced as dealing with "Modern
Science and Religious Beliefs."[113] His previous writings on evolution,
society, and politics and his Harvard Divinity School lecture on the
natural history of morality had already revealed his preliminary efforts
to delineate man's place in the natural order,[114] which made him a
natural choice for the series. *The Interpretation of Nature*, as the published version was entitled, encapsulated Shaler's current integration
of the transcendental and the empirical, or, in his wife's words, "poetic
imagination and scientific research." For Shaler, this integration was a
way of seeing that grew to full maturity in such later works as
Domesticated Animals, written, according to one reviewer, "in a far
more philosophical manner than is customary in such a subject,"[115]
and *The Individual, The Citizen,* and *The Neighbor,* a social trilogy
issued in the early years of the new century. Not surprisingly, when
Harvard conferred on him an honorary LL.D. in 1903, it was for his
contributions as a "naturalist and humanist." Of course this effort to
reconcile nature, humanity, and God, while increasingly evident in his
latter years, had been a lifelong concern and was reflected in that tinge
of cosmic sadness that his wife could detect even in his youthful
writings:

It was not only when his mind travelled into the realm of poetry that time
and eternity, the vast and the mysterious, were present with him. These
phenomena ever underlay his thought, and because of this solemn undertone
often there was a note of pathos in what he wrote—markedly so in early
manhood; and later his pen continued to obey the dictation of a subconscious

sadness, a sadness which in a measure was lost in the world of activity, where he practised a self-sufficing stoicism, but which, in the world of reflection, showed itself continually.[116]

These reflective moods, however, were interspersed with periods of intense activity, and while he held the office of dean, Shaler devoted himself to numerous public services. An active supporter of the Democratic party at both local and national levels, he was an enthusiastic worker on its behalf among voters and delegates alike.[117] At various times too he served as commissioner of agriculture for the state of Massachusetts; as a member of the Topographical Survey Commission, the Gypsy Moth Commission, and the Massachusetts Highway Commission; and as a vice president of the Immigration Restriction League of Boston and of the Massachusetts Society for the Promotion of Good Citizenship.[118] These latter activities in particular highlight Shaler's anxiety over the apparent deterioration of America's national efficiency during the heaviest immigration in the country's history. Thus the genial Harvard professor who welcomed every freshman to his table, regardless of nationality, was among the first to call for the lowering of the sluice gates on the flow of American immigration.[119] Not that this position implied any abdication of humanitarian principles. With rationalizations buttressed by contemporary social and scientific orthodoxy, he saw, like many other northerners, no incompatibility between the politics of racism and subscription to the creeds of democracy.

By now Shaler had acquired an almost patriarchal status at Harvard. The lecture hall was packed to capacity when he spoke; invitations to speak at colleges, schools, societies, and clubs flooded in; every type of organization, from the International Congress of Religions to the Military Historical Society of Massachusetts, appealed for addresses. In addition he served as president of the Geological Society of America in 1895 and was again hosted by the Lowell Institute in 1902–3 as guest lecturer on the topic "Dynamic Geology." Small wonder that William James wrote to him in 1900, "I have thought of you almost daily since July 15th, 1899, when I left home, and longed for your inspiring presence, as the myriad-minded and multiple-personalitied embodiment of all academic and extra-academic *Kenntnisse* and *Gemüths bewegagen* [*sic*]."[120]

The combination of this heavy, though largely self-imposed, workload and a quarter of a century's uninterrupted teaching, prompted Shaler in 1903 to lay plans for what was to be his last trip to Europe. But while the Mediterranean held out hopes of refreshment and rejuvenation, his vacation absence was to bring one of the most bitter disappointments of his life. Departing in January 1904, Shaler visited Egypt, Greece, Corfu, Sicily, and many other centers of Mediterranean history and culture. But the trip had to be cut short when he heard of the situation that had developed at the scientific school.

In May 1904, Eliot had received a proposal from President Pritchett of MIT that technical education in New England could be achieved more rationally by combining the efforts of the two institutions. With opposition to the Lawrence Scientific School now quite strong within the Harvard faculty, a plan to divert the McKay funds to MIT was devised, ironically with support from Abbott Lawrence Lowell, grandson of the scientific school's founder. Shaler was surprised, bitterly disappointed—and angry. He vehemently embarked on a campaign of support for the school, in which he castigated the immorality of the proposed misuse of the McKay endowment and vigorously disputed the suggestion that there was any unnecessary duplication of courses. But despite the efforts of Shaler and his friends, the Corporation of MIT and Harvard's Board of Overseers voted in favor of the merger. Shaler was embittered. And although legal technicalities prevented the amalgamation from ever taking place, the days of the school were numbered. The whole affair precipitated an inquiry into educational reorganization within the university, and after prolonged negotiations, Harvard's faculty eventually approved a new scheme of science instruction in March 1906 that brought about the establishment of the Graduate School of Applied Science. Shaler's final contribution to education at Harvard was his involvement in formulating the details of that project.

Shortly after the final approval of the plan, Shaler, who had been suffering what he thought was a bout of indigestion, walked one afternoon to Corey Hill in Brookline to visit a sick friend. As he left for home, the newly fallen snow had already hardened into an icy

Shaler in his middle years. Courtesy of the Harvard University Archives.

crust. When he reached the Harvard Yard he was in a state of exhaustion. The doctor was called immediately, and the following morning, March 24, an operation for appendicitis was performed. At first everything seemed to go well, but toward the end of the week pneumonia set in quite suddenly and unexpectedly. Another week passed, and he seemed to be improving. But by the following Monday the pneumonia had spread to his other lung, and he died on the Tuesday afternoon of 10 April 1906. Writing the next week in the *Boston Transcript*, W. M. Davis reflected, "It was as if all Cambridge held its breath when the news came of the last strategy through which death attacked him, but he fought on, even after the doctors had lost hope. And so we lost him, overcome in mortal combat, but Shaler to the end."[121]

PART TWO

Underpinnings

Evolution: Foundation and Framework

3

> He was a favorite pupil of Agassiz, to whom throughout his life he was devoted, although after graduation he soon broke away from his preceptor's anti-Darwinianism to uphold the theory of evolution.
>
> —Merrill and Dobson, "Shaler"

Nathaniel Shaler was a prominent member of the first generation of Americans to face the challenge of the Darwinian revolution. Like many of his peers he found Darwin's mechanism of natural selection compelling—just as compelling as the idealist natural history practiced by his teacher and mentor Louis Agassiz. So it is not surprising that neo-Lamarckism's idealist construal of the evolutionary *Weltbild* best suited Shaler's scientific needs.

A contextual examination of Shaler's evolutionism is important, I believe, for two reasons. First, because it will bring to light some of those specific difficulties in Darwinism that prompted many Americans to reinvoke the spirit of Lamarck's evolutionary system. To that extent Shaler provides a useful focal point for assessing the impact of evolution on conventional thought in late nineteenth-century America. And second, because Shaler was one of American geography's earliest forebears to espouse the new biology, his synthesis illustrates the way by which American geography received its initiation into the idea of evolution as a fundamental methodological precept. Shaler's evolutionism in turn may help to explain why it was the Lamarckian version, with its strongly environmentalist bias, that was implicitly assumed, on occasion explicitly adopted, by many geographers until well into the twentieth century.[1]

Shaler's friendship with Agassiz, as we have seen, was both deep and lasting, and it served to reinforce the earlier idealism he had imbibed from Escher. Through Agassiz in America and Owen in Britain the metaphysical idealism of the German Naturphilosophen Schelling, Oken, and Döllinger was most conspicuously translated into scientific practice. In different ways they sought to relocate divine design in nature's orderly relationships rather than in the individual adaptations of Paley's utilitarian natural theology. But while Owen did concede that the pattern of species might be understood in terms of natural law involving transmutation, Agassiz sternly set his face against all evolutionary interpretations.[2] Certainly Agassiz could detect a progression in the fossil record, but its true significance, he insisted, could only be understood in the light of embryological development. Just as the goal-directed path of the embryo through a hierarchical series of classes ultimately led to a perfect finale, so the pinnacle of nature was achieved in the creation of the human species. The processes and patterns of embryology were thus mirrored in the sequence of fossil forms. For Agassiz, as Bowler has put it, "Progress is not a response to changing conditions, nor is adaptation the best indication of the Creator's intelligence. Progress is a transcendental symbol of man's unique position in the world, showing him that he is the final goal of a carefully preordained and harmoniously structured plan of creation."[3]

Darwin's theory of evolution by natural selection posed a massive challenge to Agassiz's system of nature, as indeed it also did to the progressivist evolutionary speculations of Hegel, Comte, Goethe, Lamarck, and his own grandfather Erasmus Darwin. For an integral component of the Darwinian revolution, according to Ernst Mayr, was its rejection of any inevitable or automatic upward evolution. "Every evolutionist before Darwin," Mayr writes, "had taken it for granted that there was a steady progress of perfection in the living world. . . . Darwin's conclusion . . . was that evolutionary change through adaptation and specialization by no means necessitated continuous betterment."[4] Of course, as Robert Young has shown,[5] Darwin could on occasion be rather ambivalent about progress; the *Origin of Species*, for example, concludes by affirming that, "as natural selection works solely by and for the good of each being, all corporeal and mental endowments will tend to progress towards perfection."[6] Nevertheless, where any idea of progress appears in Darwin's writings, it seems to occupy the role of a secondary effect rather than an evolutionary goal, which suggests that the conflation of the ideas of evolution and progress in the contemporary consciousness owes a good deal more to the polemics of Spencer than to Darwin's incidental asides.[7] The overwhelming array of empirical data that Darwin had marshaled into a coherent framework of theoretical analysis thus distanced his efforts from the more purely speculative versions of his forebears, even if it was on occasion interpreted as legitimizing trends already present in the social thought of the period.

Darwin's work, it must be stressed, did not immediately demolish the entire fabric of idealist natural history in the Agassiz vein. For while Agassiz found in German romanticism grounds for repudiating Darwinism, many of his students who came to accept evolution incorporated into it elements derived from their teacher's idealist philosophy. Most conspicuously, through the continued fascination of American neo-Lamarckians with the analogy between evolution and embryology, the idealist legacy persisted in post-Darwinian natural science.[8]

Because Shaler found both the idealism of Agassiz and the evolution of Darwin attractive, his first presentation to the Boston Society of Natural History, reporting his research on brachiopoda, completely satisfied neither his teacher nor his own nascent disposition for Darwinism. The paper was originally intended to disprove the Darwinian theory by showing that several of the species had developed features that, far from having any utility or survival value, were actually disadvantageous. The specific point at issue concerned the prime importance of bilateral symmetry to the brachiopoda phylum. In his analysis of two of the orders, Shaler pointed out that any possible transition between them would require a series of intermediate forms, each negating this fundamental principle. This conclusion pressed him to a final rhetorical question: "And must we not, therefore, conclude that the series which united these two orders is a series of thought, which is in itself connected, though manifested by two structures which have no genetic relation?"[9]

At this point Shaler's very vocabulary exudes the spirit of his mentor's transcendental idealism, and it is not surprising that, taken over all, the study greatly pleased Agassiz. Nevertheless he was less than enthusiastic about Shaler's treatment of the Darwinian element, for while doubts had certainly been expressed over the efficacy of natural selection to explain the patterns, "there was a suspicion of heresy in the way the matter was treated." Doubtless this was symptomatic of Shaler's imminent conversion to evolution. Shortly after graduation he symbolically sought out Asa Gray; as he later recalled, "My grateful acceptance of Darwinism was a bond between us."[10] As I have already suggested, Shaler's evolution was, however, far from purely Darwinian. His version of the theory therefore cannot be contemplated in isolation from the prevailing socioscientific milieu in post–Civil War New England. In particular, the powerful "American School" of biology, associated with several of Shaler's peers, profoundly influenced his thinking on organic evolution and its application at the mental, moral, and social levels. To these we will now turn.

Organic Evolution

By the end of the nineteenth century, evolution as a fact of life was rarely doubted within the world of scientific scholarship. If he had achieved nothing else, Darwin at least had presented a superabundance of empirical evidence to support the evolutionary case. The mechanism he had advanced to explain evolutionary history, however, was quite another matter. Specific theoretical and empirical difficulties in the formula soon came to light. These have been examined in detail elsewhere and need only be summarily itemized here.[11] Among the most serious was Lord Kelvin's challenge to Darwin's assumption that the length of geological time available for organic change was almost limitless. Kelvin's much shorter physical estimates for the age of the earth, later supported by Fleeming Jenkin, presented Darwin with what he confessed was perhaps the gravest criticism advanced against his theory. Jenkin was also associated with another challenge. Given the current state of knowledge about heredity, he confidently pointed out that any new feature developed by an organism would be "swamped" within a very few generations by being blended into the surrounding stock. Then too the signal absence of intermediate fossil forms in the paleontological record was a rather embarrassing obstacle to Darwin's idea of gradual species transformation. And perhaps most startling of all was the suspicion that, despite the title of Darwin's book, it really gave no explanation for the *origin* of variations, much less of species. Natural selection could certainly account for the survival of a new feature once it had developed, but that was not the same as explaining why or how variations arose in the first place. In the United States this point was pressed with some tenacity by, for example, William North Rice and Francis Bowen. "Natural selection can originate no new function," quipped Rice, thereby supporting Bowen's earlier affirmation that "natural selection can operate only upon races previously brought into being by other causes."[12] Darwin certainly felt the sting of this censure and confessed to Huxley that he was "greatly troubled" by it. If, he pondered, "external conditions produce little *direct* effect, what the devil determines each particular variation?"[13]

Suddenly the new field of evolutionary theory was wide open to all comers. Alternative proposals soon proliferated in the attempt to out-Darwin Darwin. And undoubtedly one of the most significant was the neo-Lamarckian version forthcoming from what is now conventionally called the American School. The leadership of the movement is generally traced to Alpheus Hyatt, Edward Drinker Cope, and Alpheus Packard.[14] Hyatt and Packard were students of Agassiz, and all three reflected the pervasive influence of Agassiz's recapitulationist analogies. Through their editorial care of the *American Naturalist*, they provided a forum for publicizing their distinct brand of evolutionism. Initially what united these naturalists, however, was not their predilection for Lamarckism but their simultaneous discovery of the embryological law of acceleration of growth. Using embryological development as a model for promoting a linear conception of evolution, their law explained evolutionary change as the addition of new stages onto individual growth. When the normal process of growth was accelerated—so that some characteristics of adults became embryonic in the next generation—there was the possibility of new features being developed. Their search for the cause of such "terminal addition" led them toward the Lamarckian formula of the inheritance of acquired characteristics. Their original proposals, however, did not envisage any adaptive element in the formation of new structures. At this point they remained thoroughly loyal to the old idealist conception of orderly, predetermined development. But gradually they came to allow more and more scope to the determining role of the environment in inducing hereditary modifications.

In Hyatt's case the principle emerged from his work on fossil shellfish. While he certainly followed the path toward a Lamarckian account of adaptive evolution, he remained convinced that Agassiz's recapitulationist theory had been given evolutionary expression in Haeckel's law of biogenesis.[15] His own theory of racial senescence, which involved the idea of inevitable degeneration, introduced a more somber note of pessimism than is usually associated with the movement. Cope came to the topic via the study of fossil amphibians. In the first instance he was prepared to leave the ultimate cause of evolution in the hands of the Creator and therefore saw his first paper as a contribution to theistic evolution. But he soon came to accept the Lamarckian process of use-inheritance in response to environmental challenge as the key mechanism. Thus, while Cope never quite exorcised his fascination with the idea of predetermined sequences, his work did accord prime importance to the influence of geographical environment. His *Primary Factors of Evolution* (1896), for example, drew extensively on the work of Joel Allen, who explained ornithological variation in North America as direct responses to the continent's regional geographies. This approach, of course, was never intended to deny the role of natural selection, for, as Cope summarized his position in 1872, "intelligent choice, taking advantage of the successive evolution of physical conditions, may be regarded as the *originator of the fittest,* while natural selection is the tribunal to which all the results of accelerated growth are submitted. This preserves or destroys them, and determines the new points of departure on which accelerated growth shall build."[16] As for Packard, his experience as a field naturalist impelled him along the route from Agassiz-type "laws of growth" to a rather more environmental construal of Lamarckism. More than any of his peers, he consciously set out to revive, if only selectively,

Lamarck's own views, authoring a full-scale biography of Lamarck himself in 1901 and coining the term "neo-Lamarckism."

From these beginnings neo-Lamarckism rapidly spread among several distinguished American naturalists, among them the paleontologist and explorer William Dall, geologists Clarence King and Joseph Le Conte, the geneticist Henry Fairfield Osborn, the embryologist John A. Ryder, as well as Joel A. Allen, curator of birds at Harvard's Museum of Comparative Zoology, and John Wesley Powell, director of the United States Geological Survey.[17] Certainly they all exploited the basic Lamarckian principle in diverging ways, but whatever the internal differences within the fledgling movement, they united in their appeal to the doctrine of the inheritance of acquired characteristics. Fundamentally they attacked natural selection as the sole, or even the primary, factor in evolution, arguing that Darwin had overemphasized its power and had overlooked the more basic question as to the origin of variations per se.[18] Natural selection was a secondary mechanism that came into operation after some new feature had developed through direct environmental modification.

The prevalence of American neo-Lamarckism at the end of the nineteenth century may therefore be explained in several ways. Primarily it seemed to solve some of the most intractable problems facing the classical Darwinian proposal. The inheritance of acquired characteristics, for example, necessarily entailed a more rapid evolutionary tempo than orthodox Darwinism and therefore seemed more compatible with Kelvin's physics. The swamping difficulty could also be circumvented by emphasizing the direct environmental impress on the generality of local organic populations. It had no need, moreover, to posit those "chance variations," on whose prior existence evolution by natural selection relied entirely.[19] And the fact that some of Darwin's later revisions could readily be given a Lamarckian gloss only seemed to reinforce the point.[20] Besides, Lamarckism was wholly compatible with the more general tenets of developmentalist thinking; organicism, organic change through time, the plasticity of species, and organic adaptation to natural milieu were all as intrinsic to Lamarckism as to rival evolutionary systems. On the religious front the implicit teleology inherent in the linear Lamarckian version, when construed in terms of idealist natural theology, could help obviate the sharpest conflict between religion and secular evolutionism.[21] Culturally, too, the spirit of Lamarckian biology could be harnessed in the cause of social progress and therefore could provide the force behind various schemes for improving the condition of humanity, whether by education or by environmental management.[22]

Shaler's evolutionary thinking places him unequivocally within the orbit of the Lamarckian revival. Conventionally he had been seen as one of America's first scientists to accept Darwinism and as the source of its contagion at the Museum of Comparative Zoology after Agassiz's death.[23] But he always had quite radical reservations about the universal efficacy of natural selection. Thus, in his 1872 article "The Rattlesnake and Natural Selection," published appropriately in the Lamarckian-dominated *American Naturalist,* he set out to evaluate the typically Darwinian interpretation of the phenomenon. Contrary to his earlier beliefs he now conceded that the rattle did have survival value and that it could not be used to mount a case against natural selection. Nevertheless, he insisted, his readers should "not suppose that because I have boldly followed the lead of the most advanced of the champions of natural selection that I am convinced of its sufficiency as an explanation of the great diversities which exist among animals or of its being sufficient basis for an explanation of the snake's rattle." Admittedly he confessed to "having been driven step by step from a decided opposition of the whole theory and compelled to accept it as a *vera causa*, though . . . one much more limited in its action among animals than Mr. Darwin believes."[24] At most this was a heavily qualified endorsement of Darwinism. Shaler's treatment, incidentally, did not go unnoticed by Darwin himself. In *The Expression of Emotions in Man and Animals,* he distanced himself from Shaler's theory that the rattle had been developed "by the aid of natural selection, for the sake of producing sounds which may deceive and attract birds, so that they may serve as prey to the snake." It could certainly serve this purpose on occasion, but Darwin was more inclined to see the rattle "as a warning to would-be devourers."[25]

In another 1872 contribution, this one on the origin of the domestic cat, Shaler rather more clearly displayed his Lamarckian sympathies. In order to locate the geographical origin of the species, Shaler cataloged a number of color changes that suggested reversion to the *Felis catus* L. of western Asia and eastern Europe rather than to the *Felis maniculata* Rüppel of North Africa. In tracing this lineage, he incidentally rejected a natural selectionist interpretation in favor of the environmental mechanism associated with Lamarckism. "It is more reasonable," he argued, "to regard the color as correlated with a general physical condition which adapts the whole organization to great exposure in a vigorous climate" than to ascribe it to the "action of natural selection operating by giving these creatures an advantage in the pursuit of prey."[26]

Any lingering tentativeness as to the viability of the Lamarckian model was soon dispelled. In 1893, for example, a typically Lamarckian insistence on direct organic adaptation, the inheritance of acquired characteristics, and a progressivist construal of evolutionary history characterize what may be taken as the locus classicus of Shaler's evolutionism: "They [organisms] adapt themselves in an immediate manner to the peculiarities of their environment. Those conditions which surround them make an impression on their bodies which is transmitted to their progeny, and these influences, accumulating from age to age, become the precious store of influences which lead organisms ever upward to higher planes of existence."[27] As with the other neo-Lamarckians, of course, Shaler's views here represent in no sense an uncritical return to Lamarck's original scheme. So when Shaler conceded in 1902 that Lamarck's own theory "was evidently not applicable to plants and to animals, if at all, only in a very small measure," we need to be clear that his objection was to the supposition "that species change their shapes because of a continual effort on the part of living creatures to adopt new habits of life."[28] He had no doubts about the inheritance of acquired characteristics and indeed paraded his unashamed neo-Lamarckism two years later when specifying the essential difference between the organic and inorganic realms. Inorganic matter, he maintained, remained unaffected by environmental conditions. By contrast, "Because they inherit their experience with environment from their ancestors and are much affected by that experience, organic individuals are in their generational succession in constant process of modification."[29]

If Shaler's evolutionary outlook betrays the infiltration of American School Lamarckism, it equally displays a number of independent characteristics. The idea of embryological recapitulation, for example, so crucial to the American School's project, is conspicuous by its absence. A flavor of the "law of acceleration of growth" might seem reflected in Shaler's belief that the pace of evolutionary change increased with every stage of development. "The higher the grade of a creature's development," he affirmed, "the more sensitive it becomes to the conditions of climate and the other circumstances which affect its existence."[30] Or again, his incidental observation that schoolteachers deal "with minds and bodies which have . . . a vastly greater inheritance from brute and savage than from civilized life" suggests the merest hint of the social recapitulationist avowal that primitive races were in an early stage of evolutionary development, rather like that of an arrested child.[31] But the significance of such passing asides may not be overpressed. Certainly the idea of recapitulation may well have been among Shaler's undeclared assumptions, but there is no coherent, conscious, or systematic exposition of the theme throughout his entire writings.

More curious still was Shaler's ready acceptance of the work of Francis Galton, whom he had met during his trip to England in the winter of 1872–73. For the contribution of Galton, rigorously restated by Weismann, brought perhaps the greatest challenge to the Lamarckian mechanism. During the later 1850s Galton—already a well-known geographer—turned from the study of the human environment to human biology.[32] In his early efforts to formulate a theory of heredity, he set out to tabulate the contribution of ancestors to their progeny in terms of physical characteristics and mental aptitudes.[33] The result was the first, though primitive, enunciation of the "law of ancestral heredity." A simple corollary to the long-held principle of "blending inheritance"—that is, "the hereditary mixing of paternal and maternal elements so that characters in the offspring would be mid-way between those in the parents"—this law quantified ancestral influences on individuals in a series of geometric relationships.[34] So far, there was no necessary incompatibility with the Lamarckian mechanism. But as Galton moved toward the construction of his theory of "stirp," the term he coined for "the sum total of the germs, gemmules, or whatever they may be called, which are to be found . . . in the newly fertilized ovum,"[35] he became inexorably anti-Lamarckian: the stirp, or "germ plasm," in Weismann's expanded formula, could not be affected in any way by the environment.[36] It came to be regarded as an almost mystical, immortal substance. So the new theory implied, for example, that "bad" germ plasm could never be improved by a better environment because hereditary features were almost fatally fixed. As a result hereditarianism became vastly more important than before, despite the assaults of Spencer and other neo-Lamarckians,[37] and fueled the fires of a eugenic crusade to ensure the best animal and human pedigrees by the enforcement of a controlled breeding program.

Shaler's laudatory reference to the work of Galton in 1893 and his support for "stirpiculture"—the improvement of the human form by selective breeding—are clear evidence of his enthusiasm for hereditarianism.[38] So the question arises, how could he advocate Galton's eugenics and yet retain an unflinching faith in neo-Lamarckism? How, if at all, did he try to reconcile such diverse views? The only plausible answer to the first question seems to be that he adopted hereditarianism for ideological reasons. Only when discussing the problems of ethnicity and immigration does Shaler reveal his support for Galton's laws of inheritance, which suggests that in the new science of eugenics he found "scientific" justification for his racial policies. As to reconciling the respective influences of environment and heredity, some kind of fusion seems implied in his assertion that the American environment was truly suitable just for members of the Teutonic race, that they alone could benefit from "the stimulus of American air."[39] In other words, only certain races had the appropriate innate biological capacity for adaptive evolution. Such a resolution, of course, would sacrifice cardinal principles in both Lamarckian and hereditarian schemes, and although evidence could be pieced together from his writings to support some such reconciliation, there is little to suggest that Shaler was even aware of the inherent tensions in his own position.

Hereditarianism also displayed itself in Shaler's account of biological reversion, that is, the tendency of organisms on occasion to exhibit atavistic characteristics.[40] In keeping with a broadly progressivist linear conception of evolution, he regarded these regressive or degradational forms as exceptional. But while in general terms he believed that evolutionary "advancement or retrogression appears to be determined by the conditions of the environment," he hesitated to account for these reversionary forms in conventional Lamarckian terms.[41] "Inheritances are impulses derived from the ancestral experience of the organic form," he insisted. "We have to imagine them to be essentially separate motives handed down from generation to generation, after remaining latent for great periods, to become suddenly manifested under conditions the nature of which is not yet discernible."[42] Here, I think, Shaler is claiming that the now regressive trait had originally developed in the ancestor through classical Lamarckian mechanisms; but its reappearance after generations of quiescence is quite a different matter, and Shaler, although pleading ignorance as to the precise cause, is looking to some hereditarian resolution.

It might be expected, moreover, that only retrogression would be explained on the basis of ancestral inheritance, advancement occurring by direct heritable adaptation. Not so. Certainly his explanation of the occurrence of supernumerary digits was thoroughly ancestrian, explicable, as he put it, "on the principle of reversion to the character of a lower ancestral." But, as the following gobbet makes plain, adaptive variation could apparently take place by hereditarian means as well:

We may conceive the organic form gaining from experience and transmitting to its successors a body of divers impulses, of tendencies to variety of form and action which are ever on the watch for chance to manifest themselves. If these acted singly each for itself, the tendency would be to produce mere reversions to ancestral conditions. But, if they operate interactively, if they combine their motives in any way, it may well come about that the changes in structure or function which they cause would be in the line of advancement.[43]

For Shaler, it would then seem, evolution normally follows the Lamarckian path of the inheritance of acquired characteristics; but at least on some occasions, advantageous features may arise by other means. Indeed, in 1905, just the year before his death, he wrote of current evidence to suggest that new forms arose, not by the accumulation of slight variations, but in single, widespread, large-scale mutations. In language reminiscent of Hugo De Vries's recently publicized mutation theory, he proposed that it was "through the occurrence of sudden and great alterations, *mutations* as they are termed," that new species and subspecies evolve. Thereafter, he went on, "they are subjected to the selective process; if they are suited to the environment they abide; if they are unsuited thereto they disappear."[44]

It is interesting that Shaler ultimately felt drawn to this particular theory, for it shows his continued hesitancy over classical Darwinism. In a period when Darwin was in eclipse, De Vries's proposals were widely regarded as a viable means of transcending both Darwinism and Lamarckism. His theory seemed to explain both the *origin* of new species and, just as important, their *survival*, without the isolating mechanisms essential to the Darwinian model.[45] Still, whatever the exceptions, Shaler was sure that the normal course of evolutionary history was by way of Lamarckian adaptation. Whatever the physiological mechanisms of inheritance (he was rather unhappy about Darwin's theory of pangenesis), he remained convinced that the hereditary "units in the protoplasmic mass, can effectively contain and transmit the important elements of experience acquired by millions of ancestors."[46] Organic species, he subsequently confirmed, were in the process of continual modification because they were, so to speak, "educated by their environment."[47]

The tensions and uncertainties apparent in Shaler's treatment of organic evolution would seem to stem from a variety of sources. The lingering influence of Agassiz's natural history, Darwin's own equivocation over the significance of the Lamarckian mechanism, the diverse hypotheses advanced in the pre–Mendelian period to account for the observed phenomena of inheritance, and the apparent ideological implications of Galton's eugenics for national policy doubtless all contributed. Nevertheless his fundamental confidence in the viability of the Lamarckian alternative reveals his own attempt to balance the new science of evolution with the teleological natural history he had learned from Agassiz. It provided the foundation and framework for his interpretation of mental, social, and moral evolution.

Mind and Evolution

It was only to be expected that Darwin's theory of evolution would give renewed impetus to the study of a subject that was already fascinating biologists, namely, the brain, that most mysterious of organs.[48] Darwin himself had foreseen this in his notebook of 1837, and a later observer recalled how, as "evolution came to be the reigning hypothesis among men of science, it was to be anticipated that its central problem, the origin of the human mind, would demand consideration."[49] In the United States one of the most coherent early efforts to provide a thoroughly Darwinian, naturalistic account of the development of human consciousness from elementary antecedents in animal life was made by the Harvard philosopher Chauncey Wright. In his 1873 essay, "The Evolution of Self-Consciousness," he made the case for reflective consciousness as the mere extension of animal instinct and memory, concluding that it was both learned and functional, and therefore not intuitive or metaphysical or divinely created.[50] Yet for many post-Darwinians whose evolutionism was social as well as biological, the Lamarckian version provided the most direct link between intellectual progress and the organic evolution of the mind.[51] This link existed simply because, as Herbert Spencer put it, "Hereditary transmission applies to psychical peculiarities as well as to physical peculiarities."[52]

Whatever the precise mechanisms involved, many naturalists came to accept the idea of a hierarchy of consciousness linking animal and human life together in a single evolutionary system. George John Romanes's empirical investigations of animal intelligence in the early 1880s—despite its anthropocentrism—set the study of animal psychology on a new scientific footing. His theory of instinct, which proposed that animal instincts were at once reflexive and conscious, was subsequently subjected to the criticism and elaboration of C. Lloyd Morgan, who pushed on toward a formal biology of behavior consistent with neo-Darwinian principles.[53]

Alongside these developments in psychology, the notion of a hierarchy of consciousness was also adopted by evolutionary anthropologists, especially those influenced by Spencerian presuppositions. Thus physical anthropologists used anthropometric measurements of crania, brain weight, and so on to support hierarchical interpretations of racial mentality, while those more interested in cultural anthropology came to regard social institutions as indexes of intellectual progress and therefore of mental evolution.[54] For as W J McGee argued in 1899, any mental evolution accompanying society's transition from savagery through barbarism to civilization would be reflected in its technology, philology, sociology, and esthetology.[55] Not surprisingly, language became a new focus of ethnological research; on the one hand, it could be used to ascertain the degree of continuity between human and animal intelligence, and on the other, as evidence of mental evolution. Romanes, for instance, had found the germ of the linguistic faculty in the lower animals;[56] Darwin had suggested that language was partially artistic, partially instinctual, and that its continued use induced heritable modifications of the brain;[57] Ernst Haeckel even maintained that the "remarkable clicking sounds" he had heard made by apes were still to be found in the language of the Bushmen;[58] and John Wesley Powell, also fascinated by the origins of language, found the Lamarckian transmission of culture central to the biological evolution of mind.[59]

In typically Lamarckian vein Shaler sought the mainspring of mental evolution in the broader reciprocal relations of mind and environment. The minds of animals, he explained, were more variable than their bodies, and "by these variations the creatures are impelled in some undetermined measure to new habits of life, which in diverse ways, under the control of natural selection, brings about changes in form and structure."[60] The problem was to identify particular cases where the organic and mental functions clearly interpenetrated to foster new types of environmental behavior. Shaler suggested that physical inefficiency was quite crucial to the development of the human mind. Because the body of the bee, for example, was so perfectly adapted to its needs, its "mind operates in the automatic manner which we term instinctive." By contrast, human beings were forced to use the hand precisely because they had to overcome their poorer physical adaptation to environment. In this way, somatological and psychological development were inextricably bound together: "In a word, if man had been as amply provided with instruments suited to his needs as is the bee, there is reason to believe that the quality of his intelligence would have been no more rational than that of the insects. He has, indeed, won to his high estate because with a vigorous nervous system and corresponding will his ancestry denied him other than the most limited instruments for accomplishing his desires."[61] Shaler's original statement of this proposal dates from his 1872 discussion "On the

Effects of the Vertical Position in Man," presented at the Boston Society of Natural History. But it was left to Powell's protégé, W J McGee, to pursue systematically the theory that organic-intellectual progress was externalized in the functional diversity of the human hand, in what he called the "manifestations of manual dexterity among culti-vated men."[62]

For Shaler, however, this intimate association of bodily structure and mental endowment was not restricted to humans. A particularly close parallel was to be found in the elephant, whose intelligence Shaler attributed to "the fact that the creature possesses in its trunk an instrument which is admirably contrived to execute the behests of an intelligent will."[63] Plainly this view implied that the same laws con-joining physical features and intellectual development governed the evolution equally of animal and of human intelligence. It was a short step to sensing antecedents of the human mind in animal instinct. And so Shaler relentlessly pursued the task of identifying elements of in-stinctual behavior in humans while at the same time detecting traces of consciousness in the subhuman realm. Much of his evidence, of course, was thoroughly anthropomorphic. The behavior of a dog in encountering a long-absent friend, he pointed out, "is so thoroughly human-like, that even the naturalist, the professional doubter, is forced to believe that the dog's mind works substantially like his own, and that the feelings connected with the action are essentially the same."[64] Still, however anecdotal, such declarations do serve to redraw attention to the indissoluble link Shaler perceived between animal and human consciousness.

Given Shaler's belief in this unbroken line of intellectual continuity, it was only to be expected that the evolution of animal and human minds would share other common characteristics. His idea of mental plasticity or elasticity was one such feature. As a metaphor suggesting malleability, adaptability, and responsiveness, it simply implied that the more plastic the mind of any organism, the greater the possibility of intellectual progress. Thus, in the animal kingdom, Shaler could use this criterion to rank the dog much higher than the horse in intellectual development, while at the level of human society he could argue that the "aim of our academic culture at the present time is to make a man of varied, elastic mind, who can readily turn himself to any of the multifarious duties of ordinary life."[65] Mental plasticity was thus a condition of mental evolution and, at the same time, its objective. Again, just as the sophistication of human language mirrored organic mental progress, so, Shaler believed, the quality of a dog's intelligence could be gauged by the complexity and variety of its bark.[66] And then, too, the extent of mental evolution was manifest in the degree of group solidarity, whether animal or human. As we shall see, Shaler's reconstruction of social evolution was closely bound up with the natural history of sympathy and altruism. Mind and morality were evidently interlocked in the history of organic evolution.

For Shaler, it is now clear, the processes of mental evolution that bound together animal and human consciousness were to be accounted for by the interaction of mind, body, and environment. This thoroughly Lamarckian interpretation was thus the antithesis of William James's consistently Darwinian emphasis on the functional value of "lucky variations" in the evolution of the mind.[67] And yet, despite the close parallel, indeed the linear progression, between animal and human mentality, in one crucial respect Shaler believed there was a wide chasm between them. Mankind had attained such a level of intellectual independence that people could now begin to participate in, even guide, the future course of evolutionary history. The idea of mind-directed evolution occupies an important place in Shaler's evolutionary outlook.

While Shaler would doubtless have agreed with T. H. Huxley's insistence on "the impossibility of erecting any cerebral barrier between man and the apes," he was just as certain that the advent of the human mind introduced a quite radical break in the evolutionary process.[68] Thereafter, the human physical form was no longer subject to the workings of natural selection. If a conceptual source for this evolutionary dualism is to be sought, the most likely candidate is Alfred Russel Wallace. He had turned to the question of mind and evolution in 1863 at a meeting of the London Anthropological Society, where he argued that, from the time that "the social and sympathetic feelings came into active operation, and the intellectual and moral faculties became fairly developed, man would cease to be influenced by 'natural selection' in his physical form and structure."[69] For the first time in history, an organism had emerged that did not have to change physically in order to survive in a changing environment, an organism, indeed, that could actually modify its own environment.[70]

Shaler was well acquainted with Wallace's work, and the two-stage evolutionary process Shaler envisaged was supported by reference to him.[71] "There can be no doubt," he insisted in his 1879 article for the *Atlantic Monthly*, entitled "The Natural History of Politics," that "the view suggested by Mr. Wallace, that man has by his civilization in a great measure emancipated himself from the action of natural selection, is essentially true."[72] He departed from Wallace, however, in his resort to Lamarckian principles to account for human evolution, always with the proviso that the inheritance of acquired characteristics now governed intellectual rather than physical adaptations. "All we know of the animal man," he confirmed, "is against the supposition that he is ready to vary in his qualities through the direct influence of his surroundings." Any degree of physical plasticity once possessed by the human species had largely disappeared. "If natural selection played its ancient part," he explained, "environment might perhaps be more efficient in its action on mankind than we find it to be, but this agency has long since ceased to be of distinct value among civilized peoples."[73] To demonstrate the progressive demise of direct environmental influence and natural selection as evolutionary factors in human evolution was perhaps a rather easier project than to elaborate a convincing Lamarckian account of mental evolution. Nevertheless Shaler did do his best to present his thoughts on the central problem, namely, the mechanisms of mental inheritance. His conclusion, like that of Lester Frank Ward, was that it was the *capacity* for knowledge that was passed from generation to generation rather than knowledge per se. So, striving to avoid any reductionist rendering of the relationship between mind and body, he outlined his views as follows:

At first sight the transmission of anything like thought may seem to be essentially more difficult than that of structures alone. We have, however, to believe that the brain, which is the instrument of our thinking, is, to the utmost of our details, determined by inheritance; at least until the individual life has begun to shape it. And even when this independent personality has gone far to give his brain a peculiar stamp, the inherited features must greatly preponderate. Conceiving then that the production of thought depends upon the action of cells or other elements of the cerebrum, it does not appear to be improbable that they may, because of their shape or condition, afford the way to thought such as was the product of the ancestral forms on which they are moulded. It is not necessary to suppose that thought is a mere secretion of the brain cells in order to hold the view just above suggested. We need to do no more than recognize the fact that there is some immediate connection between the state of the mechanism and the thought that proceeds from it. Come whence the thought may, if its coming be in any way the result of the condition of the brain, a particular state of that organ, such as may be altogether due to ancestral influences, may, indeed must, be conceived as giving rise to a definite mental process.[74]

The mind–body evolutionary dichotomy once established, Shaler then proceeded to argue for the power of the mind to direct the future course of evolution. By this means human progress would be both assured and accelerated. The reason was simply that in the "place of the selection of nature will come. . . the selections of intelligence."[75] This idea, of course, was quite in keeping with Lamarck's own belief that, since the human race had acquired supremacy over the rest of the animal world, it would ultimately establish a wide gulf between itself and its nearest animal ancestors.[76] Besides, the general emphasis on will and mind in evolution was promoted by contemporary Lamarckians like Cope, who, in antimaterialist tones, turned to consciousness as the guiding force of organic development,[77] and like Powell, who announced in 1888 that "the genius of man" had transformed the brutal, cruel law of survival "into a beneficent agency for his own improvement."[78]

Now by the late nineteenth century Shaler believed that the most efficient means by which the human mind could wrest the evolutionary initiative from nature was by the practice of the new science of eugenics. Here, through selective breeding, there was a wonderful opportunity of retaining, even improving, human mental quality. A new moral duty was evidently laid on contemporary society to prevent the dissipation of mankind's mental endowment; "the moral and intellectual accomplishments of man afford the most precious heritage which it is the privilege and duty of each generation to transmit to its successors."[79] Not, of course, that such intellectual progress was either immediate or automatic. For one thing, "intellectual selection" was a recent evolutionary mechanism, attained only at the level of civiliza-

tion, and it therefore only gradually replaced the older selective processes. Checks to progress, like starvation and combat, would only
gradually be eliminated.[80] Moreover there was the omnipresent danger
of social evolutionary atavism, most conspicuously manifest in war.
So while such struggle had selective value in the early stages of social
evolution, it was now detrimental to mankind's future progress. "In
noting the fact that natural selection . . . has almost ceased to act in
human society," he told his readers, "the reservation should have been
made that in returning to the primitive conditions of war man once
again encounters those conditions which act there in a peculiar and
most effective way." His hope was that the recent advances in understanding psychological evolution would "afford the occasion for rem-
edying [this] greatest of all human ills, which has been left untouched
by all the benefits which our age has won."[81]

In Shaler's thinking, then, the genetic *continuity* of animal and human
intelligence was balanced by a corresponding *discontinuity* between the
respective selection processes. In human society natural selection was
superseded by intelligence selection, a new evolutionary mechanism
whose powers could eventually be harnessed by a careful monitoring
of future eugenic mixtures. For Shaler, of course, the history of mental
evolution was part of a larger human drama, and the emergence of
the moral sense constituted a vital chapter in that story.

Evolution and Ethics: The Natural History of Sympathy

On the surface, at least, the ethical ideals of humanity seemed to run
counter to the struggle for survival implied in Darwin's theory of
evolution. If the human species was in fact the product of a "nature
red in tooth and claw," then the evident instances of selflessness and
altruism were a mystery. This apparent dilemma soon became the
subject of many, often subtle, treatises on moral evolution. After all,
Darwin himself had confirmed the importance of the subject when
he agreed with "those writers who maintain that of all the differences
between man and the lower animals, the moral sense or conscience is
by far the most important."[82] Among those naturalists determined to
provide an evolutionary rendering of this ethical impulse, the advent
of morality was typically associated with such themes as the devel-

opment of self-consciousness and reason, the survival value of group solidarity, and the emergence of the tribe as a distinct social unit.

In Shaler's case, he was convinced that the problem of the origin of ethics should be located within the more general context of the development of emotional expression. Foremost among the affective inheritances that humanity had derived from its evolutionary predecessors was hatred—an emotion, he pointed out, present in all invertebrates that exhibited what he termed "any distinct mental capacity." It, of course, could readily be explained in Darwinian terms, for it clearly conferred selective advantage in the "unending game of chase and flight" that governed early evolutionary history.[83] With every advance in rational power, primitive men and women came to regard all about them as inimical to the fulfillment of their desires and thus to view their fellows as rivals and beasts as foes. Truly humanity had undergone a long education in the ways of hatred.

Altruism, by contrast, posed a much greater problem, since it involved actions that were the antithesis of personal gratification. Nevertheless, dissociating himself from the presumption of Kant and the pessimism of Schopenhauer, Shaler determined to root the ethical sentiment in the plain, humdrum processes of naturalistic evolution. His conviction that the "motive" could be traced in subhuman species pushed him toward this course; the expression of sympathy was "no peculiar property of the vertebrate animals, in which man belongs, but . . . originates in other series of animals, which have no blood kinship with man."[84]

By contending for this continuous line of sympathetic union, Shaler thereby distanced himself from T. H. Huxley, whose *Evolution and Ethics* was designed to juxtapose the ethical principle and the cosmic process, to displace the human race from the natural order that had produced it.[85] Shaler's standpoint was far more consonant with that of Peter Kropotkin and John Wesley Powell—both Lamarckians— who found sociability and cooperation among the very mechanisms of evolutionary progress.[86] At the same time, he was not prepared to reduce altruism to gregariousness—a social trait interpretable as a mere self-regarding sentiment. Gregariousness was a condition of altruism, not its cause. And its cause, he believed, lay in those formative psychophysical relationships between parent and offspring. In the conscious parental care of their young—a behavioral pattern contingent upon the separation of the sexes—Shaler found "the basis of altruism in all its moods." From this seedling, altruism grew by extension from the immediate family to other members of the species. Among vertebrates it achieved its greatest flowering because here the protraction of parental devotion was necessitated by the prolongation of infancy. "Thus the parental relation," he concluded, "in place of being a momentary affair, as in the lower forms, becomes an element in the emotions of the mother for the greater part of her life. The public owes to John Fiske the first statement of this important truth."[87]

In his *Outlines of Cosmic Philosophy* of 1874, Fiske had displayed his faith in the theory of evolution to supply a stronger theoretical foundation for morality than any other available alternative. To him, morality was a kind of by-product of mental evolution. Like Wallace he proposed that, after the biological development of the brain, selection only remained active in the intellectual realm. Progress in this sphere, he continued, was most efficiently achieved by prolonging human infancy for sufficient time to allow cerebral organization to be perfected before the individual had to fend for itself. During this period, conditions conducive to the fostering of social relationships within the family were brought into being. In their state of helplessness, offspring clearly required parental protection, which in turn necessitated the subordination of personal desires in the interest of others. Egoism would thus progressively give way to altruism. And so, for Fiske, prolonged infancy explained the origin of morality, family, society, and religion.[88]

In Shaler's account of moral evolution, he supplemented and extended Fiske's prolongation-of-infancy principle in a number of ways. First, he did not restrict the mechanism to human behavior. Among other vertebrates a greater duration of infancy meant that the offspring of one year was scarcely weaned before the birth of succeeding progeny. In the case of birds, for example, "The sympathetic motive has its origin in the family, which, in turn, rests on the nesting and incubating habits and on the extended care of the brood after it is hatched."[89] Second, he outlined several accompanying processes that contributed to the extension of altruistic sentiments. The evolutionary significance of cooperation among certain organisms was one that has already attracted our attention. The practice of monogamy was another. Here the parental role and affection of the male was defined and concentrated.[90] Third, at the social evolutionary level of the tribe, other factors, particularly religion, entered both to reinforce and to extend the range of sympathy.

Just as tribal sympathy later gave rise to racial solidarity, so, Shaler believed, in the religious realm "the motive of love towards the gods grows in intensity as well as purity, and is finally gathered into the intense, life-absorbing devotion of the higher monotheism." The extension of the principle did not cease even here. In the most recent past, he continued, "there has been added to this set of affections a love of nature, an affection for the whole range of phenomena, which, though in its beginning, and as yet weak, promises to become, through inheritance and habit, one of the most important elements in the structure of the mind."[91] Throughout these several spheres, moreover, sympathy with fellow creatures, with the Infinite, and with nature was supported first by religious ideals and then by secular state legislation. Clearly, while the origins of altruism might be naturalistically interpreted, there was some need for a religious ethic to be imposed on the natural order if morality was to evolve into a universal principle. Thus any comprehensive historical account of human altruism would need to do justice to the interpenetration of ethics and religion and to the ways in which these were enshrined in the structures of institutional law:

In place of the blind following of the emotions which characterizes the sympathetic movements of the lower animals, we find that even among the most primitive and lowly savages rules of conduct are instituted which serve to direct the ways in which the individual shall act with regard to his fellows. In almost all cases these rules are much intermingled with the religion of the people. . . . As time goes on and the folk attain the stage of records, these rules of conduct become definite laws which at first are based on religious ordinances; but in time they are, in the latest stage of social growth, brought into the state of ordinary statutes which, while they may have some religious sanction, are supported by the machinery of the secular government.[92]

Running through Shaler's account of biology and morality, selection and sympathy, and evolution and ethics were two convictions—one negative and one positive—that enabled him to steer his own course between reductionist perspectives on the one hand and purely super-natural schemes on the other. Primarily Shaler remained quite certain that moral evolution could not be accounted for solely by reference to natural selection. He certainly admitted that the offspring of unions where parental sympathy had emerged had a better chance of surviving than competitors; beyond that, however, particularly at the tribal level, other factors inevitably came into play.[93] Otherwise, he pointed out in an argument by reductio ad absurdum, it would be equivalent to saying that the most self-sacrificing individuals were the most likely to survive in battle! "Such reasoning," he maintained, "reduces the valuable hypothesis of selection to the level of cycles and epicycles."[94]

More positively, Shaler came to believe that the key to the question of moral development might be found in a kind of emergent evolutionary interpretation of biological history. By analogy with the quantitative and qualitative shifts evident in organic and mental evolution, Shaler suggested that such sudden, unpredictable, yet beneficial adaptations could be justifiably extended to the moral realm. This position, of course, did not imply any automatic progress toward some ethical utopia. "In the moral as well as in the physical world," he reminded his readers, "we may see these hidden seeds of ancestral impulse, when no longer overshadowed by the newer and therefore stronger motives, spring into activity, and win the creature back to a lower estate."[95] Still, cautiously optimistic in his faith for the future moral advance of mankind, Shaler ultimately found in his Lamarckian reading of evolution the grounds for an authentically natural history of sympathy. Such progress would be effected through the processes of social evolution.

Evolution and Society

Throughout the nineteenth century, evolutionary accounts of the emergence and development of society were commonplace, even in the pre-Darwinian era. Thus Herbert Spencer, for instance, insisted—as Comte had done earlier—that society had evolved in accordance with a universal law of development. Besides, in comparative law, prehistoric archaeology, and cultural anthropology, parallel evolutionary models were advanced by men like Henry Sumner Maine, Boucher de Perthes, Lane Fox (Pitt-Rivers), Lewis Henry Morgan, John Lubbock (Lord Avebury), and Edward B. Tylor.[96]

The application of pre-Darwinian evolutionary biology to the study of human society and the existence in any case of a substantial body of social evolutionary writings conceived quite independently of biology call into question the legitimacy of the term "Social Darwinism," which, following Richard Hofstadter's classic study, has come to be applied to almost any evolutionary model of society.[97] In its most vulgar form, Social Darwinism has been portrayed as an attempt to justify the cutthroat ethics of late nineteenth-century capitalism, buttressing a conservative ideology in economics and politics alike. Thus the laissez-faire radicalism of Spencer and his two devotees, Edward Livingston Youmans and Andrew Carnegie, and the materialistic determinism of William Graham Sumner that left no room for egalitarian sentiment have been seen as typical of social evolutionism.[98] Such discussion, it is now clear, has obscured the fact that really very few Europeans and Americans adopted social policies modeled directly on the Darwinian struggle; furthermore, social thought could be "biologized" in other ways.[99] One of the most significant of the latter was what could be called Social Lamarckism. In contrast to the non-interventionism of Social Darwinism, Social Lamarckians held that laissez-faire was tantamount to a denial of human creativity—a travesty of past experience and a fatalistic rejection of future possibilities. Civilization to them, as Lester Frank Ward articulated it, was the cumulation of useful social acquisitions by cultural transmission.[100]

Like Ward, Shaler conducted his social theorizing, at least to his own satisfaction, on the principle that the organic and social evolutionary processes should not be conflated, that history could not be reduced to nature. Confessedly chary of directly transferring physical laws to the social realm, Shaler did his best to dissociate himself from both monistic and positivist modes of explanation:

Certain words and phrases which were invented by naturalists to denote matters connected with the history of organic species have been taken over by the students of society, and used to denote actions which they presume to be essentially like those to which the words were first applied. In this way such terms as natural selection, environment, atavism, etc., have become current in fields where their significance is doubtful. In this as in other cases there is always a danger that the imported word becomes what Bentham well called "a question-begging epithet," for the reason that it has a connotation which, though true in its original use, is false in its secondary application.[101]

At the same time, this did not prevent Shaler from plotting his own evolutionary map of human society. Two distinct though related issues particularly engaged his attention, namely, the emergence of society and the subsequent development of social organization. Because of humanity's "liberation from the old selective forces," Shaler saw the evolution of society as grounded not in any mere continuation of organic natural selection but in those reciprocal processes of mental and moral evolution that we have just examined.[102] With a comparatively weak physical constitution, mankind's real strength lay "in his intelligence, his capacity to plan actions in relation to his needs, in his quick sympathy with his fellows, and his understanding of the world about him."[103] From these particular seeds the various forms of social organization eventually propagated themselves.

Intellectual attainment and the expansion of the sympathetic sentiment were thus the twin pivots of social evolution and found expression in two fundamental institutions—property and family. On intellectual capacity and the power of invention depended the manufacture of weapons and utensils. "These and other similar resources which nearly all savages command," Shaler went on, "begin the institution of property, and on this possession of wealth rests, in a large measure, the development of all the subsequent growth of society."[104] Property ownership had long been used by developmentalists (like Adam Ferguson in 1767, for example) as a social indicator of the transition from savagery to barbarism,[105] and Shaler easily incorporated it into a neo-Lamarckian understanding of social and intellectual evolution as mutually reinforcing processes. The development of the family was the other vital factor in society's emergence, since it was the basic unit by which social life was regulated. With the extension of familial solidarity beyond kith and kin to the clan came the second phase of social evolution—"the tribal stage of sympathetic growth"—which Shaler believed had been attained "by all the several species of men."[106] Again, moral attainment was reflected in patterns of social behavior.

These mental and moral antecedents of social evolution, however, were of themselves insufficient to explain the emergence and survival of human society. One crucial factor—inexorably biological in inspiration—that shaped this chapter of the historical record was the role of geographical isolation. Its importance in organic evolution had been recognized not only by Darwin himself but more especially by Moritz Wagner, who saw the migration of new colonies and their subsequent isolation from the main population as prerequisites for speciation.[107] To Shaler the influence of geographical isolation was most marked in the early history of human society. It was a mechanism that operated in two distinct ways.

Initially Shaler was certain that geographical isolation, because of its socioecological role, was indispensable to the emergence of civilization from the tribal stage of social development. The absence of isolation encouraged intertribal warfare and a nomadic way of life that was too precarious for the establishment of settled civilization. Thus in the Americas, for example, the lowest grades of barbarism were only superseded in those isolated tracts of the Cordilleran chain enjoying the seclusion and therefore protection "necessary for the dawn of any culture whatsoever."[108] Indeed, the very "geographic unity" of the United States had prevented it from becoming the "cradle-place" of any civilization—a marked contrast to northwest Europe. The complexion of the isolation mechanism, however, changed quite dramatically once the social level of civilization had been attained. Now it became a severe handicap, for it prevented the cultural interaction necessary to further progress.[109] The isolation of European peasant stocks, for instance, had made it "impossible for them to develop any political quality whatsoever."[110] So the very features of North American geography that had hindered development in the early stages of human evolution became the qualities now promoting social advance.

Fundamental to Shaler's conception of the subsequent evolution of civilization was his emphatic rejection, now perfectly predictable, of natural selection. If, at the lowest stages of barbarism, it could be viewed as preserving stronger over against weaker tribes,[111] in the final stages of social evolution, competition was thoroughly supplanted by cooperation. Where "societies were small and not domiciled by agriculture, the struggles between them led by a process of natural selection to the survival of those which most effectively developed the war-like motive." But beyond that, war had an "inverse selectional effect." The simple reason was that, in the modern state, unlike the primeval tribe, only a portion of the population participated in military campaigns. Since those engaged in armed combat constituted, in large measure, the most physically and intellectually vigorous of the generation, war was certainly "a very perfect system of selection, but a system that makes for the decay of the society's strength and even more for the extinction of its valor." It lowered the eugenic quality of a population (by its waste of germ plasm); it checked its rate of growth; and it brought both educational and social deprivation. So whereas thinkers like Bagehot unilaterally praised the evolutionary value of conflict,[112] Shaler saw "the insanity of war" not as something to be glorified but as a shameful relic of savagery, an obnoxious exhibit of social atavism. As he summed up his reflections on the natural history of war in 1903:

So long as war was an agent of natural selection, and there was a survival of the tribes that were the most valiant and the most skillful in the use of arms, it served the same rude purpose as like struggles in the lower life. It left the strongest in possession of the field. As soon, however, as the tribal stage is passed, when in place of the little social unit that may be readily swept away we have the larger enduring state, such as may suffer from war, but if vanquished is not swept away, all or nearly all the possible benefits arising from these contentions disappear, and the action, though selective, leads not to advancement but to the degradation of man.[113]

Like other Lamarckian advocates of solidarism and cooperation— notably Ward, Fiske, and Powell—Shaler was critical of the laissez-faire policies advanced by some Social Darwinian publicists. By contrast, he regarded the preservation of "the weak with the strong, the vicious with the virtuous, the fool with the philosopher" as one of society's most noble practices.[114] Again, he turned to education as the most effective mechanism for ensuring future progress; it was able to select the fittest in a much less wasteful way than nature and at the same time to transmit its benefits to succeeding generations.[115]

Besides these, there were other currents of thought influential in Shaler's social theorizing. These are most conspicuous in his evolutionary portrayal of state legislation, which reflected the joint influence of post-Enlightenment progressivism and the American pragmatism of Royce, Peirce, and James. To the eighteenth-century progressivists, with their naturalistic bias, the practice of law was rooted not in the authority of the imperial order but in the social realities of everyday life. Its mainspring was the coming of property ownership, since laws were primarily enacted to protect life and private possessions. So only when the emergence of a pastoral economy fostered ownership of stock and when, with settled agriculture, the disposal of land by private testament displaced the earlier practice of usufruct were the institutions of civil law established.[116] This functional interpretation was later pursued by the pragmatic evolutionists for whom legal procedures were "the cumulative social product of practical decisions."[117] To them the law was simply a societal tool for resolving conflicting aspirations in the human struggle for existence. Consistent with their philosophical stance, they showed how the processes of cultural and biological evolution could be integrated by viewing society as an organism engaged in environmental adaptation. By regarding the law as the progressive institutionalization of established social traditions, they laid the basis for modern legal realism and, in keeping with the historical studies of Henry Sumner Maine, saw it in dynamic-functional terms as an evolving body of customs.[118]

Shaler's most protracted discussion of the origins and evolution of legal practice and civil government appeared in his 1904 book *The Citizen: A Study of the Individual and the Government.* As we have seen, Shaler, like the progressivists, traced such social forms to the institution of property. The purely personal ownership of weapons, clothing, and ornaments within the tribe was enlarged, with the coming of agriculture, to individual possession of land, flocks, and herds. These acquisitions in turn raised questions of property rights and soon spawned a body of legal customs dealing with inheritance, individual liberty, and personal injury. The sophistication of a society's laws, of course, reflected the level of its evolutionary attainment, for behind Shaler's functionalist conception of legal development lurked the Lamarckian assumption that social institutions "directly contribute to the moral and intellectual nurture of the young by giving to them the experience and acquisitions of their predecessors."[119] Like education, the law was a medium through which the cultural benefits and conventional wisdom of the past could be mediated to each new generation.

Shaler's thoroughly instrumental conception of civil law as the mere codification of precedent had direct ramifications for social practice. However pragmatic and naturalistic its origin, he nonetheless felt that some judicial system was necessary if only to curb the deviant behavior of those few individuals who "turned back to the place of the primitive savage."[120] But he quite consistently rejected the retributive view of punishment in favor of the reformatory because he regarded criminals as unfortunate victims of heredity or environment simply needing their behavior compassionately regulated. Reminiscent of the pragmatist legal philosopher O. W. Holmes, Jr., who showed that criminal law had evolved from the personal desire for vengeance to the more objective need for public safety,[121] Shaler looked forward to the abolition of the death penalty. "Of old," he observed, "the theory was that the penalties applied to prisoners were for punishment; in our better day we look to the treatment of malefactors as a means of reforming the unhappy man."[122]

At the same time, Shaler was well aware that the supernatural authority of the moral law had been, if not abolished, certainly curtailed in his naturalistic reconstruction of legal evolution. "When, from the idea of laws being made by superior beings, we come to the democratic view of the matter," he admitted, "it was to be expected that the ancient reverence for these institutions would in part be lost." But with typical Lamarckian optimism, he looked to education to preserve the fabric of society in the face of its now secular identity. Such confidence, of course, could only be sustained by a thoroughgoing humanitarian faith, of which Shaler himself was a virtual incarnation. For ultimately, he believed, the behavior of citizens was guided, not so much by the precepts embedded in legislation, but by the innate morality acquired in the very processes of evolutionary history. "It is the unwritten law of every realm that makes human society possible; it was that body of sympathetic usage which laid the foundations of civilization."[123]

The theory of evolution, it is now abundantly clear, was a permeating influence on Shaler's work and thought. As a foundation and framework for organizing research, it dominated his thinking on natural and social history alike. As we shall presently discover it became a kind of teleological tool for reconciling nature, humanity, and God. Certainly many of his contributions to evolutionary theory were derivative and synthetic; but his eclectic synthesis furnishes a not unrepresentative picture of the impact of Darwinism on America's postbellum scientific culture. More specifically, the way in which the American geographical tradition received its baptism in evolution may be traced, in large measure, to the pervasive influence of Shaler on several of the discipline's founding fathers.[124] A deep sense of change through time, of ecological interdependence, and of biosocial adaptation were as much a part of Shaler's vision as of subsequent geographical practice. Moreover, the way in which Shaler was able to squeeze the neo-Lamarckian version, with its pragmatic, environmental emphasis, into the long-established tradition of British natural theology highlights some of the key conceptual maneuvers in the secularizing shift toward the new scientific theodicy. To scrutinize these efforts will be our next concern.

4   Vestiges of Natural Theology

A young man of the new school thought to trick him into an
inconsistency, and pointed out that evolution must unsettle the
beliefs of a man who accepted Christianity. Professor Shaler
turned to him and said: "I believe in Christianity, and I believe
in this evolution." Then he quoted from the book of Job, "Touch-
ing the Almighty, we cannot find him out."
—Anonymous, "Dean Shaler"

Nathaniel Shaler was, in many ways, a latter-day exponent of natural
theology. In an era when the whole fabric of natural religion, no less
than natural history, was under Darwinian reconstruction, he retained
a strong conceptual affinity with the post-Reformation tradition that
interpreted the world of nature as a place of purpose and design
incarnating the plan of a beneficent Providence. An ardent evolutionist, however, he was no mere apologist for John Ray, William Derham,
William Paley, or the *Bridgewater Treatises*; nor did he advocate an
uncritical return to the spirit of eighteenth-century physicotheology.
Only by evolutionizing traditional religion's teleology and ethics, he
believed, could nature, humanity, and God be prevented from a col-
lapse into chaotic naturalism. The reinvigoration of the old doctrine
of Providence was thus the first stage in the casting of a grand, new
evolutionary theology.

It was certainly not the first time that the doctrine of Providence had been imported into nature to make science safe for theology. To account for the transition from the mathematical geography of Ptolemy to the physio- and anthropogeography of the middle sixteenth century, for example, Manfred Büttner isolated the replacement of *Creatio* (God the Creator) by *Providentia* (God as Providence) as the key doctrinal ingredient.[1] Luther's Providence had invaded the god-forsaken world of distant Deism and had taken up the reins of its government for the benefit of humanity. To study geography, therefore, was second only to divinity because it provided access to the active God of nature; in geography the proofs of divine existence could be established, and the suitability of the earth as the home of humanity could be attested.

The impact of these Reformation currents, however, was registered not only in geography but in the practice of science more generally. As long ago as 1874, John W. Draper, himself champion of the conflict interpretation of science and religion, recognized that there was some sort of correlation; "Modern science," he confessed, "is the legitimate sister—indeed, it is the twin-sister—of the Reformation. They were begotten together and were born together."[2] And indeed, to Calvin, neither the deductions of the ancient Greeks nor the biblicist speculations of "Mosaic science" furnished adequate credentials for the genuine pursuit of scientific knowledge.[3]

The task of elucidating the precise nature of these relationships, however, has proved to be an infernally stubborn problem, and historians have retold their story of the rise of modern empirical science in sixteenth-century Europe and its flourishing in seventeenth-century England in many different forms. Some, for instance, have sequestered the Calvinistic doctrine of election as promoting a "this-worldly" asceticism that fostered at once modern capitalism and a Protestant ethic or ethos from which both science and technology received impetus.[4] Others have been impressed with what they see as a unique synthesis of Greek and biblical thought forms endemic to the genesis of science. To them it was the fusion of Greek order and biblical law during science's gestation period that encouraged a growing sense of an intelligible nature and repudiated the world-denying extremes of Gnosticism and asceticism.[5] Still others, plowing the theological furrow, have found particularly close relations between religious dissent and scientific progress in Restoration England.[6] At the same time, social and economic forces doubtless also contributed, and thus some have tied the scientific enterprise to the navigating needs of a maritime nation, the recent availability of simple technology, and the structural transformation of society due to the growing strength of its manufacturing class.[7]

Whatever the specific intellectual debt of the scientific enterprise to the Reformation, it seems that both social and theological factors played a part. Nevertheless, in the realms of the logic of scientific explanation, many practitioners turned to Puritan Providence for the spiritual sustenance they needed. Certainly the thesis that Puritanism nourished the nascent science in any special way has had its detractors and is therefore held rather less confidently now than in the past. Still, even if this case, most forcefully articulated by Michael Hunter,[8] is sustained, the theological strategies of those late seventeenth-century Christians who were enthusiastic about scientific pursuits remain crucial to interpreting subsequent developments. Their revitalization of the doctrines of design and purpose was intended to relieve theology of the worst burdens of a mechanistic universe and to keep science uncontaminated by the poison of worldly wisdom.

The impulse science had received from providential metaphysics, however, soon proved to be a double-edged sword. It could, and for many did, cement the natural and supernatural worlds together. In this spirit the Puritan geographer Nathanael Carpenter, for example, demonstrated his commitment to divine revelation both in scripture and in nature.[9] But when succeeding generations of Reformed theologians further refined Calvin's teaching on Providence, other implications could be drawn. More purely soteriological concerns were designated to the domain of *Providentia specialis* while *Providentia generalis* was allocated the interpretation of God's direction of the natural world. Distinctions, of course, all too easily widen into oppositions. Thus the seventeenth-century polymath Bartholomäus Keckermann, for instance, used it to urge the emancipation of geography from theology.[10] And his natural philosophy later influenced Immanuel Kant, who, it is well known, subjected the teleological argument to philosophical scrutiny and concluded that, since "things-in-themselves" are forever inaccessible to the human mind, the fundamental idea of causality only applied to perceptual experience. God was banished to the outer fringes of a shadowy noumenon, and it was therefore futile to hunt for him along the course of riverbeds, behind the laws of mountain building, or in the ebb and flow of tides. Conclusions about divine existence and purpose could never be reached by the pursuit of science.[11] The Kantian chasm between the empirical and the transcendental thus insulated the world of science from the realms of metaphysics.

Despite these and other detailed philosophical criticisms, notably by Hume and Spinoza, natural theology remained the love child that kept the marriage of science and religion alive during the eighteenth and nineteenth centuries. Ray and Derham further consolidated the foundations on which Paley's monumental superstructure was erected and, between them, bequeathed to history a vision of nature as the functioning revelation of divine purpose.[12] Here science and theology were sides of the same coin, and to speak of a conflict between them would thus have seemed as senseless as civil war. Of course the tradition did not always manifest itself in precisely the same way. In the wake of Laplace's nebular hypothesis, for example, astronomy witnessed a shift toward the identification of Providence with natural law.[13] Then an idealist rendition of the patterns of organic adaptation, to which we shall presently return, was already competing with Paleyan utility before the Darwinian onslaught.[14] These and other diverging theological interpretations reveal just how incoherent the natural theology endeavor really could be.[15] And yet, however contested its internal logistics, the vocabulary of natural theology constituted much of the style and rhetoric of scientific communication in the pre-Darwinian period.[16]

Nowhere are the teleological assumptions of natural theology more conspicuous than in the geographical tradition. I have explored this relationship elsewhere,[17] but in the present context it will suffice to underscore H. R. Mill's historical diagnosis of 1929:

Teleology or the argument from design had become a favourite form of reasoning among Christian theologians and, as worked out by Paley in his *Natural Theology*, it served the useful purpose of emphasizing the fitness which exists between all the inhabitants of the earth and their physical environment. It was held that the earth had been created so as to fit the wants of man in every particular. This argument was tacitly accepted or explicitly avowed by almost every writer on the theory of geography, and Carl Ritter distinctly recognized and adopted it as the unifying principle of his system.[18]

Ritter had drawn inspiration from Herder's philosophy of history and, in *Die Erdkunde,* found that the purpose and wisdom of God could be traced in the patterns of geographical distribution.[19] Through Arnold Guyot, who had come to the United States from Switzerland in 1848, the New World was introduced to the Ritterian cosmology, and in his Lowell Lectures on the "new geography," subsequently published as *The Earth and Man* (1849), he refashioned the study of humanity and nature on a new ecological model.

To this tradition, I believe, Shaler ultimately belongs. But to understand fully the nuances of his own recension of it, two further factors need to be taken into account. The first, the pervasive influence of Agassiz, has already attracted our attention. To recapitulate, his idealist construal of natural history and natural theology evoked a vision of the world as the thought of a Creator who had drafted a timeless, all-encompassing plan for the universe. Science's raison d'être, therefore, lay in its efforts to uncover the meaning and order implanted in nature by its divine author.[20]

Second, the so-called Darwinian challenge had to be taken into account. The impact of this challenge, however, has had to be reassessed in the light of recent reevaluations. Among the squad of historical revisionists who have finally dismantled the crude conflict model, mention must be made of J. R. Lucas's recent revaluation of the Huxley-Wilberforce fracas, Sheridan Gilley and Ann Loades's efforts to delve into the "bulldog" psychology of Huxley, and James Moore's remaking of the story of evolution and religion.[21] Alongside these, Himmelfarb's[22] depiction of the Darwinian revolution as a "conservative revolution" because of its bolstering well-established Victorian values and rationalizing the earlier ethical revolt against Christian morality now has its admirers.[23] Working-class defection from institutional religion and interdenominational feuding are now also seen as vitally important in the incubating of Victorian religious doubt. And there is, too, good evidence to suggest that the professionalization of science was fundamental to the breakup of the religious framework and to the transfer of cultural power from the ecclesiastical establishment to a thrusting scientific elite.[24] This latter interpretation has much to commend it, and some now regard the replacement of the sacred theodicy by a scientific one, in which the ideological justification of the social order was transferred from divine Providence to the laws of nature, as the key to understanding the misnamed "Victorian crisis of faith." Geddes's substitution of Darwin for Paley, Huxley's craving for a molecular teleology, and Galton's hankering after a "scientific priesthood" certainly invite such exegesis.[25] And Shaler's wielding of science in the cause of ideology, as we shall see, suggests a similar transatlantic intellectual drift.

Yet Darwinism ultimately played an undeniably important role in the acceleration of Victorian naturalism. Darwin's scheme certainly demonstrated how the myriad living things, so finely adapted to environment, could have developed without recourse to the purposive workings of any higher power, in a purely naturalistic cause-and-effect system. And while he did not overthrow the natural theology tradition,[26] its apologists had to modify their tactics in order to retain faith in a providential world order.[27] Alternatives, I have already suggested, were readily available, and Owen's idealist version proved to be among the most amenable to evolutionary restatement. As one publicist put it in the *Bibliotheca Sacra* as late as 1916, "I do not see why Darwin's treatise should have troubled any one who could look from the beginning down through life and see that it was all in a *plan,* where type, order, family, genus, species, and variety were always registered and executed."[28] By locating design in the harmonious relationships that linked together species in an overarching plan, these natural theologians distanced their project from Paley's efforts to catalog the utility of features adapting creatures to their environment. Not that these two versions—the teleological and the homological as they were styled—were always opposed; in many cases natural theologians were only too happy to fuel their fires from any available source.[29] Still, as the ripples of Darwinian discontent radiated outward, idealist conceptions of design achieved greater prominence, and that encounter produced a more constructive response. Its advocates, as Bowler tells us, "did their best to search out new lines of evidence to back up their belief that nature was designed in more than the merely utilitarian sense so obviously threatened by Darwin."[30] They worked to spell out the limitations of the natural selection mechanism, to identify organs that had no survival value, and to show that variation was not random but followed orderly, predesigned sequences.[31] So as Darwin marshaled the forces of evolutionary naturalism, many thinkers recast the design argument in an idealist mold, shifted it from creation to natural law, and adjusted their conception of Providence accordingly.[32]

Shaler and the Argument from Design

Throughout his life Shaler retained a deep-seated belief in a designed world. Like generations of natural theologians he sought to unify nature, humanity, and God by taking Providence with him on his scientific pilgrimage. But his concern to preserve intellectual continuity with the post–Reformation teleological tradition did not obscure the need to revitalize it in the light of current science, and so the obligation of divinizing evolution was high on his agenda.

Shaler first embarked on his reconstruction of the design argument in his 1889 discussion "Chance or Design" for the readers of the *Andover Review.* Since ancient times, he began, human societies had explained the regularity of seasons, astronomical phenomena, and the historical succession of plants and animals by reference to the presiding control of a superior intelligence. This sense of purpose in nature should be taken seriously, if only because it was part of the very fabric of human intellectual history. Moreover, Shaler added, it was philosophically essential for mankind's initial efforts to construct a coherent picture of the world and to understanding the place of humanity in the scheme of things. The idea of design so united the realms of nature and supernature that it was absurd to separate cosmological from epistemological questions. With the advent of empirical science, however, "a certain antagonism between the philosophical and religious conceptions of the world" began to mount. As this rift widened into a chasm, various schemes of reconciling science and religion had been advanced. But Shaler drew back from them all because of a general failure to acknowledge the fundamentally different methods of the two spheres. Reminiscent of the Kantian divorce between the empirical and the transcendental, Shaler suggested that when they "endeavor to invade the field of the moralist or the theologian, the votaries of science find that their methods cease to be of any use."[33]

Nevertheless, Shaler hesitated to banish science and religion to eternally separate dominions. While he felt confident that a reanimation of the teleological argument could effect a conciliation, he was just as insistent that a slavish return to the methods of Paley and the *Bridgewater Treatises* was not the right policy. Indeed, that tradition had perjured itself through logical indulgence and special pleading; it had consistently mistaken superabundant testimony for quality evidence. Still, despite both theological and scientific disenchantment with natural theology, even before the evolutionary threat, Shaler suspected that a complete abandonment of Paleyan aspirations would be too precipitate. Faults, even fundamental ones, in that system did not necessarily invalidate the whole outlook. "A somewhat careful study of the problem discussed by the followers of Paley," he mused, "has convinced me that it is not quite reasonable to dismiss the methods of their inquiry in the summary fashion in which they have been cast aside, that it is worth while to give more discussion to the problem in the light of our better knowledge."[34]

From the outset Shaler made it clear that the personal nature of human knowledge meant that observations and analyses were inevitably distorted by the individual scientist's often unconscious framework of interpretation. Advocates and opponents of design alike were thus inclined to interpret the same evidence in accordance with prior convictions. With more than just a hint of the modish Neo-Kantian creed currently captivating fin de siècle philosophers, Shaler admitted that "we are obliged to interpret the visible universe on the basis of our mental organization."[35] Accordingly, the idea of design was as legitimate a presupposition as the theory that the universe was the product of chance, and those who regarded it as a priori incoherent— they were often motivated by antireligious rather than scientific sentiments—were in their own way as unscientific as some advocates of the older natural theology.

This ground-clearing exercise, as it were, gave Shaler the philosophical imprimatur to embark on his own exposition of the argument. The perfect adaptation of organic features to environment, he recalled, was inadmissible evidence for design because evolutionary biology had naturalized its origin in the prosaic laws of cause and effect. So Shaler turned to a more idealist paraphrase of the argument, the locus of which was to be found in the overall pattern of creation. His suggestion was that the very laws of nature were themselves the expression of a governing intelligence and that evolution was itself a fully creative agency with its own internal telos that could be identified with divine power. During the first decades of the twentieth century, it should perhaps be remembered, the supplanting of the older idea of *creatio ex nihilo* by this conception of evolutionary creativity became the focal point for the process philosophy of such men as Samuel Alexander, Henri Bergson, Charles Hartshorne, and Alfred North Whitehead. Their vision was of nature as "a creative emergence within a temporalistic, teleological reality guided and directed in different degrees by a God whose very being is involved with that of the world."[36] Doubtless Shaler's close identification of natural law with divine Providence (although he stopped short of conflating them) brought his work to the attention of Bergson, who later made use of *The Individual* and *The Interpretation of Nature* in his own celebration of evolutionary metaphysics.[37]

Shaler realized, of course, that the bones of his argument needed to be fleshed out by compelling instances of design in nature. He attacked the problem on two separate fronts. First, he was impressed by the fact that living beings could only subsist within a peculiarly fine adjustment of a wide variety of terrestrial conditions. The narrow range of temperature, the delicate balance of atmospheric conditions, and the relative distribution of sea and land all suggested that "this film of life" was the product of a marvelous design. Thus life's very precariousness, not to mention its complexity and diversity, militated against interpretations resorting to blind chance, as he polemically styled alternative viewpoints. A second means of reinstating what he called "the ideal in the universe" was by reflecting on the evolutionary emergence of the human species, that crowning glory of creation. As he saw it, the human animal could have ascended from the evolutionary process only by a single route, and this implied that the tree of life was itself the expression of purpose and plan. "If at any point in the succession of steps," he told his readers, "the accidents of this complicated world had destroyed the group which held within itself the possibilities of our being, the chain would have been broken, and man would have become an impossibility."[38]

Shaler was well aware that his account of holistic design represented a more or less radical departure from the traditional Paleyite prototype.

But his own "method of presenting the problem of design," he promised, "puts the idealist into a better position as regards his method of treating facts than that which was secured by the method adopted by the school of Paley." Clearly, utilitarian design was not the only cord by which nature, humanity, and God could be bound together. So instead of working piecemeal with this variation and that adaptation, Shaler preferred to suffuse the entire gamut of evolutionary history with the pulsating power of dynamic design. Human beings, with all their intellectual sophistication and cultural complexity, he offered as a final gambit, were either "the result of a fortuitous concatenation of unadjusted impulses dependent on one chance in a practically infinite number of possibilities, or . . . the product of control."[39] With such statistical odds, the design argument was in a more healthy state, notwithstanding the Darwinian challenge, than even in its heyday under the monopoly of Paley and his devotees.

Shaler's program of teleological reclamation was just one of the new brands of natural theology that, at least for a time, helped stem the tide toward a more radical naturalism. The long Post-Reformation tradition was thus able to reincarnate itself in evolutionary guises and remain insulated from the nastiest implications of Darwinian materialism. Henry Ward Beecher, for example, America's foremost postbellum pulpiteer, carved out of Kant's philosophy, the Gospel of Christ, and Spencer's evolution a synthetic creed that was romantic in temper, religious in taste, and scientific in tone. If single acts of creation could reveal design, Beecher pondered, how much more would a vast universe, self-manufactured by inherent laws? "Design by wholesale," he concluded, "is grander than design by retail."[40] Among the many other propagandists who, in various ways, absorbed divine power into evolutionary nature, were the Lowell lecturer Henry Drummond;[41] John Bascom, author of *Evolution and Religion* (1897);[42] America's senior geologist, Joseph Le Conte;[43] the Baptist protofundamentalist Augustus Hopkins Strong;[44] Minot Judson Savage, one of the first clergymen in America to accept evolution;[45] and John Fiske, lay philosopher and popular historian.[46] In Britain, the pre-Darwinian ventures of Baden Powell and Richard Owen found continued expression in the evolutionary teleologies advanced by George Henslow, Frederick Temple, the Duke of Argyll, and George Matheson, while an even "wider teleology" dominated the outlook of those Anglicans—Aubrey Moore, J.R. Illingworth, and Charles Gore—who contributed to what Don Cupitt calls "the 'Lux Mundi' synthesis."[47]

The following year, 1890, Shaler returned to the theme, this time in the pages of the *Unitarian Review*. His aim here was to enlarge the idea of natural law so as to leave room for "the unexpected, for the suddenly appearing, in the realm of nature." Because he conceived of the Creator as intimately bound up with the laws of nature, Shaler hoped to safeguard the concept of design without invoking any direct, miraculous intervention of a deity. He wanted to find some way of keeping catastrophism within the bounds of nature's uniformity, and assuming what Hooykaas terms "the multiplication of small effects,"[48] he found it possible to argue that "by successive gradual changes, each infinitely small, we may pass in the end a critical point, leading to consequences which are infinitely great."[49] By analogy with liquid whose form quite suddenly changed to either solid or gas, he believed he could encase "catastrophes" within an actualist, even gradualist mold.[50] To account for such dramatic breaks in terms of the normal working of natural law, however, did nothing to diminish their value as teleological apologetic. Traces of the immanent designer could, for example, be detected in the delicate balance between, say, the world's range of temperature and those points, so critical to organic life, where water changed to ice or steam. If its solidifying point, for instance, had happened to be at the temperature for boiling water, "life would be impossible." Thus Shaler's world was "a place of surprises which take place under natural law," but surprises that were "quite as revolutionary as if they were the product of chance, or as a result from the immediate intervention of a Supreme Power."[51]

In such circumstances, to be forced to choose between "chance" and "miracle," it was plain to Shaler, was to be caught on the horns of a false dichotomy. So on the one hand, he resisted the idea of special divine intervention and therefore demythologized, for example, the circumstances surrounding the Mosaic lawgiving; by urging that the geology of Mount Sinai suggested past volcanic activity, he provided a naturalistic reading of the Exodus narrative. On the other, his resolute faith in a purposeful world perpetuated, albeit in a modified form, that long post-Reformation geographical tradition best represented by Ritter and Guyot. Like Charles Babbage, who "analogized" about calculating machines constructed specially to produce seemingly random results, and Edward Hitchcock, who felt compelled to posit a special divine law for miracles,[52] Shaler found room for nature's critical dislocations by wrapping natural law in a providential robe. His position here, I think, confirms Hooykaas's claim that many catastrophists were in reality acknowledging "the reign of Law . . . not only as a scientific but also as a metaphysical supposition."[53]

The providential invasion of the world order is nowhere more clearly visible than in Shaler's celebrated *Nature and Man in America* of 1891. Appropriately, perhaps, this work was the most strongly environmental determinist of all his writings and therefore suggests his transitional stance in the theodicean shift from natural theology to natural law as the source of societal validation. Unhappily, he began, modern science appeared to be in conflict with the religious traditions of the American people, since naturalistic accounts of human origins—now almost universally accepted, he claimed—had undermined the orthodox interpretation of the Old Testament chronicles of creation and of the first man instantly created in the image of God. Nevertheless, with cautious optimism, he felt sure that "we are now approaching the time when our knowledge will reaffirm the old belief which our fathers had in the essential control of a beneficent Providence." There was reason indeed to hope that the faith of the future would resemble the faith of the fathers but would transcend it because it would "rest on the firm foundations of our own knowledge, rather than on the trust in the opinions of our elders." In the struggle for the sources of knowledge, with all their accompanying authority and cultural power, dogmatic theology had to give way before the claims of scientific experience. In this volume, therefore, Shaler would present evidence to show that the evolutionary laws of nature had conspired to ensure both progressive organic development and the emergence and improvement of human society.

Nature, he again proclaimed, was not a place of chance and chaos but rather of such harmony and grandeur that only the vocabulary of providential piety could adequately capture its charm. To convey this sublime vision to succeeding generations should therefore be the task of every true naturalist. Every opportunity should be grasped to lay out the scientific justification for faith in the benevolent intentions of the creative power at work in nature. "By so doing," Shaler went on, the naturalist "may hope to help himself and his fellow students to escape from the perplexity which has been brought about through the revolution in the opinions of men which modern science has induced." So this book was offered to the public as Shaler's own testimony that the "geographic changes and the consequent revolutions of the climate which our earth has undergone, though rude and in a way destructive, have nevertheless served the best uses of life, driving organic creatures by the whips of necessity upward and onward toward the higher planes of being."[54]

That a celebrated Harvard geologist could promote a scientific faith fusing divine Providence and evolutionary nature may explain why the faculty of Andover Theological Seminary invited Shaler to deliver the Winkley Foundation Lectures in 1891. According to the seminary prospectus, these lectures were "established for the presentation of sociological, scientific, philosophical, and historical subjects which have a relation to the doctrines or the practical work of the Christian church." Quite fittingly, Shaler's course was advertised as dealing with "Modern Science and Religious Beliefs."[55]

Shaler's receipt of this invitation reflects not only the standing he now enjoyed as an authority on science and religion but also the change taking place in New England theology during the last decades of the nineteenth century, particularly at Andover, as it was accommodated to the prevailing scientific ethos. Indeed Shaler's disquisition was presented during the years of controversy at Andover, when the new theology of "progressive orthodoxy" was being forged, and his presence there in 1891 symbolized the changing climate of theological opinion. Established in 1808 to exorcise the ghost of New England liberal theology in the shape of Harvard Unitarianism, Andover Seminary remained dogmatically Calvinist for three quarters of a century. The appointment of George Harris to the faculty in 1883, however, heralded a new phase in Andover theology. Sensitive to the major currents of thought in the period—German idealism, Emersonian transcendentalism, biblical higher criticism, and Darwinian evolution—he helped to establish the *Andover Review* in 1884. From its inception, the journal's contributors—philosophers, theologians, and scientists—were specially exercised about the implications of evolution for teleology, ethics, and belief in God, and much of the polemic was directed against the imposing naturalistic system of Herbert Spencer. Andover's progressive orthodoxy was the offspring of the union between biology and theology, and it continued to thrive, notwithstanding six-year-long heresy trials instigated in 1886 by a board of visitors unnerved by creeping liberalism.[56] Symptomatic of Andover's new credenda was Harris's portrayal of salvation as "moral transformation"—a kind of individual recapitulation of humanity's moral evolution. Indeed he was so keen to cultivate Darwinian divinity that he subjected theology to the evolutionary laws of survival; "the theology which gains currency, at any time," he proposed, "is that conception of Christianity which commends itself to the reason and experience of the most enlightened and spiritual Christians."[57] At heart, then, Andover's progressive orthodoxy lay in its rechristening of secular evolutionism and in the sacralizing of the idea of progress in order to guarantee society both moral improvement and material prosperity.

If the Andover progressives sought encouragement for their theological ventures from a member of the rising scientific establishment, they certainly could not have done better than turn to Shaler. On one point after another, *The Interpretation of Nature* (the version of Shaler's lectures published in 1893) confirmed the expediency of credal reformation. Their philosophically synthetic inclinations, for example, were mirrored in Shaler's own journey toward the rediscovery of a dynamic nature. This insight had helped dispel those mechanistic images that in his youth had waylaid him from the path of Christianity. But now, in a blending of post-Kantian idealism, evolutionary progress, and a rehabituated Providence, a new appreciation of the world environment could be achieved. Again, in Shaler the Andover fraternity found an enthusiast to reinforce their own fascination with evolutionary dissemination. Shaler, as we have already seen, had satisfied himself on the direct links between instinctive animal behavior and human understanding. So it is no surprise to find him reconstructing the history of metaphysical speculation along evolutionary lines.

Human curiosity about the causes behind natural events was the fertile soil in which religious sentiments were initially propagated. Animism in its various forms was its first yield, but this soon gave way to the idea that the spirits of the departed ruled the natural order from the world beyond. It was a short step to the postulation of a whole pantheon of divine agents, and only latterly was this "great conception . . . completed in the unique idea of a Supreme Being."[58] Shaler's retelling the story of religious awareness in evolutionary style, of course, fitted snugly into the spirit of Andover's developmental reconstruction of dogmatic theology. But the most recent thrust of the historical process Shaler hoped to resist, namely, the secularization of the causal principle in the idea of natural law, and he hoped to circumvent the new sense of cosmic rootlessness by impregnating the reign of law with Supreme Power. Currents of thought were now moving in this direction, he told the seminarians, for "naturalists are being driven step by step to hypothecate the presence in the universe of conditions which are best explained by the supposition that the direction of affairs is in the control of something like our own intelligence. . . . In other words, it seems to me that the naturalist is most likely to approach the position of the philosophical theologian by paths which at first seemed to be far apart."[59] However different the subject matter and methods of science and religion in general, it now seemed that, courtesy of the idea of design, natural law and natural theology remained opposed only in their choice of vocabulary. So "the two methods of interpreting nature which were originally united, then long separated, are again to be conjoined."[60]

To refurbish the tools of natural theology, however, Shaler made it clear, did not exonerate the Christian church from its long-standing repression of scientific pursuits. The brilliance of Greek science, which had flourished pari passu as animism was replaced first by polytheism and later by the Platonic Forms, was a case in point. It was shattered with the advent of Roman Christianity, a religion that, Shaler was convinced, was no less energetic in its denigration of empirical science as in its protest against evils of pagan culture. This antipathy stemmed from the traditional Hebrew resort to the sovereign reign of Yahweh as the all-sufficient explanation of natural phenomena. Thus for one and a half millennia Western science was stultified, and it was only through the "Graeco-Christian Renaissance" in the sixteenth century that the door to scientific rehabilitation began to open. But even then, the Catholic church vigorously resisted any move to liberate science from ecclesiastical surveillance, and only now in the latter part of the nineteenth century could the cessation of hostilities between science and religion be expected. The discovery of Providence at the head-waters of natural law promised a mutually enriching future for science and theology; indeed there already was "really little of moment to separate the men who approach the unknown through the old ways of the imagination and those who find their path into the depths from the newer avenues of science."[61] If theologians could only reject the unnecessary dualism entailed in the belief that events in nature could occur miraculously outside the realm of law, the last obstacle to substantial reconciliation would be removed. Instead of hankering after the theophanic certainties of a past age, theologians needed to recognize that religion was rooted in the moral nature of humanity and "on the sense of the depth of the universe"—a sentiment not unlike Rudolph Otto's idea of the holy.[62] Once again this proposal would have been congenial to the Andover reformers, with their stress on the foundational nature of moral character and on the exploration of Christian consciousness. To see the ethical impulse as the bedrock from which religion was quarried soon became a common article of the liberal creed; Shaler's views on this subject again mirror the changing face of New England theology at the turn of the century.

Science, Sympathy, and the Human Family

Just as Shaler's renovation of the argument from design echoed the sounds of contemporary theology, so his driving religion back to innate moral feelings reflected the naturalistic trend toward a religion without revelation. For a natural scientist such a humanitarian faith was doubly attractive; on the one hand its history could be disentangled by the methods of empirical science, and on the other its origin could be explained in the language of evolutionary functionalism. At the same time, "altruism," "sympathy," and "the brotherhood of man" were the new vogue words among those enlightened New Englanders who spearheaded the cerebral and rather esoteric religion of humanity. So in that vital period of transition from Puritanism to naturalism via Unitarianism, Shaler's mind registers many symptoms of the secularist leavening of the New England tradition.

While Jonathan Edwards would doubtless have detested Ralph Waldo Emerson's *Nature* (1836) from cover to cover, he would nevertheless have recognized in him an equally passionate lover of nature. And herein was genuine divine encounter, for in the New England religious mind from Edwards to Emerson, from Puritanism to transcendentalism, the love of God's creation was itself a spiritual experience. Under Illuminist influence Edwards saw religion as fundamentally an affair of the heart that needed emotional forms to express adequately its intense love of Perfect Being. But whereas Edwards was far too sensitive to the nuances of historical theology to fuse God and nature into a single divine entity, Emerson—no Edwards redivivus—had no concern to hold in check any such mystical or pantheistic tendencies. By universalizing the privileges of Calvin's elect, he eradicated all traces of original sin and human depravity. "The ecstasy and vision which Calvinists knew only in the moment of vocation, the passing of which left them agonizingly aware of depravity and sin," Perry Miller comments, "could become the permanent joy of those who had put aside the conception of depravity. . . . Unitarianism had stripped off the dogmas, and Emerson was free to celebrate purely and simply the presence of God in the soul and in nature, the pure metaphysical essence of the New England tradition."[63] If Edwards sought the God behind nature, Emerson looked for god in Nature, convinced that there a spontaneous fusion of universal existence could be achieved.

Extracting the spiritual wedge driven by the Puritans between saint and sinner, Unitarians and transcendentalists alike rejoiced in mankind's common moral nature. Here at last was a solid foundation on which to ground experimental religion—an earthly Christianity that rejected miracle and historical revelation, emphasized divine immanence in nature, and hailed the ethics of Jesus as "the absolute religion."[64] It was a monumental reversal of the Calvinist soteriological drama that would lead many to a solely ethical religion of duty. Now a bowdlerized New Testament could be read as a record merely of Christian ethical experience, with nasty themes like judgment and hell suitably expurgated. The doctrines of "fall" and "regeneration" no longer had significance either, for free religionists saw "progress" as the great universal law of history, in which the "Divine Idea" perpetually revealed itself in the unfolding world order. Not surprisingly, the theory of evolution in its Spencerian garb was welcomed as the latest product of rationalistic thought. And even then the glow of Enlightenment progressivism was sufficient for Emerson, Parker, and Thoreau to keep out the chills of Darwin's darker materialism.[65]

In this age of doctrinal restatement, when almost every article of traditional Christianity was rejected by one scholar or another, New England radicals happily turned to morality as the last stronghold of religious certainty. But the secularizing drift of the intellectual tide merely attests to the ultimate bankruptcy of their accommodationist maneuvers, blandly designed to confirm the most sanguine assumptions of human perfectibility. However optimistic was the faith of free religion in humanity's innate goodness, in society's ethical virtues, and in individual altruism, its final legacy, as Stow Persons tellingly wrote over three and a half decades ago, was "the transformation of Unitarianism from a Christocentric religion to a pragmatic, humanistic theism, retaining the Christian name but actually being Christian only in the sense of recognizing its dependence upon the religious patterns of Western culture."[66] Yet, if free religion's strategy proved *theologically* moribund, it was certainly *culturally* opportunistic. For by eliding the natural and supernatural worlds, it created the impression that immanent values were really transcendental; and by sacralizing the language of a purely mundane morality, it eased the transition from a religious to a scientific theodicy.

The humanistic ethos of New England's religious culture toward the end of the century found full expression in the pragmatic thrust of Shaler's ethical reformulation of the Christian heritage. Religion, after all, was simply the highest of the social motives, and behavior accordingly had precedence over belief, ethics over dogmatics. As we have seen, Shaler happily subjected the altruistic drive to evolutionary exegesis. Indeed he was prepared to go further: evolution had actually put ethics on a far stronger basis than the traditional creation-based morality. Witness, for example, his locating society's responsibility for nature in the very earthiness of human evolution:

So long as men considered themselves to be accidents on the earth, imposed upon it by the will of a Supreme Being, but in nowise related in origin and history to the creatures amid which they dwelt, it was natural that they should exercise a careless and despotic power over their subjects. Now that it has been made perfectly clear that we have come forth from the maze of the lower life, that all these tenants of the wilderness are sharers in the order which has brought us to our estate. . . we can no longer keep the old careless attitude.[67]

Shaler freely admitted that this position represented a reversal of the traditional claim that ethics were based on divine command. No matter how "formalized" or "concealed by the superstructure they derive from the accidents of the mind," it was certain that altruistic sentiments were "the foundation of religion."[68] But Shaler soon sensed that to naturalize morality could be a socially risky business. "Experience," he reflected, "shows that the soundness of a government such as our own depends upon the existence of religious motives among the people. It is clearly the main source of patriotism, for it leads to those relations between men, and to that sense of duty, which are the foundations of true democracy." Religion, simply, was needed to reinforce the sense of moral duty, and unapologetically manipulative, Shaler insisted that, "from the point of view of the citizen, it is in the highest measure desirable that Christianity should prosper and that it should profoundly affect the conduct of every person in the state."[69] Clearly, society was in no position as yet to respond to the purely naturalistic ethic of the scientific theodicy. Such social prescriptions, moreover, were resonant with the strains of William James's pragmatism. For had not James told his Gifford hearers a couple of years earlier that the essence of religion lay, not in abstract theological formulas, but in the promise of richer and more satisfying lives people enjoyed when assured of its truth?[70]

So far our perusals have established the idea of sympathy as the linchpin in Shaler's evolutionary depiction of both society and religion. They were, nevertheless, mutually reinforcing agencies. In three particular spheres, I think, this process is specially discernible: the sense of empathy with nature; self-consciousness, alienation, and psychic development; and altruism as the criterion for judging comparative religious claims. To reflect on these will further highlight the conformist yet optimistic spirit of Shaler's New England religion in its encounter with scientific culture.

For a large part of human history, Shaler maintained, religion had encouraged love of God and humanity at the expense of empathy with the natural world. Only with the emergence of monotheism from its polytheistic precursors could the fearful attitude to nature give way to a more enthusiastic affirmation. The idea of a Creator intimately bound up with the stuff of creation fostered at once a greater affinity with the material world and "the sense of ordered control in nature, which is the breath of all science."[71] Nature was no longer capricious; religion had pronounced it a place of regularity and harmony. Not that this was a mere restatement of a static natural theology. Evolution had dynamically transformed nature and theology alike. Indeed sympathy with nature was no less religious than biological in origin. Still, however secularized in the principle of natural law, this revolution in the attitude to nature continued to undergird the whole scientific enterprise.

Self-consciousness was another human attribute born of altruistic evolution. The extension of maternal love beyond the family to the tribe was umbilically tied to the emerging sense of individual identity. For a mutual recognition of each other's particularity was plainly at the heart of any sympathetic relationship. Yet here, Shaler believed, in the psychological evolution of self-consciousness were the seeds of humanity's experience of isolation, lostness, and alienation. Now "for the first time the soul feels itself naked and alone in the world." Surely there was no one, he went on, who had not felt "the almost mortal sickness which sometimes comes from the many varying moods of self-consciousness. Men seek some lightening of the burden whenever they can find it. They give themselves back to the unconsciousness that blessed their animal ancestors, by means of alcohol or opium."[72] And yet, paradoxically, if altruism played midwife to despair, it still could liberate the individual from personal preoccupation and introverted introspection. If evolution had revolutionized society's understanding of its own ethical history, it did nothing to undermine its crucial role in the life of the individual.

Judging by the Gilded Age standards of ethical excellence, Shaler applauded the great Christian command to unselfishly love God and neighbor because it "carried man farther out of the prison of self than all the other teachings that have come to him." Altruism plainly could act as a kind of hermeneutical principle for discriminating the virtues of competing religions, and not surprisingly, he concluded that "the Christian doctrine, looked at purely from the point of view of natural science, has the merit of setting the altruistic motives on a wider foundation than any other form of religion."[73] Still, this praise was not unqualified, for the sympathetic principle had an edge sharp enough to cut a deep gorge between the respective teachings of the Old and New Testaments. Shaler's unhesitating concurrence with the moral excellence of Christ's teaching did not prevent his discrediting much of the Old Testament on the very same grounds. So when he told the readers of the *North American Review* in 1889 that he hoped world peace would be the last but greatest gift of the nineteenth century, he castigated the Old Testament for its "praise of battle."[74] Indeed this intertestamental disjunction forced him in 1904 to cut off the biblical "allegory" of creation from mainstream New Testament teaching. The writings of Moses, scholars could confirm, had little or no connection with the teaching of Christ.[75]

In keeping with the temper of New England religious opinion, then, Shaler saw in the ethics of sympathy the integration of religious and scientific views of the world. As a means of escape from the prison of selfishness, sympathy foreshadowed "that real absorption into the Infinite, that true *nirvana,* which nature offers by the way of the sympathies"; it pointed the way for the resolution of human alienation, and it became the focal point for the synthesis of evolution and Christianity. Convinced that "the doctrine of Christ is the summit and crown of the organic series," Shaler found in the idea of sympathy a fusion of biology and theology surprisingly akin to the "Omega Point" more recently celebrated by Teilhard de Chardin.[76]

Toward a Scientific Theology

Shaler's passion to bring scientific scrutiny to bear on religious instinct was not confined to unifying evolution and ethics. Consider, for example, his periodic preoccupation with personal immortality. It had fascinated him since childhood, but not until his Andover lectures did he give public expression to his feelings on the subject, though he soon discussed the topic further in *The Individual: A Study of Life and Death*. Because, as he candidly admitted, the question of immortality lay beyond the domain of scientific methods, naturalists had been known to deny authoritatively its reality. Such materialist sentiments, however, were now being assailed, as the inadequacy of wholesale reductionism became more and more apparent to the scientific mind. The advent of Darwinism in particular promised to shed light on the landscape of the mysterious psychic world. By contrast with the theology of special creation, the principles of holistic evolution revealed the continuity of all organic life, while the new science of heredity held out hopes of isolating the mechanisms of mental inheritance. The evolutionary recasting of biology and psychology thus convinced Shaler of the mind as an independent entity, of a continuous mental genealogy among organisms, and of the human psyche as irreducible to the mere functions of a nervous system. And all this suggested that the disembodied survival of the mind after death was an entirely coherent supposition. Evolution had apparently taught individuals that, since they were but parts of a "gigantic organization," a cosmic totality moving forward under the inexorable control of "Design," they could abandon themselves to the "Power" controlling "Nature" in the confidence that it worked for the benefit of all.[77]

At the same time, if Shaler concluded his final Andover lecture on natural science and immortality with a confident flourish, it was just a shade out of character with the tone of the lecture as a whole. Indeed it was his tentativeness and uncertainty that made him write to Horace Scudder of Appleton's Publishing House on 6 August 1892 about the manuscript for this last chapter. "The fact is," he confessed, "the subject is one of exceeding difficulty. It is in fact a little beyond the confines of my understanding. I said the little which I did, for the reason that I found it would not do to leave the ground quite untouched. I rather regret the essay, though it has good, and somewhat novel features."[78]

Whatever his hesitancy, however, it did not stop Shaler from again putting pen to paper on the subject. For in 1900 he issued *The Individual*, the first installment of a trilogy that also dealt with the neighbor and the citizen, and he looked on this latest offering as a small appendix to Royce's great contribution on the Individual.[79] If only in its self-confessed idealism, the book was thus typical of those that looked back to Kant to shield religion from Darwin's metaphysical pessimism. This stance was certainly consistent with the philosophical tone of the

Cambridge intelligentsia in general and the Harvard pragmatists in particular.[80] But Shaler did sense the need to go beyond an a priori defense of divinity by locating some point where science and religion converged on the horizons of human experience.

Modern science, it seemed, had subverted the conventional wisdom about humanity's place in nature, and so the main aim of Shaler's new book was to reexamine the meaning of the individual's life in society in the light of this revolution. On the problem agenda of the new evolutionary confession of faith, one enigma above all needed to be solved: what was the reason for the brevity of life, for that transience that seemed to make mockery of the strivings of men and women and to render their very existence meaningless? The claims of traditional Christian eschatology, of course, would no longer suffice. And so, while Shaler insisted that his statements were not in the least directed "against Christianity in its native form," he made it clear that "the curious remnants of primitive religions" that had crept into the Christian church had turned "its blessings into curses and its light into darkness, thereby laying a burden upon human life and hope in ways even more grievous than those of pagan times."[81] It remained for him to spell out the superiority of a natural history solution to the question of the meaning of life.

As Shaler reflected on the problem from the evolutionist's standpoint, it struck him that the consciousness of death could be used to identify the point where the human species appeared on the evolutionary continuum. The realization that death was a mystery to be confronted was a purely human phenomenon and denoted the moment when the species passed beyond the realm of the mere animal.[82] If it was the presence of death that questioned the meaning of life, however, that in itself did not solve the knotty problem of what the purpose of human existence actually was. Evolution, when narrowly construed as natural selection, was plainly powerless to provide any satisfactory explanation. Whatever its origin, "the need of death," Shaler wrote, "is established by influences which lie outside of the selection field." A more broadly conceived evolutionism that focused on the reciprocal relations between individual and society, however, might illuminate the dark enigma of life's brevity. Supposing, after all, that the purpose of evolution lay, not in individual selection, but in the progress of a community, that might suggest that the individual's raison d'être was to be found in the responsibility of transmitting to the future the store of gains procured by predecessors. Certainly such a corporate Lamarckism would run counter to Shaler's earlier individualist inclinations, but it would nevertheless make sense of the brief duration of life for the period most consistent with the attainment of maturity and the care of offspring. A rediscovery of the interdependence of the generations, moreover, would counterbalance an overweening Western individualism and would encourage "a sense that man is not here for himself, but that he is part of a vast order which he is bound to serve; that his individuality has dignity and beauty only in so far as it recognizes this order and intelligently shares in the work thereof."[83]

Without recourse to transcendental values, the modern scientific theodicy could show that human existence had an immanent meaning within the structure of nature. It was, plainly, a thoroughly conformist prescription underscoring the necessity of society's acquiescence to the laws of nature. But Shaler was not prepared to leave the matter there. His belief in personal immortality, as we have seen, was too deeply ingrained for that. Now he expanded his Andover speculations in two ways. First, Shaler wanted to extract the venom with which paganism had infected Christianity. In an impassioned rejection of eternal retribution, he lamented the survival of spirits and devils in the minds of modern Christians; such unyielding demonology merely showed how deep rooted was the ancient, inherited fear of death. In the most scathing of tones, he denounced the scaremongering tactics of those who had used religion to aggravate the torment rather than to relieve it:

One of the best things that can be said of the century that is drawing to its close is that it has seen the end, or at least the promise of the end, of the ancient demon-worship. The physical hell, the personal devil, his imps of all degrees, the fiery furnaces, and all other agents of torment are passing away from the imaginations of men. There is probably not an educated clergyman who believes in them. . . . The idea of suffering for evil done is still firmly rooted in the minds of all men of sound moral nature: suffering in this or any other world until it has accomplished its fit work; but the old conception is now being purged from our religion, which it has so long disgraced.[84]

Second, whereas Shaler believed there were good grounds in an evolutionary conception of the world to sustain a faith in immortality, he was just as sure that the truth of personal survival would not be proved by scientific methods. Indeed he remained thoroughly skeptical of those parapsychologists claiming empirical evidence for life after death. For one thing, the fact that many psychic investigations had been bedeviled by cases of fraud or mental disturbance did little to compel scientific respect. Then there was the distasteful nature of occult practices, which led Shaler to conclude that any reports forthcoming from such quarters were worthless compared with belief grounded in the canonical faith, the cumulative judgment of religious history, and the prophetic insights of the church's seers. Certainly none of these sources was scientific, but they seemed to Shaler as valid as the "revelations" of spiritualism. So while advocating his scientific religion, Shaler wanted to prevent the wholesale absorption of theology into natural science and to leave room for alternative supplies of knowledge beyond the frontiers of empirical exploration. Nothing better illustrates Shaler's position, I believe, than his trenchant critique of the psychic research pioneered by Frederic Myers.[85]

No doubt because death—by cholera, typhoid, or childbirth—was one of the few unifying experiences in Victorian England's highly stratified society, Myers became increasingly fascinated by it and by the question of human survival. Death, after all, had robbed him of an intense platonic friendship with Annie Marshall, and this loss propelled him toward the joint goal of reaching her departed spirit and of finding empirical proof of immortality. He had long been convinced that orthodox Christianity no longer satisfied spiritual needs in the age of science and what he termed the "New Nature," and so he shared Henry Sidgwick's conviction that such ultimate questions must be confronted scientifically.[86] Accordingly he set up and presided over the Society for Psychical Research, and the fruits of his personal investigations were posthumously published in *Human Personality and Its Survival of Bodily Death*.[87] Steeped in the thought forms of the English romantic poets, he remained staunchly antimaterialist in philosophy and opposed Spencer's physiological psychology, which seemed to reduce love, life, and emotion to matter and energy. His efforts to elucidate the unexplored natural laws of the psychic universe, he hoped, would provide the building blocks of a new scientific religion.[88]

In his critical review of *Human Personality* for the *New York Independent* in 1903, Shaler castigated Myers for what he saw as a fundamental misconception of the nature and limitations of scientific method. "He is affected by the common delusion that this method of inquiry is applicable to anything under and above the sun, and that a pinch of it will serve to clear away the darkness that hides the most inscrutable parts of the realm," Shaler urged. "The truth is that natural science is but a limited resource available as a means of determination to those fields of phenomena where the occurrence can be accurately observed."[89] By definition, Myers's interrogation of his "subliminal man" by means of hypnotism must be literally unscientific. Besides, the realization that scientific findings were inherently provisional, as attested by the rejection, modification, or replacement of theories, had done much to moderate its authority, especially in matters beyond the observable. So for all Myers's expeditions into the territory of spiritual consciousness, Shaler felt that faith in the personal survival of death had still not been advanced beyond the primeval stage of human experience. For Shaler its truth lay in the intuitive, not scientific, realm, and therefore "to demand that the evidence be scientific is as idle as it would be to require scientific proof of a mother's love for her child." In short, Myers's *Human Personality* stood as a monument to misdirected effort and misapplied toil.[90]

It is now clear that Shaler's desire for a scientifically informed theology led him to question, reject, or reinterpret many articles of the traditional Christian creed. Where science could teach religious lessons, Shaler was more than willing to learn. But where scientific methods were inapplicable, he preferred to rely on intuition, imagination, and insight rather than to jettison long-established convictions that had sustained the faith of many generations. Just as he railed against Myers's blind faith in science, so, in a review of the *Life and Letters of Thomas Henry Huxley* for the *Critic* in 1901, he chastised Huxley for "his worship of tangible facts" and for his intolerance of all beliefs not supported by empirical evidence.[91] Plainly Shaler yearned for a genuinely synthetic scientific theology in which neither science nor theology solely managed society's moral and metaphysical affairs.

Reactions to Shaler's credo, especially as presented in *The Individual,* were diverse. Some found it compelling; others were less convinced. Among its admirers, no one was more enthusiastic than the anonymous woman who expressed her exhilaration in the warmest of tones. "You can regard this merely as a word of appreciation, of gratitude for the light and comfort I have got from your book 'The Individual,' " she wrote. "I wanted to talk with you and ask you questions, feeling that one, to use your words, who sees so far on dark ways must be able to penetrate the mysteries of life and death, and make the Universe plain. . . . I have much comfort in the truth that what we see is not all that is to be known." A second correspondent was no less sparing in adulation: "I've just finished reading your exceedingly interesting 'Individual.' I feel sure that I, in a measure at least, appreciate and 'absorb' it. Will not the 'solidarity' of organized society in this century apply the peaceful, fraternal teaching of the Christ to their daily life? What a vast theme it all is! I feel cheered, animated, exalted and more full of myself, as capable of being uttered. Most heartily do I thank you for this work."[92] Satisfying though such congratulation may have been, Shaler himself was surprised at the "consolation" that some of his readers had found in the book and mentioned this to his old friend Charles Eliot: "I believe I gave you a copy of 'the Individual.'. . . The book has aroused a curious interest . . . among clergymen and doctors. Both classes have taken it well. I am surprised that the ministers should find the consolation in it that they appear to."[93]

Among dissenters the most telling criticisms came from William James. To James, incidentally, Shaler had disclosed his idealist aspirations for the volume. "Since you left, a book on The Individual. . . has come out," he wrote to James, currently in Britain. "If you have read Royce's great contribution on the Individual, you might take mine as a small appendix."[94] In fact, Royce's Gifford Lectures of 1899–1900, published as *The World and the Individual,* represented the high-water mark of metaphysical idealism. But despite their brilliance and the formidable defenses of F. H. Bradley, the philosophical tides were running too strongly in the opposite direction. Still, however moribund idealism generally, Shaler was dazzled by Royce's metaphysical speculations. He had already expressed enthusiasm about Royce's "marvelous thinking" to Eliot and had told him that Royce was "far and away the ablest intelligence of his species" he had ever known.[95] Moreover, Royce's belief that objects were merely ideas of the absolute was redolent with Agassiz's scientific idealism,[96] while his ethics, grounded primarily in the virtue of loyalty to the community, would have reinforced the role Shaler accorded to sympathy in social evolution.[97] So it seems symbolic that Royce was the philosopher to whom Shaler was most drawn, for both were seeking to defend a vision of the universe that the next generation would substantially discard.

William James in his turn found that Shaler's book left "a good taste behind, in fact a sort of *haunting* flavor due to its individuality" and vowed that he would have to quote it as a powerful case of the "religion of healthy-mindedness." But he was dissatisfied with Shaler's efforts to reinvigorate the argument from design. To James it seemed that Shaler had far too easily made the leap from the apparent regularity of observed phenomena to a purposive intelligence operating analogously to the human mind. For one thing, James found the statistical-probability argument quite unconvincing. Only one world had actually evolved—the world we live in—and it was therefore impossible to know if another form of existence might not have emerged by a different evolutionary route. In any case, James had the suspicion that the semblance of an overall purpose might be no more than a series of short opportunistic "purposes" compatible with the strictest principles of Darwinian orthodoxy.[98]

Between Religion and Science

The vestiges of natural theology in Shaler's outlook would seem to have been the product of a tension between the older teleological tradition and an emerging secular evolutionism. That he was increasingly at variance with the naturalistic temper of the new century's scientific ethos is certainly clear. But just as clear is his mediating role between the two theodicean systems. James Moore has recently spoken of the process by which "the locus of the sacral moved from the noumenal towards the phenomenal, from the eternal towards the temporal, from another world towards this world."[99] There seems, in other words, to have been an ideological and institutional displacement of one religion by another. It was not so much, then, the abandonment of religious ideals but rather their naturalistic translation into the vocabulary of secular science. This restatement, to my mind, is perhaps most pointedly illustrated in the thinking of W. M. Davis, Shaler's most celebrated pupil. In two unfinished manuscripts Davis set out to distill a system of ethics from the Christian heritage, but one divorced from its metaphysical underpinnings. Pressing Shaler's emphasis on behavior rather than belief to the limits, he issued his call, in a manuscript entitled "The Faith We Need," to a scientific faith that would help man "realize his highest ideals." Supernaturalist beliefs of any kind must be held tentatively, he urged, so that if modern science should undermine them totally—as he himself was quite sure would happen—disillusionment would not follow. To Davis religion was rooted, not in abstract ideas about divine revelation, miracle, or the nature of Deity, but rather in upright behavior and moral practice.[100] In this vein, he concluded his second manuscript, "The Greater Influence of the Brotherhood":

If the layman of today were asked to advise his minsters as to the best way of increasing their influence, the reply would be:—Look around you; learn what elements of religion are common to all beneficient [*sic*] religions, and teach them. Learn what elements are peculiar to certain denominations and absent from others of equally good standing; do not preach them. . . . Go lightly on metaphysical abstractions, like the immanence of God; for the thoughtl [*sic*] listener in your congregation—and there are many such—will know that your metaphysical argument proceeds from postulates which neither you nor he can substantiate.[101]

Davis further pursued this theme in the second Hector Maiben Lecture presented to the American Association for the Advancement of Science in 1933, significantly entitled "The Faith of Reverent Science." Here he waxed warm in his celebration of the role science had played in liberating humanity from its subjection to ancient dogma. But now he was concerned with how to preserve society's moral sense in the face of its religious emancipation. And here the radical continuity between the ideologies of religion and science was tellingly exposed.

Our declaration of spiritual independence can be best celebrated by a long festival of cooperation between the organized forces of religion and science— the priesthood and the professorhood—directed to the object which the priesthood has always, but the professorhood has not always held in view; namely, the betterment of humanity; . . . the festival of cooperation should enlist all those members of the priesthood who, recognizing the victory of modern science over ancient theology, desire to replace a good share of their study of theological apologetics by a scientific study of the nature of modern man and of the methods by which he can be ethically moved.

Evolution, of course, had restructured both religious belief and scientific knowledge, while Andrew Dickson White had charted the stultifying tactics of the religious resistance movement against science. So, less cautious than Shaler, Davis happily called for the overthrowing of the religion of the nation's fathers. But, as is equally clear, he wanted to retain traditional ethical ideals and so, like Shaler, he turned to evolutionary naturalism in the confidence that morality—now fortified by science—could stand in independence of its original doctrinal foundation. "For my own part," he concluded, "the ground for that forward-looking confidence is an optimism which is based on a study of the past and which springs from a firm belief in the philosophy of evolution and the faith of reverent science."[102]

In many ways, then, Shaler represents an important transitional strand in the American intellectual response to what has been called, somewhat dubiously I think, the Victorian crisis of faith.[103] Belonging to an era stranded between two ages, he wanted to inject the scientific study of nature with transcendental purpose and, indeed, went as far as to suggest, like his post-Reformation predecessors, that the details of the world environment provided testimony to the overriding control of a Supreme Providence. Reminiscent of Emerson's spiritual journey through Nature up to Nature's God, Shaler felt sure that "whoever deals with a realm of the actual in a proper way finds himself persuaded that he has to do with the infinite."[104] Equally sensitive to the speculations of free religionists and transcendentalists, he firmly anchored his own metaphysical convictions to evolutionary moorings. For, like many other late nineteenth-century intellectuals, Shaler believed that altruism had given birth to religion—not the reverse—and that, in

Shaler in 1894. Reprinted from *The Autobiography of Nathaniel Southgate Shaler* (Boston: Houghton Mifflin, 1909), p. 372.

the ethical perfection of "pure" Christianity, it had finally come of age.[105]

Daily rubbing shoulders with such men as Josiah Royce and William James, giants in the golden age of American philosophy, Shaler would have found much in their conversation to justify a broadly religious conception of the universe. Like these philosophers, Shaler was left to evolve his own reconciliation of science and faith. In one sense it was a theological enterprise doomed to failure. By expecting to integrate the empirical and the transcendental and yet radically distinguishing them and all the while rejoicing in the promise of a liberated rational faith, Shaler represents that Gilded Age mood that created its own "crisis of faith" by displaying an optimistic hope that it could resolve what, in its own terms, was irreconcilable. But as an ideological undertaking it certainly aided the delivery of a naturalistic ethic to society by investing it with all the aura of transcendental authority.

PART THREE

Applications

Science and Society: A Racial Ideology

5

One of the most comprehensive efforts to deal with racial-national questions in biological evolutionary terms was made by Nathaniel Southgate Shaler.

—Persons, *American Minds*

Previously vague and romantic notions of Anglo-Saxon peoplehood, combined with general ethnocentrism, rudimentary wisps of genetics, selected tidbits of evolutionary theory, and naive assumptions from an early and crude imported anthropology produced the doctrine that the English, Germans, and others of the "old immigration" constituted a superior race of tall, blonde, blue-eyed "Nordics" or "Aryans," whereas the peoples of eastern or southern Europe made up the darker Alpines or Mediterraneans—both "inferior" breeds whose presence in America threatened, either by intermixture or supplementation, the traditional American stock and culture.

—Gordon, "Assimilation in America"

To Bostonians of the Gilded Age, New England society seemed the acme of human evolution. With an optimism that reflected Enlightenment rationalism and material progress, they interpreted the history of human society as an advance from savagery through barbarism to civilization and perceived the vigor of their own American nation as the sturdy outgrowth of Anglo-Saxon racial heritage. Despite this progressivist perspective and the humanitarian implications of a democratic ideology, however, there were signs of a nascent sense of national insecurity. Disillusionment with the politics of Reconstruction in the aftermath of the Civil War and the feeling of vulnerability arising from an unprecedented influx of foreigners reawakened the nativist tradition that expressed a xenophobic opposition to internal minorities on the grounds of their un-American ways. Anti-Catholicism, fear of foreign radicals, and Anglo-Saxon infatuation were chief among the stylized themes in the tradition. In New England especially, the self-named Brahmins, for whom Boston was the hub of the universe, self-consciously appealed to descent from English forebears and easily found historical and biological fodder on which to feed their racial ideology. Soon the nativists shared the growing conviction that the physical and mental qualities embodied in a finite number of human types might be diluted through intermixture with other inferior racial strains. In justifying the ways of the nation to the alien in its midst and foreigner on its doorstep, the scientific theodicy could be relied upon to preserve America in the face of an incipient crisis of national identity.[1]

As a son of the South and a long-established Harvard professor, Shaler seemed to possess that qualification to pronounce on race that comes from having the view both ways. On the one hand, his own mother had come from a prominent slaveholding Virginia family, and he, on the other, was a member of a vigorous intellectual circle at Harvard devoted to reforming society by scientific regulation. Both socially and academically, therefore, Shaler felt equipped to reevaluate the black's place in postbellum America, to furnish a comprehensive account of ethnic diversity from the perspective of natural history, and to offer a solution to the nation's immigration "problem." By doing justice to the findings of science and religion, his analysis promised to be truly eclectic. "In many ways," as John Haller has written, Shaler epitomized "the most 'scientifically' accepted attitudes of the late nineteenth century on the Negro, the immigrant, and the so-called 'inferior races.' "[2] Ready-made concepts in a variety of emerging academic disciplines were certainly to hand; evolutionary biology, physical anthropology, the Teutonic school of history, and environmentalist human geography all supplied ammunition for the armory of Shaler's scientific racism. To unravel the scientific sources of his ethnonational policies, therefore, will highlight both the eclecticism of his own cultured racism and, at the same time, the rootedness of scientific theory in social practice.

Evolutionary Biology

Suffused with the assumptions of biological naturalism, late nineteenth-century social theory paraded its fascination with the problems arising from supposed racial characteristics. In his own diagnosis Shaler, like many other naturalists, fastened on the biological factors in the origin and evolution of humanity as of primary importance in finding an effective treatment for society's racial diseases. Whether the human species was of monogenetic or polygenetic origin and whether its progress was by hereditarian or environmentalist means were the two questions uppermost in the minds of contemporary ethnic theorists.

The monogenist faith in the common ancestry of the human race certainly enjoyed the corroboration of the biblical narrative. As chief among its chosen theological sources, the Book of Genesis taught that all the world's inhabitants were propagated from an original pair.[3] Paradoxically enough, however, monogenists had to explain human differences as the result of some evolutionary process, for only in this way could the humanity of exotic peoples be accepted in the face of their failure to conform to the European physical form. Challenging this conventional biblical anthropology, the pre-Darwinian writings of S. G. Morton and his student J. C. Nott set the style for American polygenism,[4] a movement that received the fullest endorsement of Louis Agassiz, who was convinced that the different human races had been specially created for particular geographical regions.[5] Soon the polygenists also numbered among their ranks Paul Broca, Henry H. Kames, and Karl Vogt.[6] At the same time, polygenist racial thought was far from monolithic. Traditionalists, for example, tried to preserve both the strictures of a Mosaic cosmology and the time scale of a short earth history by proposing several separate, distinct creations of the different races. Others, only marginally less conservative, construed the biblical record in such a way as to allow for the existence of pre-Adamite human groups or for the modification of original types by subsequent acts of creation. A third group, drawing inspiration from Lamarckian biology, urged that, since geographical barriers were so formidable as to prevent migration from a single center of creation, races must be regarded as the environmental products of specific physiographical zones.

Darwin's theory of evolution did not, as might have been expected, put an end to the monogenist-polygenist controversy. After all, Darwin himself had confessed that a naturalist confronted with a Negro, a Hottentot, an Australian, and a Mongolian would "assuredly declare that they were as good species as many to which he had been in the habit of affixing specific names";[7] Huxley, meanwhile, hastily commended his theory for "reconciling and combining all that is good in the Monogenestic and Polygenestic schools."[8] Those searching for an interim merger, moreover, needed to look no further than A. R. Wallace, who set forth his own integration in an 1864 address to the Anthropological Society of London.[9] For Wallace, humanity certainly had a common root. But the period of single ancestry was so remote that, by the time the species had acquired those intellectual and spiritual qualities that made it truly human, the various races had already been differentiated by natural selection. Wallace's proposal, in other words, amounted to physical monogenism and moral polygenism.

There are clear signs in Shaler's writings that the polygenist alternative most appealed to him. After all, he was a student of Agassiz and an admirer of Wallace. And however vague in his definitions[10] he had nevertheless persuaded himself that the genus *Homo* could be subdivided into several distinct living species.[11] This position Shaler saw as entirely defensible from a scientific standpoint, and it had the added attraction of addressing those thorny social issues that the assumption of a common human ancestry had recently raised. Polygenism therefore provided an anthropological option that intelligent and humane scientists like Shaler could and did reasonably embrace.

In Shaler's mind there were both cultural and biological reasons for subdividing the human race into a number of separate species. He attached much importance, for example, to the alacrity with which races safeguarded their own customs and ways of life. And he rejoiced that the world's canon of sacred writings could be called as witness to his own words. "The Hebrew Bible and all similar harvests of knowledge is full of these ideas as to the fixedness of racial attributes," he told his readers. "Investigators have only extended the conception by showing that the varieties of men, following a common original law, hold fast to the ways of their forefathers, and that the moral as well as the physical characteristics of a race are to a greater or less degree indelible, whether the given kind belong to the human or to lower creatures."[12] Shaler's prescriptions were even more compelling in that they could be buttressed by the vocabulary of biological science. The interbreeding of fertile offspring had long been taken as prima facie evidence that animals belonged to the same species, and it was patently obvious that it provided an acid test for polygenism. On the face of it, the polygenist case seemed hopeless because all human races were plainly interfertile. But it soon began to be rumored that, in certain cases at least, such interfertility might be more apparent than real. Paul Broca, for instance, had pronounced on the inferiority of human hybrids,[13] and Shaler gladly joined forces, professing that "the off-spring of a union between pure black and white parents is, on the average, much shorter lived and much less fertile than the race of either parent."[14]

Soon a whole range of socioscientific strategies were elaborated to relieve the Bostonian elite of the anxiety over racial purity that all the talk of human hybridization had helped trigger. The distinction between "eugenesic" and "dysgenesic" race mixtures was frequently made, and Shaler tried to implement it by encouraging the crossbreeding of superior races and by opposing miscegenation. He did not hesitate to inform his public that "the hybridization of groups so far apart that they may be termed specifically distinct species is almost always disadvantageous, for the progeny of such unions are more or less sterile and usually have not the vitality of either parent."[15] At the same time, he was just as certain that the English, French, Germans, and Italians could "mingle their blood and their motives in a common race, which may be as strong, or even stronger, for the blending of these diversities."[16] Evidently the old American "melting pot" had served its purpose well. But now, with the changing source of immigration, it needed to be replaced by a new, sanitized crucible in which impurities could be purged away as its racial ingredients were more carefully sifted.

If the assumptions of polygenism went some way to easing racist consciences, there still lurked the suspicion that, given enough time and the right environment, differences between the human species might eventually be obliterated. Whether environmental or hereditarian forces lay behind human evolution was thus another scientific problem impinging on social policy. Initially, at least, Lamarckian evolution seemed particularly suited to the needs of those early social theorists who saw the forms of society as the function of its response to environmental conditions.[17] Most conspicuously, social Lamarckians displayed their evolutionary colors in their racial theorizing, for they believed that the formation of the various races could be traced to environmental adaptation and the transmission of acquisitions by physical and, just as important, by cultural inheritance. No doubt this conflation of biological and social processes aided the casual and widespread misapplication of the term "race" to specific national groupings. Under the influence of Lamarckians like Spencer, many nineteenth-century social scientists thus found that the inheritance of acquired characteristics provided a formula that explained the origin of racial differences and afforded direct links between organic mental evolution and sociocultural progress.[18]

With the advent of the neo-Darwinian model promoted by such figures as Galton and Weismann, Lamarckism began its slow, but not relentless, decline into disrepute. Assumptions of a heritable hierarchy of physical and mental talent—quite independent of environmental experience and therefore of fatally fixed racial attributes—plainly conflicted with Lamarckian biology and were vigorously contested by representatives from that quarter.[19] But just as common were those Lamarckian-inclined anthropologists who tried to resolve tensions by dividing humanity into "historical" and "true" races as distinct species. The basic idea here was that some primitive races had simply remained in evolutionary infancy and were therefore incapable of cultivating civilized patterns of behavior. The Lamarckians had to concede that they might improve over a long period of time through adaptation and use-inheritance, but their inherent inferiority ultimately disqualified them from attaining racial parity with the superior "true" race. In this way Lamarckian evolution and hereditarian biology could be fused to account for the dominance of Caucasian racial character.[20]

Shaler's casual efforts to harmonize these diverging evolutionary traditions have already attracted our attention. Particularly significant in this context is his using his own rapprochement both as a lever to pry biological and cultural processes apart and then immediately as a bond to reunite them. The separate human species had come into being by the conventional operation of Lamarckian mechanisms prior to the time when mankind's intellectual achievements placed it beyond the sphere of natural selection. From that point in the evolutionary past, hereditarian processes had apparently been engaged, and so, with Galton, Shaler could insist that hereditary talent must not be diluted by the crossbreeding of superior and inferior racial strains. His claims were, plainly, riddled with inconsistencies. Not least of the difficulties was the untidy fact that on numerous occasions he had expressed his confidence in the applicability of Lamarckism to mental evolution. But these conceptual maneuvers did nevertheless serve the purpose of reinforcing his polygenist inclinations and of legitimizing his recourse to biological and social criteria for race discrimination.

If human species were to be defined in both biological and cultural terms, it was only to be expected that biosocial signposts would guide Shaler back along the course of human evolution. Within the genus *Homo* the racial groups from which the different human species arose he termed "tribes." These had emerged from an earlier stage of "savagery" where violence and antagonism to the "neighbor of all degrees" had characterized interpersonal relationships.[21] By contrast, the tribal stage—a stage corresponding incidentally with Morgan's "barbarism" phase—represented an advance of profound biological and cultural significance.[22] The tribe, for example, was a distinct biological unit: its members shared a common blood and preserved it by outlawing exogamous marriage. Culturally, social relationships had now developed to the point where hostilities within the tribal group had given way to sympathy. Indeed during this stage the cultural began to gain ascendancy over the biological, as social cooperation at last began to triumph over the primitive struggle for survival. "The fact that in all the several species of men the tribal stage of sympathetic growth is readily attained, that in which all within the group are regarded as friends while all without are enemies, and that very few peoples go further in the enlargement," he was convinced, "is of the utmost importance for our inquiry."[23] Only the white species in fact had managed to press on beyond the tribal stage to modern civilization—

a social evolutionary innovation still on probation. By emphasizing human rights and equality of citizenship, it was, at heart, a hopeful if tentative expression of Christ's religion of universal brotherhood. To Shaler this shed light on the contemporary problems of racial conflict. Civilization, after all, was in essence an enlarged tribe, and attempts to recapture tribal solidarity often failed to take seriously enough the very real differences between human groups. Scientific analyses of the various races were therefore urgently needed if the future success of the civilization experiment was to be ensured. Biology had contributed its share; now physical anthropology must play its role.[24]

Physical Anthropology

If scientific scrutiny was to be brought to bear on the study of racial groups, Shaler knew only too well that the methods of physical anthropology as developed in Britain and France must yield rich returns. He himself, however, had little time to devote to this particular exercise, and his contributions remained largely programmatic. Certainly he did resort to the medical statistics available from military records as an anthropometric source when discoursing on the eugenic excellence of the American male. But overall he was content to outline the areas of anthropological research relevant for the study of racial discontinuities. At the same time, Shaler was undoubtedly familiar with mainstream currents of thought in the discipline, and the corpus of anthropological theory on which he could readily draw throws further light on the interweaving—whether inevitable or artful—of science and ideology in his racial recommendations.

If the institutional beginnings are to be taken as a guide, physical anthropology as an organized field of study emerged around the time of the publication of *The Origin of Species* with the formation of the Société d'Anthropologie de Paris by Paul Broca in 1859 and with the breaking away of the Anthropological Society of London in 1863 from the Ethnological Society (itself founded in 1843) under the influence of James Hunt.[25] But the methodology and apparatus of physical anthropology were rooted in the assumptions of post-Enlightenment developmentalism, and at least for a few years, the project continued to be conducted with pre-Darwinian categories of interpretation.[26] This orientation prevailed also in North America. Here there was no anthropological organization comparable to either the London or Paris societies, and so the Americans tended to draw on the older preconceptions and systems of analysis employed by their European counterparts. Besides, Jeffries Wyman, described by Packard at the time of his death in 1874 as "indisputably the leading anthropologist of America," was always a reluctant evolutionist.[27] It was, therefore, all the easier to turn a blind eye to any monogenetic implications in Darwinism and to remain obsessed by the apparent need to catalog as exhaustively as possible all racial differences. Indeed Hallowell has suggested that the discipline's scientific hallmark in the Victorian period lay in its attempt to formulate a standard index to discriminate the as yet unnumbered races.[28] Specially designed instruments to measure living bodies and dry bones soon became the insignia of the anthropological professional.

The dramatic widening of anthropological horizons arising from the voyages of exploration and discovery had prompted physical anthropologists to turn to anthropometry or somatometry as the focal point of their science.[29] Linnaeus, for example, used skin color to differentiate the varieties of *Homo sapiens,* while Blumenbach, in 1781, used a combination of skin color, hair pile, and skull and facial characteristics in his designation of five human races. Methodological innovations came thick and fast in the nineteenth century, as a host of more sophisticated instruments were invented to analyze the same basic criteria. Consider, for example, the study of crania. Broca, for instance, displayed his technological skills in his invention of the occipital crochet, the goniometer, and the steriograph to analyze the human skull, while on the theoretical front Petrus Camper advanced the theory of the facial angle, Samuel G. Morton and Josiah C. Nott worked on skull circumference, and William Z. Ripley resorted to the "cephalic index."[30] So the coming of the Darwinian theory merely reinforced a movement that was already the chief interest of most nineteenth-century physical anthropologists—the measurement of racial differences. And yet, for all this undoubted craft and toil, no uncontested schema emerged: Linnaeus identified four races, Blumenbach five, Cuvier three, Burke sixty-three, Geoffroy Saint-Hilaire eleven, Haeckel thirty-four, Huxley four, and Deniker seventeen, with thirty different types.[31]

Already convinced of the biological reality of different human species, not to speak of races, Shaler did not need to enlist anthropometry in the service of racial division. But he was well aware of this anthropological tradition, for he took skin color, hair pile, facial profile, and skeletal structure into account in his efforts to delineate the ethnic strains among American Negro stock. Besides, he called for a regional anthropometric survey of the United States so that comparisons could be made of different races in a variety of environmental settings. Such a reconnaissance would seek to determine, for example, the effects of a vertical sun on skin color, comparative racial liability to disease, and the range of cranial dimensions. And it would also include autopsies, like those proposed by Stanford B. Hunt,[32] to quantify environmental influence on brain weight; such research would quickly resolve the problem of "whether the brain of the American African is larger than that of his African prototypes."[33]

At least one anthropometric registry of sorts already existed, namely, the vital statistics of United States volunteers collected during the Civil War by the Sanitary Commission. Here was a ready-made source of actuarial data, and Shaler did not hesitate to use it for observations on subregional differences of population. None of the American volunteers, he noted, was inferior to those born in England, Scotland, Ireland, or Germany in terms of average height, and the statistical differences—where they existed—in mean weight, chest circumference, and head measurement were really insignificant. Indeed Shaler remarked with some pride that, in almost every category, his own Kentucky ranked highest for anthropometric excellence—an achievement no doubt reflecting the strong Anglo-Scottish element in the population![34] Besides, as B. A. Gould had demonstrated, Americans enjoyed a greater capacity for recovery from war wounds than Europeans. The implications of this study of course were plain. Scientific anthropometry had vindicated the claim that the North American continent was an eminently suitable environment for the settlement of the white race.[35]

Teutonic History

Among the other strands in Shaler's racial thinking, the Teutonic theory of American history figures prominently. It suited his historical needs perfectly, for it affirmed the essential Anglo-Saxonism of the United States by tracing the nation's political institutions through the English shire to the primitive folkmoot of the ancient German forests.[36] From among the pages of a host of ethnological studies by European scholars marched the fabled Anglo-Saxon, that archetype of Teutonic racial excellence. In the 1870s, for example, Edward A. Freeman traced nearly every feature of English civilization to Teutonic forebears, and it was therefore easy for Henry Adams at Harvard in 1873–74 to outline the Anglo-Saxon origins of its lusty American offshoot.[37] By the 1880s, when Herbert Baxter Adams of Johns Hopkins gave his support to this new scenario, American history had become a "scientific" quest for the Teutonic antecedents of American institutions. Intoxicated with Old World values, it was almost inevitable that nativists would soon transform such academic history into ideological practice. Indeed Richard Hofstadter was of the opinion that the "Anglo-Saxon dogma became the chief element in American racism in the imperial era."[38] It was a vision that many found exhilarating. Josiah Strong, Congregationalist clergyman and author of *Our Country* (1885), for example, saw the Anglo-Saxon race as the most vital force on earth because it was the bearer of both freedom and faith, while the liberal Woodrow Wilson announced to the readers of the *Atlantic Monthly* in 1889 that American democracy was a unique expression of historic Teutonism.[39]

Some patrician intellectuals, of course, wanted to go even further and, driven by the social construal of Spencer's "survival of the fittest," they acclaimed the American people as the finest branch of the Aryan race.[40] So just as John W. Burgess found vindication in the Darwinian struggle for Anglo-Saxon, and therefore American, imperialism, so John Fiske portrayed the United States as in the vanguard of evolutionary history.[41] "Race" and "nation" were now rapidly becoming interchangeable terms, and by conflating biological constitution, social identity, and cultural heritage, nativist vocabulary helped reinforce the confusion between natural history and national history. By the end of the century, nations were widely thought of as species engaged in a desperate struggle for survival, and it was therefore all the easier to turn the Anglo-Saxon creed into an overt attack on immigrants of a supposedly inferior constitution and culture.[42]

Shaler was, in many ways, a reluctant nativist. He always remained a bit touchy, for example, about Brahmin exclusiveness, and while welcome in Boston homes, he never felt quite at one with the society of which he had become a part. Ever since the time when he "had been a stranger in a strange land and had somehow got the impression that a man from outside of New England did not count, indeed that the rest of the country was in a way superfluous," he felt he could sympathize with the outsider and the exile.[43] But he *was* sure all the while of his ancestral pedigree and social credentials, and these admirably fitted him for membership in the Boston circle. Awed by the achievements of modern Teutonic society, he soon came to share the belief of his Brahmin peers that American institutional democracy rested on Anglo-Saxon foundations, and so he looked to northwest Europe for points of resemblance with American race character.

Shaler's highly popular *Nature and Man in America* (1891) thus naturally contained an extended excursus on European backgrounds. "Europe concerns us almost as much [as North America]," he affirmed, "because it is the cradle of our people, the place of nurture where our race came by its motives and learned how to act its parts in the new theatre of the Western world."[44] Put simply, the vast majority of Americans belonged "to a stock which was nurtured in northwestern Europe and acquired its civilized character in several states of that continent."[45] The inherited traditions that the early colonists brought from Europe thus constituted the raw material from which the nation's culture had been carved. That the New World's environment had induced many modifications, Shaler had no cause to deny; but there was no doubt in his mind that American civilization was nothing less than the flowering of those ancient Teutonic "seeds" carried across the Atlantic ocean.

All the progress of our people has been shaped and controlled by the inherited traditions which they brought from the Old World. This accumulation of motives which determined the spirit of individual and associated action, and had been manifested in the morals, religion, and politics of this country, is the aggregate of experience, extending over thousands of years. About the best thing we can say of our society is that it had maintained and affirmed this precious store of inheritances which are embodied in the family, the churches, and the political organizations. They have cared for these invaluable seeds in such a manner that they have all greatly prospered on the new soil.[46]

Here, then, is an unmistakable reliance on the "germ theory," which, in its most advanced form, proposed the spontaneous reproduction in the woodlands of the Atlantic seaboard of those social traditions of democratic self-government evolved in the forests of the Old World. Carried to America both directly from Germany and indirectly via England, these seeds, when germinated, replicated in the New World the self-same structures of free political life.

As to the vehicles of sociocultural inheritance, Shaler saw the family as the basic institutional unit. Since it was the channel for the transmission of moral and religious acquisitions and therefore guardian of national purity, its traditional Aryan structure had to be preserved from all forms of secular attrition. So he was glad to report that the stability of American family life had survived such abortive experiments as aggregate households, Mormon polygamy, Shaker communes, and socialist cooperatives like the one at Oneida. Not, of course, that Old World inheritance was restricted to family structures; almost every facet of American society, in theory at least, could be traced to some transatlantic antecedent. The various field sports, to take just one other instance, were apparently derived from English roots. "Even base-ball," he insisted, "which appears as a distinctively American game, is but a modification of an English form of sport, which is really of great antiquity."[47]

If, then, the American nation was in a very special sense Anglo-Saxon and its accomplishments represented the dizziest heights of Teutonic high culture, it was entirely natural for Shaler to transform historical investigation into political prescription. The Germanic theory thus provided a powerful tool in his advocacy of immigration restriction. The nation's "original population retained and in a way restored, the primitive social form of the Germanic race." And because, in his eyes, the "American commonwealth could never have been founded if the first European colonists had been of peasant stock," it was "doubtful whether it can be maintained if its preservation comes to depend upon such men."[48] So, like evolutionary biology and physical anthropology, American history in its Teutonic guise could easily be deployed in the cause of political apologetic. Shaler's personal views on the subject, indeed, undergirded as they were by a respected tradition of historical scholarship, played a significant role in the eventual emergence of Boston's Immigration Restriction League, expecially as mediated through the public activities of his student Robert DeCourcy Ward, a leading light in this most recent nativist coalition.

Geographical Influence

If race constitution was, to a substantial degree, determined by bio-social endowment, that was really only half the story. The direct effects of the physical environment, Shaler felt, had to be weighed on the other side of the racial scales. The birth and growth of the Aryan race, he had long maintained, had been induced by the physical geography of northwest Europe, while the effects of the New World's environment could be seen in the regional readjustments of colonial New England's Teutonic institutions.

The cataloging of the effects of nature on culture, of course, has been a perennial theme in the Western tradition. From the Greek writings of Hippocrates and Polybius (who affirmed that "mortals have an irresistible tendency to yield to climatic influences; and to this cause, and no other, may be traced the great distinction that prevail among us in character, physical formation, complexion")[49] to the eighteenth-century florescence of geographical determinism in John Arbuthnot's *Essay concerning the Effects of the Air on Human Bodies* (1733) and Charles Montesquieu's *De l'esprit des lois* (1746), the influence of environment on society had been continuously investigated.[50] In the nineteenth century, with the mushrooming of population-pressure studies, environmentalism did not wane; indeed Otis T. Mason's work on "ethnic environments" did much to keep alive the idea of the environmental determination of race.[51] Within the developing field of human geography, Friedrich Ratzel's two-volume *Anthropo-geographie* (1882, 1891) quickly established him as the leader of the latest environmentalist school, which was popularized ("vulgarized," according to Spate) for United States readers by Ellen C. Semple.[52] Through her writings a whole generation of professional geographers were initiated into the arcana of the Ratzelian program.

Extending to human society the biological theories of his teacher and friend Moritz Wagner (an astute critic of Darwinism), Ratzel had argued that geographically isolated groups were directly affected by the environment.[53] He found that he could now explain, courtesy of Wagner's migration theory, the origin of racial diversity. For after the diffusion of organisms over the face of the earth, geographical barriers prevented the blending of incipient species into the neighboring stocks. Ratzel's theory of the lebensraum, or culture area, thus perpetuated the major tenets of Lamarckian evolution, and when it was conjoined to his organismic conception of the state—elaborated in the *Politische Geographie* of 1897—it lent "scientific" support to expansionist political jingoism. That it served as a source for the pan-Germanic Geopolitik developed in Europe during the interwar years is now well known.[54]

Shaler's *Nature and Man in America* was fully consonant with the spirit of anthropogeography. Indeed precisely because he approved of geography's engulfing history, he felt compelled to "give the latter half of this essay to the discussion of geographic influences upon man, endeavoring to show, at least in a general way, how the development of race peculiarities has been in large part due to the conditions of the stage on which the different peoples have played their parts."[55] When it came to fleshing out this theoretical skeleton with substantive cases, he turned to climatic influence and geographical isolation as the key anthropogeographical ingredients. The conditions of climate, for example, could account for everything from biological variation to the origin of civilization. Indeed it was no accident that the Aryan race, that "most vigorous of all the varieties of man," had grown to maturity along the shores of the Baltic Sea. For here "the winters are severe, and life is a constant struggle with the long-enduring cold which winter brings." Then, casually moving from contingent history to determinist geography, he went on:

In such conditions, as long as food and fuel is abundant the human body attains its greatest vigor, size, and longevity, and the mental powers are at their best; thence southward it may retain these qualities as long as the winter season brings a moderate degree of frost. When we pass into the tropics, though the frame may with certain races exhibit a moderate development, it can not endure hardships as well as where it is bred in contact with an earth which is often snow-clad, and the mental powers appear always to be enfeebled.[56]

Even if the first men and women had evolved from tropical creatures, as Shaler in fact believed, this origin did nothing to undermine the geographical law that the genesis of civilization depended on "the stress of high latitudes, the moral and physical tonic effect of the cold."[57] Of course, the climate could be too severe: if it prevented the practice of agriculture—the social foundation on which he, like Lewis Henry Morgan, believed civilization rested—the barbarous stage of human evolution could not be passed. A rigorous and invigorating climate was one thing; a barren and desolate wilderness quite another. Indeed, precisely because social progress rode in tandem with settled agriculture, the native North Americans had failed to advance beyond the lowest grades of barbarism.

Many later geographers were also to see austere environments and agrarian practice as the twin tracks along which civilization had progressed. J. Russell Smith, for example, one of Shaler's most enthusiastic devotees, spoke of civilization as the product of adversity and insisted that it had only arisen "where nature made production possible only a part of the year, and thus made it necessary for man to work and save up for the time when he could not produce."[58] Again, Huntington was to underscore the necessity of a sterile season for the fostering of foresight and ambition, while Semple argued that the agricultural revolution went "hand in hand with civilization."[59]

Shaler's engagement of the migration-isolation theory in his account of social evolution has already attracted our attention. So far as racial matters were concerned, he believed that the regional fragmentation of northwest Europe had made it the most natural "cradle of strong peoples," not least because it allowed primitive societies to defend their moral and cultural acquisitions against the incursions of foreign conquest. Central Asia, therefore, could no longer be regarded as the birthplace of the Aryan, and it was "with satisfaction, from the naturalist's point of view . . . that in the peninsulas of Scandinavia and in the islands of the British archipelago we find the point of origin of the dominant people in the world; for there more perfectly than anywhere else is the environment adapted to making strong races."[60]

Thus, whereas the *origin* of civilization was determined by climate and agriculture, the *preservation* of cultural identity required geographical isolation.

Just as severe climates could have detrimental side effects, however, so too could regional insularity. Once essential social structures had emerged, further tribal quarantine would only stultify development and prevent expansion. Great Britain, Denmark, Sweden, and the remote valleys of the Alps certainly possessed "the geographical conditions which most favor the development of peculiar divisions of men, and which guard such cradled peoples from the destruction which so often awaits them on the plains." But once a race had "come to possess a certain body of characteristics which gives it its peculiar stamp, the importance of the original cradle passes away."[61] So while it lacked the environmental credentials for begetting civilization, North America boasted the finest landscape for the dissemination of the racial qualities evolved in Europe. Fusing Lamarckian environmentalism and racial hereditarianism, Shaler concluded that the whole Teutonic branch of the Aryan race was particularly equipped by both biological and social history to benefit from the qualities of the New World. Moreover, its geography had induced sufficient physiological improvements to allow him to speak of *Homo americanus,* a new human variant that lived longer than its European counterpart and had greater recuperative powers owing to modifications in the nervous system. For non-Aryans, of course, it was a different story. Lacking the necessary sociobiological equipment, the eastern European, for example, just simply could not respond either to the stimulus of "American air" or to the equally bracing American way of life.[62] America, in short, was only suitable for the Teutons. And reminiscent of the language of Manifest Destiny, Shaler concluded that the broad sweep of human history had been teleologically determined:

Looking back over the history of life upon the earth's surface, the physiographer is forced to the conclusion that its highest estate embodied in the moral and intellectual qualities of man has been, in the main, secured by the geographic variations which have slowly developed through the geological ages. Thus our continents and seas cannot be considered as physical accidents in which, and on which, organic beings have found an ever-perilous resting place, but as great engines operating in a determined way to secure the advance of life.[63]

Shaler was now well supplied with the academic apparatus to bring specific racial groups in the United States under the searchlight of scientific scrutiny. With the apparent backing of these new scholarly industries, he saw no reason to desist from prescribing his own socio-political solutions to the problem of the Negro,[64] the American Indian, the Jew, and the European peasant.

The American Negro

Of the many post–Civil War problems in the United States, perhaps the greatest was the challenge of the Negro and the South. Writing throughout the decades that, according to Haller, "marked the hiatus in the old New England conscience," Shaler's evaluation of the Negro as "the most serious of the questions before the South" typified the change taking place in the late nineteenth-century Bostonian mind.[65] After the fiascoes of Reconstruction, New Englanders who had formerly condemned the South's racist ideology and had fought for black emancipation acquiesced in the 1880s and 1890s to southern legislative and institutional segregation of white and black.[66] Antislavery, of course, was one thing; racial equality, quite another. Many New Englanders were only too happy to leave the ultimate resolution of the problem in the hands of the southerners. The humanitarian assumption of men like Lowell and Norton that the common humanity of all men and women should be expressed in universal suffrage began to be eroded in the light of the postbellum reassessment of the black's place in the America of the future.[67]

First and foremost Shaler believed that the black race constituted a distinct human species, "a lowly variety of man." No less than any other biological species, however, the American Negro was diverse in geographical origin, biological constitution, and intellectual capability. A small proportion of the country's original black population were Semitic Negroes or coastal East Africans from Zanzibar and Mozambique, but the largest element had come from Guinea Coast, where the inhabitants, Shaler believed, were the least capable of the African peoples. With facial expressions that were the "remnant of the ancient animal," this majority was to be distinguished from the rarer Zulu type, whose superior posture reflected a more vigorous, alert, and hardy stock. Diversity, then, there certainly was, but it was quite definitely diversity within a separate species. Surely anthropologists had established beyond all doubt that the Negroes were outcasts from evolution, the victims of a truncated biological history. As Shaler put it, "The folk for all their ethnic variety had, in their African life, come to a state of arrest in their development; they had attained to a point in that process beyond which they were not fitted to go."[68]

The assumption that the black and white races were different species did much to keep Shaler's miscegenation phobia at almost feverish pitch. The very same "scientific" arguments that had buttressed Louis Agassiz's visceral revulsion against direct contact with blacks—mulatto weakness, hybrid sterility, and biosocial dilution of the Aryan stock—found their way into Shaler's racial diagnoses.[69] Relying on blood-blending theories of heredity,[70] for example, Shaler saw the mulatto, that "third something which comes from the union of the European with the African," as inferior to both parent strains and therefore lacking the vital energy for the progress of civilization.[71] Blurring distinctions between the biological and the cultural, he fell easily into the fashion for ethnosocial typecasting, as when he alleged that the "mulatto, like the man of most mixed races, is peculiarly inflammable material. From the white he inherits a refinement unfitting him for all work which has not a certain delicacy about it"; and, "from the black, a laxity of morals which, whether it be the result of innate incapacity for certain forms of moral culture or the result of an utter want of training in this direction, is still unquestionably a negro characteristic."[72]

Already moving ingenuously between nature and society, it is scarcely surprising that such "scientific" observations about hybridity should be ideologically crystallized in the politics of racial segregation. American blacks could indeed hope for social assimilation—provided they could achieve what Milton Gordon has termed "Anglo-conformity"[73]—but not biological amalgamation; there could be personal integration but never racial unification: "The African and European races must remain distinct in blood, and at the same time they must, if possible, be kept from becoming separate castes; there must be a perfect civil union without a perfect social accord; they must both march forward with entire equality of privilege as far as the state is concerned, yet without the bond of kinship in blood to unite them in the work of life,—indeed, with a sense that it is their duty to remain apart."[74]

Expressed in the 1890s, this call to nuptial apartheid progressively hardened during the next decade into an even more stringent community segregation. By 1904, for instance, he could quite unapologetically affirm that it was "desirable to separate the blacks and whites in public conveyances" and that he could understand the offense to a southerner if federal appointments were made by a system "which gave such places of honor or profit to the inferior race and not to his own."[75]

If Shaler's progress report on the blacks was grave, it still was not unrelieved gloom. Both biologically and socially, they had displayed amazing adaptability, first in adjusting to new climatic conditions and an unfamiliar diet and then in accommodating to new customs and a strange language. The African slaves had withstood "the trials of their deportation in a marvelous way. They showed no peculiar liability to disease. Their longevity or period of usefulness was not diminished, or their fecundity obviously impaired."[76] Only in the Middle Atlantic states of Maryland and Delaware had the race failed to maintain its numbers, which merely confirmed Agassiz's conviction that northern climates were unsuited to a race created in and for an African environment.[77]

Biologically nearer to mankind's anthropoid ancestors than to other contemporary human varieties, Shaler believed that the Negro, this "relatively simple species" of humanity, was no less intellectually, morally, and socially inferior.[78] With little historical sense and less literary culture, the African had developed no conception of formal law, no organized commerce, and no education. Too often these facts had been ignored. In the New World, for example, political machination and popular myth alike had conspired to make the black a pawn of the "worst political rabble that ever cursed the land."[79] A thorough program of Americanization was needed. His own proposal was a blatant ideological incarnation of the biological speculation that the blacks had yet to attain full evolutionary potential. In a nutshell, black progress was conditional on the imitation of a superior culture: "Any further progress must depend on an imitation of a mastering race: it cannot come from the innate motives of the folk."[80] Quite simply, the race was biosocially deficient in those very qualities necessary to the creation and maintenance of social structures beyond the order of barbarism. What better illustration could there be than the current political situation in Haiti? It presented an undeniable social analogy to the biological process of reversion to ancestral type. For here, the Negro had been left unsupervised and, not surprisingly, had failed to preserve even the semblance of Aryan social structure. The lesson to be learned from this bureaucratic mess was simple: whenever "such communities have remained apart from the influence of the whites for a generation, they commonly show signs of a relapse towards their ancestral estate."[81]

Quite predictably, Shaler's assessment of black political involvement in the postwar period was full of ambiguities. It was easy, of course, to castigate the extravagance of those opposed in principle to the governmental appointment of blacks. By cosmetic touches, however, it was just as easy to temper that extreme without losing its thrust. If, as he claimed, the ballot box was "as dangerous a plaything as a gun" without some informed political expertise, it was plain that "until the negro acquires the habits of thought and action which make it an effective arm, he will be impotent to use it to any good effect."[82] The source of the problem lay buried in the race's African past. There blacks had failed to attain the stage of social evolution where monogamy superseded polygamy, and thus they had been unable to forge strong family relationships. With no family structure there was no satisfactory vehicle for the transmission and preservation of social inheritance, sexual morality, or altruistic behavior. Besides, the absence of any developed system of religion beyond that of primitive animism meant that there was nothing intrinsic to black society to encourage personal determination, business partnerships, or more general social action.[83]

Still, Shaler did remain committed, at least in theory, to the political ideals of a democratic ideology, even if his "science" was pushing him in the opposite direction. Nowhere is this tension more apparent than in his equivocation over slavery and manumission. The need for emancipation, he freely confessed, was morally inescapable. But what was best ethically might not be best ethnically. For one thing, emancipation had ultimately retarded black social evolution by separating the race from its Aryan masters—a bitter irony indeed, for the blacks' greatest asset, Shaler judged, was their "capacity for adjustment by a process of imitation, a power denied to the higher races."[84] Moreover, the social security of the slave system had been stripped away, leaving the Negro exposed to the chilly winds of ethnic struggle in an environment where the competition was far too one-sided. It was all too easy, he reflected, to forget that the parallel between the social structures of African tribal life and of the slave system had eased their transition to the New World. "The master took the place of the chief, to whom the black for immemorial ages had been accustomed to render the obedience and loyalty which fear inspires," Shaler insisted; "under this white lord's control, he was hardly more a slave than before. On the whole this lowly man gained by the change in the quality of the servitude: by the contact with the new master he gradually acquired some sense of the motives of the dominant race."[85]

By 1900 Shaler had rather tempered his wholesale praise of slavery. Now distinguishing between domestic and plantation slaveholdings, he candidly applauded the former for the cultural enrichment it had fostered through social intercourse, but the latter, associated with the rise of the cotton and sugar industries, he stigmatized for its breakup of families, cruelty, educational privation, and inevitable moral reversion. And yet even plantation slavery had advantages, for, as an agent of natural selection, it weeded out those constitutionally unfitted for domesticity. The two classes of slave, indeed, were as different as two races.[86]

Taken as a whole, American slavery had "lifted a savage race nearer and more rapidly towards civilization than had ever before been accomplished."[87] But it was ultimately unacceptable. Its "aristocratic" spirit, symbolic as it was of the disjunction between northern and southern political temperaments, was thoroughly inconsistent with democracy. Besides, on moral grounds, the social benefits of domestic servitude had been vastly outweighed through the abuses associated with the rise of massed industries between 1820 and 1860—a period during which, as Shaler put it, "the educational value of slavery to the black was in good part lost."[88]

For the black race, then, Shaler remained certain that the only escape route from the prison house of the past was by way of education. In his vision, Spencer's theory of primitive mentality, which implied that educational aims must run in double harness with mental evolution, furnished a much superior model to the old Lockean tabula rasa.[89] Foremost among its curricular implications was the fatuity of exposing blacks to the higher scholastic forms. In 1870 Shaler even questioned the value of mechanical training for the Negro (and this despite the technical schools for blacks in Atlanta University, the Hampton Institute, and the Tougaloo Normal and Manual Labor School), although he came increasingly to support instruction in crafts, basic finance, and music. Besides, there were useful social by-products. For to provide a race with technical expertise would ultimately mean their dispersal throughout the country; to educate a people was to scatter them. And this result, Shaler rejoiced, would bring the black into contact with larger numbers of Aryans and thereby better their physical and moral standing.[90]

Shaler's thirty-year study of black America had convinced him of one thing, equivocations on policy notwithstanding: the Americanization of the Negro must be achieved without northern intervention. Federal meddling had already occasioned some of the worst evils of misapplied humanitarianism and misdirected philanthropy. It was high time the nation recognized that only the South truly understood black psychology. In pursuit of this ideal, indeed, Shaler went as far as to defend the impulsive lynching of black criminals in the southern states as essentially "American" in character and free from the cruel excesses of the Old World![91] Though later retracted, this effort at national self-vindication merely demonstrates the extremes to which New Englanders could go in renouncing the politics of Reconstruction.[92]

The North American Indian

To the degree that Shaler's writings on the Negro were profuse, if not prolix, his treatment of the North American Indian was summary, popular, and thoroughly derivative. His single account of "The Aboriginal Peoples of North America" —chapter four of *The Story of Our Continent* (1892)—was neither as subtle nor as sophisticated as Powell's essay on the subject, which Shaler was soon to publish in the three-volume *United States of America* (1894). But it did, nevertheless, synthesize many of the prevailing attitudes to the Indian, dating back to the pre–Civil War era; herein lies its significance for mapping out the territory of racial ideology in the culture of nineteenth-century America.[93]

The precise sources of Shaler's portrait of the American Indian are impossible to identify, not least because of his bibliographical abstinence. Numerous studies were certainly already available, notably those by Samuel Stanhope Smith, Lewis Cass, and Henry Schoolcraft on ethnology and those by John Pickering, Peter Stephen Duponceau, and Albert Gallatin on linguistics. But with a theoretical scaffolding constructed out of the presumed stages of social evolution, it seems more likely that Shaler would have turned to Lewis Henry Morgan's pioneering study *The League of the Iroquois* (1851) and his later classic, *Ancient Society* (1877).[94]

Fundamental to Shaler's own appraisal was his conviction that "man did not originate on this continent, but came to it from the Old World."[95] From China, he reckoned, along the Alaskan archipelago, the first settlers found their way to the west coast of the United States. To sustain this claim, moreover, he casually assured his readers that recent empirical research favored a far more recent date of colonization than many scholars had assumed. On this Powell bluntly demurred. His reading of the archaeological and linguistic evidence had persuaded him that the "Western continents" had been inhabited long before the acquisition of articulate speech and that theories of immigration from the Eastern Hemisphere were thoroughly unsatisfactory.[96] The burden of proof certainly lay with Shaler, but his overriding textbook intention was to convince his audience that no part of the aboriginal population "had ever attained to an economic organization which deserves the name of civilization," and he pressed ahead with that particular objective. In a few cases, he had to admit—notably in the region stretching from northern Arizona southward along the Cordilleras to the southern reaches of Peru—the inhabitants "had advanced beyond the savagery of the other tribes," perhaps as far as the stage of barbarism. But this achievement was the exception that merely proved the rule.

As might be expected, Shaler's account of Indian social retardation was suffused with the logic of environmentalism. Because they "dwelt in a very open land, unlike the peoples of Asia and Europe, who were nurtured in districts more or less completely shut off from the other parts of the world by decided barriers" and because the only sufficiently isolated tracts were too arid or too sterile or too tropical, their evolutionary growth had been retarded. Geographical assumptions, therefore, encouraged Shaler to see the Iroquois as atypical of American aboriginal culture; their attaining "the strongest and most enlightened savage state" merely reflected the unusual measure of protection they had enjoyed.[97] An uncongenial physical environment was thus a severe evolutionary hindrance, and the absence of animals capable of domestication made the North American world even more hostile to the genesis of civilization. So when Powell confirmed that the establishment of patriarchal societies depended "upon the domestication of animals and the possession of flocks and herds" and when Morgan umbilically tied civilization to the idea of property, Shaler could only concur.[98] For without domestication there could be neither settled agriculture nor individual property ownership, and these, as we now know, Shaler took as the mainsprings of civilization. Furthermore, Indians had never been able to cultivate mining or other mechanical arts, which made them peculiarly vulnerable to the whims of fortune, as witness the social retrogression accompanying the eastward migration of the buffalo. Not, of course, that the Indian mind was constitutionally impotent. Their inventive powers were certainly displayed in an albeit primitive technology of war, while their social skills found expression in a tribal organization akin to an extended family. But with no writing, no legal system, and no developed literature, Indian culture had remained verbal rather than literate and therefore of necessity shackled to the primitive evolutionary stage of savagery. External physical environment and internal mental endowment had merged to make the Indian the victim of a soulless evolutionary law.

The Jew

Since Shaler's racial neurosis was stimulated so directly by the endemic problems of America's ethnic minorities, it was inevitable that he would sooner or later turn his attention to the Jews. That any race could outlast some two thousand years of persecution greatly intrigued him, and in a historical review of anti-Semitism since the diaspora, he determined to identify the source of their cultural grit and determination.[99] Initially anti-Semitism was the illicit child of the union between the Roman and Christian worlds. For when the xenophobic sentiments of Cicero, Seneca, and particularly Tacitus were coupled with those of "Christian Europe," anti-Jewish feeling gained a solid foothold. Even now, their attainment of civil rights had not lightened the weight of the centuries' resentment; indeed they seemed ubiquitously more distrusted than ever, even in North America, where Shaler sensed a rejuvenation of anti-Semitism in the wake of the Dreyfus affair.[100]

Given his familiarity with the anthropometric tradition, the absence of any attempt in Shaler's writings to identify somatic criteria for the Jew would seem surprising were it not for the conspicuous failures of that project to provide any satisfactory schema.[101] In the Jewish case, Shaler was quite content to conduct his analysis on the cultural plane, and to that extent his reactions mirror the anti-Semitic feeling then gathering renewed force in Europe and America. There were, certainly, obvious differences between Europe's heavily politicized anti-Semitism and its less cohesive American counterpart.[102] But the anti-Semitic mood did progressively harden in the United States under the impetus of mass Jewish immigration and typically found psychological security in its resort to those racial stereotypes of the Jew as nouveau riche, financially shrewd, and nationally disloyal. Besides, the cultural life of the eastern European Jews—their institutions, publications, and societies—remained untouched by the American spirit, which acted to reinforce their alien image. In an era of mass strikes, social upheavals, and economic instability, a growing xenophobia manifested itself in a proliferation of patriotic societies and heralded the demise of the Christian democratic tradition that envisioned America as an asylum for the oppressed of all nations.[103]

Shaler could not agree with those who traced anti-Semitism back to religious sentiments. In his mind it was just a special case of the human psyche's instinctual allergy to ethnocultural differences. To identify just exactly what these sentiments might be, he endeavored to reconstruct his own response to particular Jews in that critical moment of personal encounter. The "certain definite repulsion" that he—and for that matter some twenty of his friends—experienced in first meeting members of the Jewish race was characteristically attributed to their shrewdness, avidity, and unfriendliness. If, as Shaler of course believed, these stereotypical judgments were thoroughly Aryan, it was understandable that this sense of psychological discomfort would continue to govern subsequent dealings with the race. Again, however, Shaler's racial estimates were not untempered. From a more strictly "scientific" standpoint, he ventured, the Semite could be seen as the highest species of humanity—"clearly the ablest folk the world has ever known."[104] Besides, their unsurpassed ability, morality, and charity had prevented them from sacrificing their own way of life to the customs of the people among whom they settled. To them acculturation was nothing but adulteration. And so their determined sense of identity had enabled them to resist assimilation and to preserve their distinctive qualities throughout two millennia of dispersion.

Like many of his generation, it is clear, Shaler's anti-Semitic prejudices were covert and cloaked by a lingering respect; as Higham has put it, "many Americans were both pro- and anti-Jewish at the same time."[105] By this attitude Shaler typified the paradox at the very heart of the nativist movement. The same populists who contrasted the "Shylocks of Europe" with the "toilers of America" were also the groups most deeply committed to those social ideals that had made the United States a beloved homeland for thousands of expatriate Jews. Writing during an era when anti-Semitism spread like wildfire, Shaler's prognosis can be seen as representative of those who were concerned to balance "realistic" racial thought with benign humanitarianism.

The European Peasant

The Jewish scare, in fact, was merely part of a more general immigration problem that, beginning in the early 1820s, reached its zenith in the 1880s. Under the sheer weight of numbers, old-stock Americans began to lose confidence in the processes of assimilation, especially when the newcomers tenaciously clung to their own folkways in disregard—defiant or defensive—of American culture.[106] Furthermore, this newest wave was heavily Roman Catholic, in marked contrast to the Protestant influx of the eighteenth century, and therefore encouraged the establishment of a clutch of anti-Catholic societies in the 1880s and 1890s: nativist fraternal orders, Masonic Lodges, and the American Protective Association. Again, the new arrivals fell foul of ethnic typecasting and inhabited the American mind in the form of nasty caricatures. Fleeing from poverty and settling in urban squalor, southern and eastern Europeans were seen as carriers of the social and economic diseases with which immigrants, nativists believed, were generally infected. As Higham has put it, "Most of the hatred of Italians, Slavs, and Jews consisted of general anti-foreign attitudes refracted through specific national stereotypes."[107] The meeting of Chicago anarchists in Haymarket Square in May 1886, at which a bomb exploded, did little to pour oil on troubled waters. Quite the reverse. It awakened antiradical strands in the nativist tradition and precipitated an outburst of manic nationalism near to epidemic proportions. The American consciousness, already haunted by ghosts with colored skins, now had to cope with the new specter of imported European anarchy.[108]

Doubtless reflecting the prevailing antiradical mood, Shaler's 1878 monograph *The Nature of Intellectual Property* warned of the imminent dangers of American communism. Sharing sentiments akin to later exponents of the closed frontier/safety valve theory, Shaler hoped that, so long as American society was "pliant" enough to allow the most disadvantaged social and material improvement, communism would remain distasteful to the nation's political palate. There were, however, incipient signs of civic vulnerability. Contemporary calls to communist ideology, "to plunder the few for the profit of the many," suggested that radical politics had already begun to take root in American soil. This new growth would have to be carefully monitored, otherwise the legislation governing real estate and chattel property, despite thousands of years of accumulated tradition, would rapidly be swept out into the ocean of vulgar populism. Meanwhile, the absence of satisfactory copyright and patent laws made "intellectual property" particularly susceptible to communist subversion. "If the proletariat spirit—that spirit which is, in fact, the return into man of the instinct of greed proper to the lower animals—overturns the protections guaranteed by the contracts of society to its inventors and authors," he insisted, "the results to the principle of property will be disastrous."[109]

If the threat of communism caused Shaler some political unease, it was nothing compared to the vexation he experienced in the face of the "new immigration." During the 1880s the countries of southern and eastern Europe had begun to be represented in large numbers for the first time, and Shaler rose to the new challenge by exercising his journalistic muscles in the pages of the recently established nativist weekly, *America*. The original European colonists, he told its readers in 1888, were made of much sterner stuff that these latter-day immigrants, who, since the Civil War, no longer composed a "pure" racial stock nor a normal social mixture. The destitute of Russia, Poland, Hungary, and Italy were simply less fit material from which to carve American citizens. As if that were not bad enough, their Catholicism bothered Shaler even more. For some fifteen centuries the Roman Catholic church had stamped an indelible mark of inferiority on the institutions and peoples of Catholic Europe by siphoning off the cream of its talent for the priesthood and other religious orders. This continual drain had seriously diluted the heritable qualities of peasant racial stock.[110]

The same racial-religious refrain echoed in Shaler's mind throughout the next decade. Consider, for example, his 1896 reflections on the Scottish element in the American people. A temperamental difference, he began, could clearly be discerned between the Lowland Scottish and the Highlander, a difference traceable to hereditary factors. Now, moving away from his earlier stress on racial purity, he suggested that the literary, political, and scientific excellence of the Scandinavian Lowlander reflected the "mixed blood" of "strong, related but varied peoples." By contrast, the obdurate conservatism of the Highlander, crystallized in a continued adherence to "the old faith of Rome" and an intransigent resistance to the civilizing virtues of Anglo-Saxon culture, reflected a purely Celtic heritage. So too with the "Celtic Irish." Like the Highlander, Shaler saw them as "an unmixed race, perhaps the purest blooded in western Europe; their geographic isolation having kept them from the intermixture due to the Germanic and other migrations."[111] In the New World, these respective psychophysical traits found the fullest cultural expression. For whereas the European peasant class clung closely in clanlike groups and displayed a thoroughly predictable immigration pattern, the Lowland Scottish exhibited all the qualities of the pioneering spirit by disseminating themselves, their culture, and their commerce throughout the whole country.

In this assessment of the European immigrant, Shaler was by no means alone. Theodore Roosevelt, for example, had written in praise of the Scotch-Irish pioneer as a "bold and hardly race."[112] Then Frederick Jackson Turner expressed the opinion that the coming of Italians, Poles, Russian Jews, and Slovaks was detrimental to the social organism of the United States; in his mind, these immigrants were "counteracting the upward tendency of wages" by "encouraging the sweatshop system."[113] Again, Francis A. Walker's 1899 statistical survey advanced the startling thesis that the population would have increased as rapidly from 1830 to 1880 if not a single newcomer had been admitted. The falling birthrate among native Americans, he felt compelled to tell the nation, was nothing less than a numerical demonstration of their disinclination to rear children in the sordid environment created by mass European immigration.[114] As with his expressed sentiments on blacks, Indians, and Jews, Shaler's judgments on the European peasant thus fitted snugly into the faddish racial science of his day.

Solutions

Given his enthusiasm for evolutionary biology no less than for natural theology, it is no surprise that Shaler's proposals for the resolution of racial tension in the United States lay in a fusion of reason and faith, science and sympathy. Eugenic control, immigration restriction, and Christian tolerance were chief among the eclectic proposals he advanced.

If the modern scientific foundations of eugenics lie in the writings of Darwin's cousin, Francis Galton, who first coined the term, its social hereditarian counterpart was already in vogue during the middle third of the nineteenth century, when its spirit of "optimism and confident manipulativeness" found favor with America's Progressive generation.[115] Sharing working assumptions about human heredity largely derived from Lamarckian science, physicians and publicists alike soon discovered in biology plausible sanctions for social action. As Nathan Allen, an enthusiastic advocate of social hereditarianism, explained in 1869, "Like begets like, is for the existing state of science an ultimate fact."[116] Still, if Galton did little more than reinforce a long-familiar tradition of social thought, the impressively empirical tenor of his *Hereditary Genius* (1869) excited many Europeans and Americans with the vision of pioneering a scientific program to raise inherent human ability. And with racial self-consciousness gathering momentum, such respected biologists as Charles Darwin, Alfred Russel Wallace (although he later repudiated it), and E. Ray Lankester all implied that evolution sanctioned a breeding policy for humanity, while Galton, himself an agnostic, found in eugenics an emotional surrogate for religion.[117]

However popular the hereditarian mood, its underlying motivations were both diverse and changeable and serve to illustrate its inherent flexibility. Those impressed by Cesare Lombroso's deterministic theories of criminality, with its scientific savor of atavism and reversion, at first stressed the need for environmental reform. Later, they held up eugenic control as the only means of coming to terms with moral degeneration. In Richard Dugdale, one of the movement's most influential postbellum expositors, both strands found expression. His famous study of the Jukes family, in which mental deficiency had persisted for several generations, carefully balanced the influence of environment and heredity, and so he was claimed by reformist and eugenicist partisans.[118] But the days of such synthesizers were numbered. August Weismann's militantly anti-Lamarckian account of the germ plasm and the hereditarian educational psychology of G. Stanley Hall and James Mark Baldwin broke through the environmental barrier to a sense of the prime importance of inheritance. The idea quickly gripped popular imagination, so that the habitual recourse to environmental optimism was unable to survive unaltered in the new climate of opinion. By the end of the 1880s, the eugenics movement had been born, if not as yet christened.[119]

Hereditarian social thought, it should be noted, was not of necessity bound up with xenophobia. But "racial realities," as they were called, seemed to indicate a disproportionate incidence of insanity, tuberculosis, criminality, and alcoholism among particular ethnic groups. So with mounting immigration and a claustrophobic feeling of closed space, a new batch of ethnoeugenic policies received an airing during the mid-1890s. Campaigns to secure miscegenation and sterilization laws and increased agitation for immigration restriction seemed to be the inevitable social implications of the young science of eugenics.[120] In the new biology both racists and restrictionists had found the reassurances they needed for rationalizing political action.

Long anxious to find scientific sanctions for national policy, Shaler saw in Galton's research the means of easing the burdens of a multiracial society. "The truth is that a man is what his ancestral experience has made him," he began. "He is but the momentary expression of the qualities bred in his race for immemorial ages. No sound national polity can afford to disregard the great truths of inheritance, which, though long known to men, have first taken a clear shape before their eyes within the limits of our own generation."[121] No less concerned for national efficiency, British eugenicists were similarly to present their cause as a patriotic one; only patriotism could provide the necessary incentive and discipline. Thus F. C. S. Schiller, for example, prophesied that the "nation which first subjects itself to a rational eugenical discipline is bound to inherit the earth."[122] Shaler happily concurred. Galton's findings, he was certain, unambiguously pointed to the need for a breeding program for the human species. After all, had not selective breeding already enjoyed a tremendous success in the animal world? Without it, domestication, and therefore civilization, would still be in a primitive state. And yet, despite this vast array of zoological evidence, the nation's most revered and august organizations were joining forces to prevent the application of eugenic science to human society. The whole future course of human development was thus at risk.

Just as eugenicists were varied in their motivations, so too were they divided in their policies. Some, like Galton himself, advocated "positive eugenics"; in the Huxley Lecture for 1901, he thus confirmed that the production of the best stock was "far more important than that of repressing . . . the worst."[123] Others were more negative in their aims. Certainly Shaler dissociated himself from the extreme measures that some had toyed with—like killing the unfit in infancy.

But his sympathy for negative eugenics is plain in his expressed concern that the weakling should not be allowed to paralyze the national organism by transmitting hereditary deficiencies:

There is reason to believe that we are now coming to a stage where the disease tax, which has hitherto mounted with the advances of culture, is to be diminished by the extirpation of maladies. This is evidently not to be accomplished by any hideous Spartan plan of destroying weak infants, but by a fitting care that such come rarely to life and that they do not send their weakness on to mar the race. We are rapidly coming to a sense that while the individual life has an absolute right to a seemly place in the world it has absolutely no inborn right to send its infirmities onward through the generations.[124]

If eugenics was to play the role in America's future that Shaler envisioned, it was plain that it needed to be supplemented by a more specific racial strategem. Immigration restriction was the natural issue of the marriage between ethnic and eugenic ideologies. By the 1890s older Brahmins were on the verge of an intellectual decision in its favor.[125] The changed character of the new immigrants now flooding in from the relatively backward agricultural communities of eastern and southern Europe, together with black emancipation, the problems of Reconstruction, and the placing of the conquered Indian with his broken culture on government reservations had all compelled New Englanders to reassess the various theories of assimilation. A restrictionist ideology was the inevitable outcome. Some rationalized it on the grounds of America's historic "Anglo-conformity." Whether in its discredited Teutonic form or based on an altogether instinctual infatuation with Anglo-Saxon culture, they used their admiration for English institutions to challenge the wisdom of allowing further unrestrained immigration. Others turned to a restatement of the "melting pot" ambition. Its earlier advocates had dreamed of the United States as a totally new blend of humanity, but that idealist vision had been warped by the suspicion that the ingredients of the "urban melting pot"—compared with its old frontier counterpart—were tainted with the worst ethnic ailments. These doubts and uncertainties, reinforced by incipient fears of losing their cultural identity, soon forced New Englanders in particular to question the assumption that all stocks could be successfully melted.[126]

Then there was Francis A. Walker's obsession with the fecund foreigner. At a time when the frontier was pronounced closed and limitless natural resources declared a myth, his dramatic immigration statistics for the 1880s (five and a quarter million) were as depressing as his report on the falling marriage rate among native Americans.[127] Via Walker, the science of demography entered the race arena just as confidently as biology had already done. Bolstered by its findings, the seaboard states of New England sought to protect themselves from the burdens of the heaviest immigration in the nation's history, and out of this mood was born the Immigration Restriction League of Boston. Posing as the guardian of respectability, morality, and standing, it readily appealed to the Boston intelligentsia. All of its founders were practical-minded intellectuals from long-established Boston families and had attended Harvard in the late 1880s, where they had come under the spell of Shaler as dean of the scientific school.[128] The alliance never really enjoyed widespread political clout, but it did achieve some intellectual prestige from its president and list of vice presidents. John Fiske, for example, was eventually persuaded to take up the presidency by Charles Warren, Robert De Courcy Ward, and Prescott Fransworth Hall, who had already approached Henry Lee, George Edmunds, and Walker himself. Most fittingly, Shaler, who had helped rationalize the very idea of restriction, joined the roster of supporters by becoming one of the ten vice presidents.[129]

Shaler's involvement with the league was only the most recent expression of a long-standing nativism. He had been concerned for many years that Anglo-Saxon Americans were losing their breeding power and was fearful that the solid yeoman backbone of the nation would be entirely swept away. He told Norton as much in 1879 when he bemoaned the fact that the residents of Lennington, Virginia, were "all that is left of the *yeoman*; simple, devout, careless of gain. They know nothing of our modern improvements in human nature but they keep some of the best the past has won."[130] What a contrast they were to the recent arrivals, who he predicted as early as 1873, would sorely "try the digestion of our New England civilization."[131] So during the social and political upheavals of the 1880s, he threw his weight behind several patriotic projects, only one of which was his assuming the vice presidency of the Massachusetts Society for Promoting Good Citizenship. His estrangement from the foreign born had intensified during the strains of these years as he came to the conclusion that the immigrants had not played fair with the nation that had shown a welcoming face to the world's refugees. They had remained permanently unassimilated both culturally and politically. To Shaler, of course, assimilation meant Americanization, which implied that any individual or group not sharing the traditions of Anglo-Saxon culture should be denied entry. Immigration restriction was not an option but a necessity, for "the laws of inheritance" had shown that these aliens would "remain laggards in the way of progress."[132] Science once again had spoken.

To achieve his objectives, Shaler pressed for educational and health tests to be applied before any immigration permit was granted. The British, Germans, and Scandinavians, of course, would pass with flying colors, while the less desirable southern and eastern Europeans would fall by the wayside. These were sentiments widely shared by Shaler's Boston peers. Henry Cabot Lodge, for instance, the league's first and most influential congressional advocate, singled out the pauper, the criminal, the diseased, the anarchist, the communist, and the polygamist for exclusion; with the support of Senator McCall of Massachusetts, he managed, after strenuous efforts, to have the literacy bill passed by Congress.[133] In their case, no less than in Shaler's, immigration restriction was the best means of protecting America from the delusion that national quality was the same as population quantity.

For all that, Shaler was reluctant to leave the race problem solely in the hands of eugenicists and restrictionists. As time passed, he more and more felt the need to integrate the facts of biology and history with the ethical ideals of the Christian heritage. A fusion of science and sympathy furnished the ultimate clue to relieving racial tension. As Persons has quite rightly said, "Shaler believed that America was fortunate that social evolutionary progress had brought the dominant Aryan race to the point where it could grasp the principles of the Christian religion and the method of scientific investigation."[134] Progressivists, we have already noted, had long been enthusiastic about sympathy. Adam Smith, for example, traced the ethics of social living to the human capacity to sympathize imaginatively with others, while

Adam Ferguson affirmed that "love" carried the "attention of the mind beyond itself" and thereby freed men and women from enslavement to selfish desires and animal appetites.[135] That his outlook was consonant both with that tradition and with the scientific spirit greatly appealed to Shaler.

The origins of racial conflict, Shaler pointed out, lay deep in humanity's evolutionary past. Competition and cooperation, hatred and sympathy were the warp and woof of the human species' natural history. Racial hostility was thus a reinstatement of hatred at the expense of sympathy and therefore little more than an atavistic return to the primeval struggle for survival. By contrast, the Christian religion had been the chief agent of altruism, sacralizing its virtues and emphasizing its universal claims. More than any other religion, it was the bedrock from which the vision of the brotherhood of man had been hewn. To Shaler, therefore, "real Citizenship consists in the application of the motives of Christianity to human relations. That religion far more than any other belief has laid for its foundation the idea of the brotherhood of men, and of the need of their living in affectionate relations with one another."[136] So when the American nation fabricated fictions to maintain tribal isolation and to foster race prejudices, it was turning its back on its own religious heritage. Was it not "evident that while Christ set his face against all the sins of the flesh, he above all opposed the motive of tribal pride and hatred"? As he went on, "With a clearness of understanding which puts him immeasurably above all other leaders he saw straight to the center of the ills that beset mankind; saw that they lay in the lack of friendliness for the neighbor of every estate. He sought the cure where we have to seek it, in the conviction that whatever be the differences between men, they are trifling compared with the identities which should unite them in universal brotherhood."[137]

For the future, a blending of science and faith was needed. If the findings of science and the dictates of faith could be expressed in personal encounter, an empathetic understanding would be aroused and a final solution to the race problem achieved. "Here at least science and enlightened faith are one," he mused. "All that learning, religious or scientific, gives us leads to the same conclusion, which is that the aim of life is to bring men to love one another."[138]

Doubtless these final comments seem strange in the light of Shaler's earlier veiled, and also overt, attacks on various ethnic groups. To try to knit together these strands into a coherent racial policy, I believe, would be profoundly misguided, not least because they illustrate some of the very ambiguities with which nativism was infused. Committed to the ideals of Christian democracy, nativists struggled to find some means of expressing that heritage in the face of economic and social problems of an altogether new order. In Shaler's case this national tension mirrored his own personal dilemma. On the one hand, the social pressures of the era jarred with his "sympathetic" temperament. On the other, his recurring concern to reconcile science and faith, the social and the individual, the scholar and the citizen, and ultimately nature, humanity, and God stemmed from a widening gulf between the ethical imperatives of religious faith and the ideology displayed in public practice.

Typical of the scientific humanitarianism of the New England mood during the Gilded Age, Shaler thus emerges as a key figure in the attempt to chart the historical configurations of racialism as a creed, a scientific doctrine, an emotional attitude, and a symptom of national insecurity. Plainly he made few original contributions to the study of race, but by distilling the relevant "findings" from a variety of sciences, his popular synthesis stands as a focal point for elucidating a dominant strain in the social thought of the period. In many ways his contribution was a response to those structural changes in late nineteenth-century American society that fostered the fin de siècle sense of crisis and depression: the advent of industrialization and urbanization, which created emotional needs that social hereditarianism could meet; the secularization inherent in the shift from theological to scientific schemes of reference as a means of legitimation; a conservatism striving to retain continuity with traditional social values; and a lingering optimism that placed confidence in biosocial science to grapple with ominous trends toward moral entropy. Perhaps ultimately Shaler's resort to racist ideology can be seen as "an adaptational tactic," to use Rosenberg's term, a tactic that helped "to create a cognitive world picture through which particular individuals could impose a consoling order upon a continually shifting reality."[139]

6 Environment and Inheritance: A Geographical Interpretation of History

Shaler—geologist, humanist, idealist—led his pupils to see, beyond and through the facts of geology or palaeontology, the larger problems and relationships of the earth and man. A geologist by training and yet a geographer in instinct.

—Dodge, "Brigham"

Shaler, one of the greatest geographers of America.

—Sauer, *Pennyroyal*

The interweaving of geography and history, the ways in which the human story had unfolded over time across the face of the earth, was a subject never far from Shaler's mind. From his early popular writings on earthquakes and his depictions of American regional physiography to his historical sketch of *Kentucky* and the celebrated *Nature and Man in America,* the influence of nature on culture and the environmental setting of social history remained central concerns. Given the increasingly environmentalist cast of geography's newest literature, it is not surprising that Shaler's contribution has been stereotyped as standard nineteenth-century geographical determinism. One writer describes him as a "geologist still seeking to explain human societies in terms of 'controls' of the physical environment, as conceived by Carl Ritter in Germany and through him by Arnold Guyot"; another represents

Nature and Man in America as a volume that "popularized the central argument of Ratzel's theory"; while a third observes that "Shaler's contribution was not only to popularize Ratzel's theories, but to identify the specific features of the American landscape. . . that 'explained' the nation's history."[1] Such designations have stood as the definitive evaluation of Shaler's geographical thought.[2]

Certainly there are strong grounds for sensing in Shaler's writings a fascination with environmental controls on human behavior. But his acknowledgment of the human species as an agent of geological change should itself caution against any facile dismissal of him as a mere "geographical determinist." In point of fact, his work was much more subtle and supple. For him society and nature form a partnership that is dynamic rather than static, mutually reactive rather than mono-deterministic. Indeed, to make a passing methodological point, the "determinist-possibilist" polarity seems a particularly inadequate tool for understanding the work of those writers whose perspective on humanity and nature was integrative and synthetic. A less restrictive set of historical categories is plainly needed to deal with the subtleties of influence that many naturalists perceived in the historical relationships between social inheritance, acquired experience, and geographical context. Now I do not mean to suggest that Shaler's writings exhibit no interest in geographical influences on human affairs. They certainly do, and to that extent the traditional "geographical interpretation of history" finds full expression in the corpus of Shaler literature.[3] But his equally firm insistence on the moderating influence of inheritance, whether biological, educational, or religious, testifies to an understanding of human geography that transcends the standard dichotomy between nature and nurture.

The Geographical Interpretation of History

From Hippocrates' "On Airs, Waters, and Places" of the fifth century B.C. to the more recent compositions of Ellsworth Huntington and Arnold J. Toynbee, the study of environmental influences on society has received sustained exposition.[4] Indeed the historical and philosophical dimensions of the subject have been so amply documented elsewhere that even a thumbnail sketch would justly incur the charge of pleonasm.[5] The rehearsal of the theme in its American setting is more germane to our purposes, though its advocates here were thoroughly tutored in the classical and medieval sources and in the works of post-Reformation and Enlightenment origin. In the case of George Bancroft, the geographical factor is little more than marginal, the first volume of his 1834 *History of the United States* displaying the faith of a historian who saw the United States as a divinely guided nation marching toward perfection.[6] Francis Parkman was rather more partisan: "Not institutions alone," he conceded, "but geographical position, climate and many other conditions unite to form the educational influences, that acting through successive generations, shape the character of nations and communities."[7]

Environmentalist doctrines, however, received fullest expression in the writings of the Swiss geographer Arnold Guyot. Building on the work of such modern pioneers as Alexander von Humboldt and Carl Ritter, Guyot's 1849 lectures to the Lowell Institute, subsequently published as *The Earth and Man,* presented what Sauer was to call "a full and simple gospel of environmental determinism."[8] Patterned on the teleology of Ritter, the work bore witness to Guyot's conviction that "nature and history, the earth and man, stand in the closest relations to each other, and form only one grand harmony."[9] This indissoluble link meant that the contrast between the unified physiography of North America and the regional diversity of the Old World was mirrored in the racial differences between the two continents. The climatic determination of human disparities was central to Guyot's program, and like Ritter and Hegel, he used it to explain the westward march of civilization from its eastern cradle *(ex oriente lux).*[10] Throughout, Guyot's was a history shaped on a geographical template. It bespoke the overriding control of Providence and thereby forged a mystic union between nature, humanity, and God that was teleological in emphasis and predestinarian in character.

Among historians writing during the middle decades of the century, Henry Thomas Buckle and John W. Draper deserve mention. Buckle, of course, was English and did not concern himself with American history especially, but Shaler did find inspiration in his *History of Civilization in England* (1857–61). As part of his vision for a scientific history, Buckle made much of the "influence exercised by physical laws over the organization of society and over the character of individuals."[11] More narrowly cast was Draper's *History of the American Civil War* (1867). But like his mentor Buckle, he sought explanations in the influence of geographical environment. His earlier study entitled *History of the Intellectual Development of Europe* (1863) had brought Draper to an awareness of the ways in which national life was directed by "uncontrollable causes." In such seemingly remote topics as the physical geography of North America, its climate, and the patterns of colonization were the ultimate sources of the Civil War to be found. Like so many others, Draper saw in climatic conditions the greatest influence on human society; his study of isothermal lines on a world scale, for instance, led him to the conclusion that the January isotherm of forty-one degrees marked the final boundary between the Catholic and Protestant peoples of Europe! The "springs of history," it therefore seemed, lay less in "the machination of statesmen or in the ambitions of kings [than] in the silent influences of Nature."[12]

To highlight the necessitarian bias in the writings of such historicist historians, while important for contextualizing Shaler's own work, is really something of a retrospective exercise, in light of the currents of thought that later came to dominate American geography. For environmental determinism, as it is thought of today,[13] owes more to the popularization of the work of the German naturalist-traveler Friedrich Ratzel. Truly the popularization of Ratzel's contributions became the programmatic outline for geographical determinism, because those environmentalist disciples who turned to his two-volume *Anthropo geographie* of 1882–91 and to the later *Politische Geographie* of 1897 did so largely at the expense of his other—arguably more substantive— offerings in, say, *Culturgeographie* published in 1880 as the second volume of *Die Vereinigten Staaten von Nord-Amerika* (1878–93).[14] Moreover, while the broad canvas of Ratzelian anthropogeography was tempered in France by a "meticulous and luminous style of regional description" and in Germany by "the more subtle chorographic analyses of *Landschaft*,"[15] a more fundamentalist environmentalism became paramount among English-speaking geographers largely through the bowdlerization of Ratzel's work in Ellen C. Semple's *Influences of Geographic Environment* (1911).[16]

Shaler and Environmental Influence

In 1922 Arthur M. Schlesinger expressed the opinion that Shaler's essay "Physiography of North America" and his *Nature and Man in America* were the first systematic attempts at a geographical reading of American history—a view later endorsed by his son, A. M. Schlesinger, Jr., and by Oscar Handlin, among others.[17] Berg's doctoral dissertation on Shaler challenged this assessment, identifying Buckle, Bancroft, and Draper in particular as precursors of Shaler's geographical reconstruction of historical development.[18] But whether or not Shaler's was the first exposition of the theme, it certainly was systematic, in contrast to many earlier treatments that were both evanescent and episodic. His writings, I believe it would be fair to say, provided an early conceptual framework for later elaborations of American history in its geographical setting. This conclusion does not mean, I hasten to reaffirm, that his work on geography and history can be subsumed in toto under the rubric of environmental determinism. His interpretation of the American frontier experience, to take just one case, shows his concern to intermesh the influence of physical environment with cultural inheritance. Even less can it be dismissed as a mere Americanization of Ratzel, if only because his earliest writings on earthquakes, published prior to that of Ratzel, reveal his propensity for tracing the influence of environment on regional character. Indeed, his providentialist rendering of world history suggests a rather greater intellectual affinity with Ritter and Guyot, to whom he directly refers, than with Ratzel, of whom he makes no mention (general lack of referencing notwithstanding) even in *Nature and Man in America*. Besides, this later contribution to the topic, I would suggest, derives more from the social construal of a pervasive Lamarckism than from Ratzelian anthropogeography, whatever the conceptual affinities between them.

As early as 1869, writing in the June issue of the *Atlantic Monthly,* Shaler called attention to the influence of earthquake belts on national temperament. The extent to which he saw his own exposition as original is evident from his footnote observation that "Buckle, in his History of Civilization, has incidentally referred to the influence of earthquakes on national character; but, so far as is known to the author, no careful effort has yet been made to determine the influence upon human development of this very important assemblage of phenomena."[19] To Shaler this neglect seemed all the more surprising because the effects of earthquake regions on their local populations were both psychological and material. If, he began, a map of the relative frequency and intensity of earthquakes could be correlated with a second depicting the distribution of superstitions, the two would closely match up.[20] Seismology and mythology, it seemed, were tightly linked together. And there were other psychological repercussions. The "constant destruction of architectural and other records" sentenced the inhabitants of seismic zones to a measure of cultural isolation between succeeding generations that inhibited social progress and stultified national growth.[21] Material culture, too, reflected a people's exposure to earthquakes. The architectural styles of Gothic Europe, for instance, were entirely unsuited to "the tremulous lands of the Latin peoples of the South," where the very elements contributing to their architectural elegance—arches, steeples, columns—would be the greatest points of weakness.[22] This line of thought, of course, fitted hand in glove with Shaler's historical Teutonism and ethnoenvironmental policies.

In the August, October, and November issues of *Atlantic Monthly* for 1869, Shaler pursued the earthquake theme, although now he seemed more interested in sketching a historical overview drawn from diary and journal records. Still, he did insist that the impact of earthquakes on societies was as great as "those agents of change which are generally included under the name of climate" and, to this extent at least, widened the scope of environmental influence beyond the traditional climatic limits.[23] Indeed environmental advantages paled into insignificance before the crippling force of the capricious earthquake. The history of the South American continent, for example, had shown that superlative scenery, healthy climate, even rich mineral resources were powerless to resist the earthquake's reign of terror. "There seems not much to be hoped for the future there," he lamented, "at least until the disturbing forces sink to rest; for how can political stability, continuous effort, or any other result of an advanced civilization be expected, where the land is as treacherous as the sea, and the forces of nature seem man's natural enemies?"[24]

Nor did the range of environmental agents stop here. The swamplands of coastal Virginia, he noted in passing the following year, 1870, were "destined to exercise a great negative influence in the development of the country." In much the same way, South Carolina's "meagre soil and wretched cultivation" had set decided environmental limits on the state's agricultural potential.[25] Again and again throughout the 1870s, periodic asides on the way in which the racial qualities of a population echoed its regional geography would creep into Shaler's picturesque travelogues presented in lyrical prose. And yet even these articles already demonstrate Shaler's distaste of an absolutist environmental determinism. The occupants of the Berkshire hills, to take just one example, preserved a social structure that reflected the institutions and traditions of their transatlantic forefathers no less than it did the local climate.[26] Another regional vignette, "Reelfoot Lake," penned for the *Atlantic Monthly* in 1878, called attention to geography's role in military history: the strategic value of the Mississippi River bluffs in western Kentucky, Shaler urged, had enticed a Confederate army to occupy the site even before Kentucky had declared its Civil War allegiance, and thus "the island of hill-land in the lowlands of the Mississippi became the means of determining the course of a State which more than any other held the key position in the great contest."[27]

This varied folio of landscape depiction provided Shaler with the tools for a more systematic survey of the environmental foundations of American history in the 1880s. Justin Winsor's invitation to set the geological scene for his four-volume *Memorial History of Boston,* published in 1880, afforded Shaler the admittedly brief opportunity (his chapter was a mere eight pages long) to pinpoint the effects of Boston's soil types on the pattern of early settlement.[28] Here he fastened on the absence of boulders as of paramount importance in explaining the growth of Boston, more important even than the positive advantage of soil fertility. Compact though Shaler's contribution was, it served the useful purpose of cementing his links with Winsor. The success of Winsor's editorial techniques on this his first cooperative venture paved the way for his prodigious eight-volume *Narrative and Critical History of America* (1884–89), for which Shaler was again invited to provide the physiographical overview. He began by presenting a sixfold regional division of the continent that synthesized climate, soils, and topography. But in the second, more substantial part of the essay, entitled "Effect of the Physiography of North America on Men of European Origin," he made his first concerted attempt to sketch the broad sweep of geography's control of American history.[29]

To be sure, this was neither the first nor the last word on the subject. Others had plowed at least part of the same furrow, and Shaler himself would return to it repeatedly during the rest of his life. Like Guyot, for example, Shaler noted that the Atlantic coast afforded the easiest means of access to the continental interior;[30] if the original colonists had first encountered the Rocky Mountain range, he mused, "the history of America would have been very different."[31] Again, in keeping with John Richard Green's elaboration of the influence of marshlands on early English history,[32] Shaler claimed that, next to deserts and mountains, swamps were the single greatest impediment to the early movement of population. Thus the difficulties of preparing the glaciated soils of New England for agriculture were offset by the relative absence of swamps, at least as compared with northern Europe. And of course, like other predecessors, Shaler isolated the climatic factor as having a crucial effect on the human species. But here his interpretation took a more independent turn. Certainly he acknowledged the direct effects of climate. But of even greater importance was its indirect influence on the symbiotic relationship between soil capability and agricultural practice. For the fertility of the land ultimately allowed the visions of the early pioneers to be translated into reality. And just as colonial settlement was umbilically tied to soil conditions, so too was America's later social and political history. Nowhere was this more obvious than in the case of Negro slavery. Indeed, to the cultivation of tobacco, "which demands much manual labor of an unskilled kind, and rewards it well," Shaler reported, "we owe the rapid development of African slavery. It is doubtful if this system of slavery would ever have flourished if America had been limited in its crops to those plants which the settlers brought from the Old World."[33]

In interpreting American society, the soil factor, whether in terms of its exhaustion or its fertility, was never far from Shaler's mind. Either way, politics and pedology were intimately bound together. In his 1892 monograph "The Origin and Nature of Soils," Shaler fleshed out the bones of his argument. First he reiterated his earlier points:

On the returns given by this industry [tobacco] the political and social culture of the central colonies of the Atlantic coast chiefly rested. To it also in the main was due the profitable and rapid extension of African slavery. In a similar manner the soils of the more southern States proved in the present century well adapted to the culture of cotton, a crop which led to the establishment of large and numerous plantations, and thus to the further diffusion and firmer establishment of the slaveholding system. Though in part due to climatic features, this system by which the descendants of Africans were held as slaves is principally to be accounted for by the characteristics of the earth in the southern States.[34]

But now he pursued the implications of this interaction between soil, agriculture, and politics even further, weaving them together in his account of the origins and state alignments of the American Civil War. His story ran along the following lines. Since the soils of the plains were suited to raising cotton and tobacco, slavery tended to be restricted to these areas rather than to mountainous terrain, where a small field size could not sustain plantation agriculture. The ensuing slave distribution was, in turn, reflected in the polarized political allegiances of the upland and lowland peoples during the Civil War era. The state of Kentucky was a paradigm case, for "a majority of the people on the richer lands where it was profitable to keep slaves were led to cast their lot with their kindred of the same class in other parts of the South, while those dwelling on poorer soils, where they knew nothing of the institution, were overwhelmingly on the Federal side of the debate."[35]

As a supplement to these comments on the pedological determination of social and political history, Shaler included a five-page summary "Effects of Soils on Health." Published in this same 1892 monograph, it coincided with the appearance of Alfred Haviland's *Geographical Distribution of Disease in Great Britain*. In Shaler's opinion the state of health was affected by the soil in three distinct but related ways: the quantity of water retained in the soil determined surface humidity; the conditions of soil water influenced the propagation of bacteria; and the quality of drinking water reflected soil characteristics. Together these influences, induced very largely by the vertical oscillation of groundwater, generated both animal and human disease. As evidence, Shaler pointed to the populations of the Dismal Swamp region in Virginia and North Carolina, Holland, and the English Fenlands, where the introduction of drainage systems stabilizing groundwater levels had brought about a substantial reduction of malarial-related illnesses.[36] These examples pressed Shaler to the conclusion that in the future "many diseases of a geographical and limited nature, the causes of which are yet unexplained, will be found to be attributable to the action of the soil in the regions where they occur."[37]

Shaler's *Kentucky*

To return to 1884, the year in which the essay "Physiography of North America" appeared in Winsor's collection, Shaler also published his *Kentucky: A Pioneer Commonwealth*. The third in the American Commonwealth series instigated by the Houghton Mifflin Publishing Company in 1873 (it was preceded by studies of Oregon and Virginia), the book rehearsed the history of its author's beloved state in its geographical context. According to the publisher's introductory press notices, these respective state portraits were not so much detailed scholarly monographs as "rapid and forcible" historical sketches, for, as the editor Horace Scudder made clear, the aim of the series was "to secure trustworthy and graphic narratives."[38] That Shaler was invited to contribute reflects the regard in which he was already held as a lay historian, as a stylist who could write in an engaging and lucid way, and as a devoted Kentuckian with an intimate knowledge and deep love of his native state.

From the outset Shaler made it clear that he saw his brief as giving the "reader a short story of the development" of Kentucky, not a definitive history. On two accounts indeed—one practical, the other methodological—he eschewed the designation "history"; first, because such a project would require many volumes and then, perhaps more significantly, because no one could "write a thoroughly unbiased account of a civil war in which he took any part whatever." To forestall speculations as to his own political allegiance, therefore, he openly declared "that the reader should know that the writer, a native of Kentucky, was a unionist during the war. But while his opinions have the color given by his political position, he believes that he has in most cases done substantial justice to his friends, the enemy of that unhappy yet glorious time."[39]

In organizing his material, Shaler's predilection for military history, in keeping with the thrust of contemporary historical scholarship, meant that considerable space was devoted to that topic. Chief among his historical landmarks were the settlement of Kentucky via Virginia, the subsequent separation of the two states, the government of Kentucky, various Indian battles, the War of 1812, the Mexican War, and the political struggles in the pre– and post–Civil War period. Besides the battle motif, however, there were two other recurring themes: the Anglo–Scottish origin of Kentucky's colonial population and institutional framework and the domestic (and therefore patriarchal) character of the state's slave system.[40] Shaler scarcely missed an opportunity to underscore Kentucky's staunchly European heritage, which suggests that, in his historical interpretations, no less than in his ideological predispositions, the inheritance factor—whether biological or social— was beginning to assume greater importance. His overall conclusion to the volume, for instance, was couched in ethnic rather than environmental terms:

Kentucky has had the good fortune to inherit a nearly pure English blood. Aside from the diminishing negro population, the blood of the people is of a singularly unmixed origin. Her success in meeting the strains of the Civil War could not have been secured if its people had not had this singular unity of race and the solidarity of motive that it brought with it. . . . The history of the commonwealth gives us one of the most encouraging chapters in the history of our English race; it shows us that its blood, entirely separated for two centuries from its parent influences, can carry on its development on the American soil, undiminished in vigor, and true to its original motives.[41]

Notwithstanding this concession to the formative influence of bio-social inheritance, Shaler's history was firmly tied to geographical roots. "Before beginning the historic account of Kentucky," he opened, "it will be necessary to examine the physical conditions of the State. This we shall be compelled to do in a somewhat extended way, for here even more than in Virginia has the physical character been effective in determining the history of the people." The topography of the state, he was convinced, had exercised its control in a range of ways: the shape of the political boundaries, the rivers, not to mention the soils, all had direct social and political ramifications. Recapitulating his argument on what he now called "the geological distribution of politics," he concluded that the "dwellers on the limestone formations, where the soil was rich, gave heavy pro-slavery majorities, while those living on the poorer sandstone soils were generally anti-slavery in their position."[42]

Nevertheless, Shaler remained chary of the turgid environmental determinism that reduced all history to nature. To understand the sources of the Civil War, the social differences between a modern industrial North and a rather more archaic feudal South had to be taken into account just as seriously. If geography had stage-managed the American drama, cultural tradition had directed it. So in his 1890 essay, "The Peculiarities of the South," he explained that the rise of a southern slavocracy had provided fertile soil in which feudalism could continue to thrive. A resurrected relic from medieval Britain, southern society and southern slavery thus mutually reinforced one another, and herein lay roots of the national clash: "If we adopt the hypothesis, as it seems reasonable to do, that Southern society is to be regarded as a survival of the same feudal-life conditions which two centuries ago still existed in the mother-country, while the new English life of the North, untrammelled by domestic slavery, had gone forward into the modern fields of thought and action, we may account for the most important differences between the two sections."[43]

Ultimately, Shaler's *Kentucky* may be seen as the effort of a natural scientist to write a history informed at once by geography and biology. I do not mean, of course, that he satisfactorily integrated the twin influences of environment and inheritance at this stage. It does show, however, that his treatment of the historical process in its broader context was substantially more sensitive than that of contemporaries like John B. McMaster and H. H. Bancroft, whose geographical observations amounted to little more than a restatement of traditional climatic determinism.[44] At the same time, it is not clear why Shaler's work has continued to be lampooned as outmoded geographical determinism. Three possible suggestions may be ventured. The subsequent use of a narrow range of Shaler's output by later writers who disseminated a naive environmentalism doubtless played a part. Then, too, the parallel advent of Ratzel's *Anthropogeographie* and its widespread popularization through Semple's writings invited invidious specifications of influence. The close intellectual affinity between the two men arising from their shared predilection for Lamarckism and migration-isolation theories of human evolution only compounded the error. But the chief reason must surely lie in the extreme popularity of Shaler's next volume, *Nature and Man in America*. Published in 1891 and reappearing in new editions until well into the twentieth century, it can legitimately be regarded as the most strongly environmentalist of all Shaler's works.

Nature and Man in America

From its inception, Shaler's new book invited comparison with Guyot's earlier *Earth and Man*. Logistically both began as a course of lectures at the Lowell Institute, both appeared as a series of journal articles, and both were subsequently published as a book. Conceptually, too, Shaler's aim of tracing "the influence of environment on organic life, and. . . the conditions of man in North America" would have received the applause of Guyot, who would likewise have relished the providentialist mold in which the work was cast. Yet all this would be to underestimate the independence of Shaler's thought and the revisionist spirit with which both his geography and his natural theology was tainted. The evolutionary fire that Shaler had breathed into the subject had transformed the quest of earlier geographers almost beyond recognition. As he insisted at the outset:

This inquiry was not possible in the time of Louis Agassiz, or with the students who held to the belief as to the nature of organic changes which he entertained. He believed that each species of animal and plant was the product of direct creative action. Mr. Darwin and Mr. Wallace have forced us to admit that the development of new species is, to a great extent, due to circumstances, to the action of the inorganic conditions upon them or the interaction of species with species in the struggle for existence. Geographic conditions may greatly affect the struggle for existence in most important ways; as, for instance, when the sea advances or retreats, the assemblages of marine and land creatures, the faunae and florae of the neighboring waters and lands, move to and fro, with the change of their domains. Such migrations lead to the death of weak species, brought about by a struggle for existence with forms with which they have not previously come into contention. Such times of migration are necessarily periods when rapid selective changes occur.[45]

While Shaler certainly did reiterate and expand many earlier ideas, the evolutionary garb in which they were dressed could not but broaden the scope of his self-appointed task. Much of the first half of the volume was devoted to explaining the origin, evolution, and influence of American physiography on the natural history of organisms. He charted, for example, the effects of orogenic processes, climatic variability, prevailing winds, and marine conditions on the plants and animals of the continent, demonstrating how environmental change and evolutionary natural selection went hand in hand. And again reminiscent of Guyot, he found that an essential harmony in the natural world could be sensed in the way that "the evolution of geographic features" had "controlled the development of organic life." But whereas Guyot had never found the courage to embrace the evolutionary cause wholeheartedly, Shaler felt that the new vision only reinforced the "sense of order and relation between the apparently rude machinery of the earth's crust and the delicate beings which are bred upon it"; for him it "inevitably led to the conviction that there is an essential unity in all the life of this sphere—the physical and organic being but parts of one great plan."[46]

The Lamarckism with which Shaler's biogeography was suffused found no less expression in his human geography. So when he now turned to influences of environment on social history, he repeated such well-worn themes as the dependence of cotton and tobacco culture on slavery, the tying of soil fertility to Civil War politics, and the significance of an undifferentiated landscape on America's aboriginal population. In this latest publication, however, he went even further. First, he made it clear that society's contemporary dealings with nature manifested the cumulative repercussions of the story of humanity's individual and social evolution. Therefore, contrary to expectation, modern civilization was actually more dependent on the physical environment than its predecessors savagery and barbarism. Why? Because the agricultural and commercial bases of advanced societies were so closely coupled with soil quality and with the spatial arrangement of the earth's continental landmasses over the face of the globe. Indeed the increasingly international character of the modern world had encouraged a greater dependence on an even more close-knit network of socioecological relationships. Thus his discussion of the economic structure of Britain came closer to environmental reductionism than most of his other work. "A bad harvest in the plains of the Upper Mississippi," he informed his readers, "means dear bread in England, fewer marriages, and shorter lives; in other words, it produces an effect on the whole social status of that country." Furthermore, "Almost every storm and every drought which affects the remotest lands and seas reacts upon that State."[47]

As a second extension, Shaler now followed the tracks of climatic influence into unexplored territory. Rigorous climatic conditions, he now suggested, precipitated the growth of Aryan civilization and, on the local scene, contributed to the "soldierly qualities" of the Iroquois Indians. Besides, climate had a direct bearing on agricultural practices, so much so that, if "the isothermals had been drawn one or two hundred miles farther north, so that the southern crops could have prospered in these States," slavery might have become so deeply embedded in the fabric of American society that it could never have been rooted out. Marine and atmospheric circulation also assumed a new importance in the latest Shaler scenario. The prevailing trade winds and ocean currents very largely explained the Spanish colonization of America's more tropical regions and, indirectly, the turning of the French and English pioneers to the more northern coastal plain.[48]

Finally, *Nature and Man in America* afforded Shaler the opportunity to develop in greater detail some of those embryonic ideas with which his earlier writings were sprinkled. The fundamental importance of geographical isolation on social evolution and the "effect of the geography of North America on its savage tribes," themes we have already discussed, found their fullest expression here. The influence of the Appalachian barrier on early colonial expansion and the conviction that "almost every feature, every river and plain, had its effect in controlling the distribution of the population in its westward march" were other topics that we will presently find crucial to understanding Shaler's exposition of American frontier history.[49]

Nature and Man in America proved to be an extremely successful volume. William Z. Ripley, for instance, thought that the ways in which natural environment controlled primitive ethnology were "ingeniously worked out by Professor Shaler in Nature and Man in America."[50] The Royal Geographical Society's editor found its discussion of the effects of American environment on population "very carefully prepared and of high geographical value."[51] Yet the book's popularity, however well deserved, ultimately served to earn Shaler a fossilized reputation as a geographical determinist. In point of fact it was the high-water mark of his writings on environmental controls; although other pieces on the theme appeared throughout the 1890s, later treatments were far more muted and much less direct.

Already by 1892 a tempering of the determinist strain is clearly evident. Now he wanted to take more seriously the accumulated enregistration of multiple environments on the history of particular peoples. The geography of a race's "cradleland" was certainly deeply imprinted on its eugenic character. But exposure to new environments could eventually erode the formative influences of even the national birthplace.[52] Besides, he now suspected that the degree of environmental influence was contingent on the stage of a race's social evolution. His 1893 United States Geological Survey report on the history of harbors is further witness to this strategic change. When he came to discuss "Harbors and Civilization"—a topic later recapitulated in *Sea and Land*—he consciously steered a middle course between the influence of cultural heritage and that of physical environment. Thus, if a shoreline with natural harbors favored "the development of sailors," it was still the maritime spirit, "transmitted by the inheritance of blood and habit, which has made our own people for two centuries successful in the exploits of war and peace upon the seas." The coastal environment, therefore, did not so much determine the maritime traditions of a nation as act as a kind of limiting factor on seafaring possibilities. Certainly the construction of harbors on "harborless coasts" was one of the miracles of modern engineering; for all its brilliance, however, it was an expensive achievement, and there was therefore "no reason to expect that any advance in the engineer's art will ever exempt a country from the disadvantages which the absence of good ports entails."[53] More muted still was his introduction to the three-volume *The United States of America* in 1894. No peculiarity of soil, climate, or natural resources, he now felt, could of themselves make a people great.[54] Moreover, within the next two years he could express the considered opinion that "the influence of environment on man, at least in his civilized estate, though considerable, is not direct, but is brought about in a secondary manner."[55]

Affirmations such as these might prompt the label "possibilist" as an appropriate designation for Shaler's geographical thought. One student indeed has urged that, as "one of the earliest adherents of possibilism in America . . . Shaler is not an idealogical precursor of Semple but instead the forerunner of Isaiah Bowman and Carl Sauer."[56] But such a designation is too precipitate. As I have said, the sheer pressure of the evidence simply overrides the determinist-possibilist polarity, rendering it historiographically redundant for understanding the work of a naturalist like Shaler. Apart from anything else it constitutes a historical framework for interpreting relations between society and nature that Shaler himself would not have recognized. Otherwise he could never have written within the pages of a single work that the "physical circumstances in the end determined that North America should become the great seat of the English race" and also that "all the progress of our people has been shaped and controlled by the inherited traditions which they brought from the Old World."[57] On the contrary, the distinctiveness of Shaler's geography is to be found in the very interplay of environment and inheritance. Nothing more clearly illustrates this interface than his treatment of the American frontier—a theme that runs like a high-voltage current through his writings on United States history coming into even sharper focus during the 1890s.

Influences and Interpretation

Before turning to Shaler's frontier theory, however, it is worth pausing to reemphasize one of the major reasons for the retrospective specification of him as an advocate of environmental determinism. If approached with a predilection for identifying geographical controls on history, there is certainly much in *Nature and Man in America* on which partisans could fasten. A selective reading of the literature could, and did, provide deterministic geographers of the next generation with the inspiration and confirmation they sought that history was governed by geography.

Consider, for example, the two works that appeared in 1903 devoted to discussing the physiographical basis of American development and welcomed by Frederick Jackson Turner as "significant both of an enlarging view of the science of geography in this country and of the changing attitude of historical students toward their own subject"—

Albert Perry Brigham's *Geographical Influences in American History* and Ellen Churchill Semple's *American History and Its Geographic Conditions*.[58] These two books represent in many ways the locus classicus of the geographical interpretation of American history. Although a student of Shaler during his geological training at Harvard, Brigham acknowledged no indebtedness to his teacher in the introduction to the book. This omission in itself would not be worthy of mention, were it not for the fact that he felt the need to inform his readers that, in studying the geographical controls on American history, "one must invent a method as he can, for models in this field can scarcely be said to exist." This disclaimer notwithstanding, Brigham actually referred to Shaler sources throughout the volume more often than to any author claimed as inspiration in his preface.[59] Besides, R. E. Dodge later observed that in "Shaler's course in palaeontology (Geology 14) Brigham caught and absorbed Shaler's philosophy which clarified and perhaps accentuated his own. At least it seems to the writer that in all his popular and some of his semi-technical writings, Brigham has reflected that philosophy."[60]

As for Semple, while her *Influences of Geographic Environment* (1911) contains only two references to Shaler (on the increased dependence of civilization on nature and on America's lack of local centers of civilization), her earlier *American History and Its Geographic Conditions* drew very substantially on his work. In particular her treatment of the role of isolation in the later stages of social evolution, the westward movement and continental development, the unsuitability of the North American environment for cradling civilization, and the relationship between soil, slavery, and southern politics were all directly rooted in Shaler's writings.[61] Both Brigham and Semple, therefore, abstracted from Shaler's work (they referred almost exclusively to *Kentucky, Nature and Man in America,* and *The United States*) those elements most susceptible to an environmentalist reading of historical development. Moreover, when Jones and Bryan later told their readers in 1924 that the "main and simple lines of American Historical Geography have been indicated once and for all in the works of Shaler, Semple, and Brigham," their praise only served to cement these authors together in an ideological trinity.[62]

References to Shaler's geographical thought in the 1920s helped to further reinforce this determinist image. W. M. Davis, for instance, scarcely did justice to the breadth of his teacher's interests or influence when he reported in 1924 on the progress of geography in the United States. "Note may here be made that Shaler, 50 years after the appearance of Guyot's 'Earth and Man,' brought out a similar book, 'Earth and Man in America' (1899) [*sic*]," he reflected. "Its interest in the present connection lies in the extension that it gave to the rational correlation of earth and man, introduced to us in Guyot's book half a century before."[63] That same year J. Russell Smith, who had studied anthropogeography under Ratzel in 1901–2 and later became one of Huntington's devotees, published his regional geography *North America,* prefacing it with that well-known tribute to Shaler, his "unseen master."[64] And when, during the following year, 1925, Franklin Thomas published his interdisciplinary review *The Environmental Basis of Society,* he too drew exclusively from *Nature and Man in America* to support his own views on civilization and geographical isolation.[65] Many other writers using Shaler's environmentalist theses could doubtless be enumerated. John Fiske, who, though no geographical determinist, maintained that it would be idle to disregard the importance of physical conditions, gleaned many ideas from his "friend, Professor Shaler"; Harlan Barrows used *Kentucky* in his historical geography lectures; Robert DeCourcy Ward, who discussed the effects of climate on human life, derived much from his teacher and colleague.[66] But invariably these individuals relied on a narrow range of Shaler's output and perpetuated an image that, as Sauer wrote of Ratzel, "was indeed one likeness of the man and his time but showed him in only one, and lesser, attribute of his person."[67]

American Frontier Historiography

In his exposition of American frontier history, Shaler came closest to a genuine synthesis of the effects of environment and inheritance. This achievement, of course, was not until the 1890s, when the earlier anticipations of the frontier melody finally broke through in full refrain. Like other contemporary theoreticians, notably Walker and Turner, Shaler's use of the term "frontier" was thoroughly American in implication; for all of them it meant "the edge of settled land," not "boundary" or "border," as in conventional European usage.[68] With them, as with Shaler, the continental interior was a place of boundless potential, and every westward step a further national advance. Such retrospective reconstructions of Western colonization, typical of late nineteenth- and early twentieth-century historiography, gave birth to the thesis that the frontier had created all that was desirable in American life and culture, that it was the mainspring of democracy and freedom, and that it had been the decisive factor in the perennial regeneration and social evolution of the American nation.

To the first European pioneers, however, the frontier was more a physical threat than an intellectual enigma. So much did survival depend on overcoming a wild environment in order to secure the most basic necessities of life that it was only appropriate that the broadax should come to be viewed as the symbol of early American attitudes towards nature.[69] Animated further by a Western tradition that envisaged untamed nature as cursed and chaotic, frontiersmen struggled with wild country in the name of nation, race, and God, while poets projected evocative and terrifying images of a "howling wilderness" inhabited by / "hellish fiends, and brutish men / That devils worshipped."[70] Yet, despite the prevalence of such adverse portrayals, few English-speaking colonists had reliable knowledge of the interior of the continent at the opening of the eighteenth century, for under the British policy, settlement in the interior was regarded very largely as a means of forestalling the French.[71] Soon, however, the picture began to change. The new vision of the American West as a place of pristine glory and potential agricultural wealth found reinforcement in the zestful nationalism of America's newfound independence. A belief in continental destiny became its principal article of faith. Thus, Freneau, for example, felt that this new empire of the future would bring agriculture to the summit of perfection and would make the nations brothers by disseminating the riches of the New World throughout the whole earth.[72] St. John de Crèvecoeur and Benjamin Franklin added their support to this Arcadian dream of America as the modern Garden of Eden, an agrarian refuge retaining the old ideals of simplicity and virtue.[73]

It was, however, Thomas Jefferson who became the intellectual father of the advance to the Pacific by sending Lewis and Clark up the

Missouri River and over the Rocky Mountains on their famous government exploration. Their reports, together with those of Pike (1806–7) and Long (1819–20), conjured up the image of a western desert to the east of the Rockies, a designation partly based on their personal observations and partly quarried from the impressions of earlier French, British, and Spanish pioneers.[74] But whereas the Spanish had spoken of a physical desert and the Puritans had imagined a demographic wilderness, these expeditions added the distinctively American concept of a desert where agricultural settlement was impossible.[75] These images dominated the American mind until the 1840s, when the explorations of Frémont gave rise to a new evaluation. The Great Plains very largely relegated the idea of a Great American Desert to the realms of myth. Frémont's portrait was to become the driving force behind westward expansionist policy and, as promulgated by William Gilpin, was eagerly promoted by railroad companies after the Civil War. Gilpin simply rejected the easterner's perception of the West by affirming its agricultural significance in superlative tones. Indeed he pressed beyond a mere recognition of the region's economic potential to a predestinarian sense of the inevitability of its development. With the rhetorical zeal of a Hebrew prophet he shared with his hearers the ecstasy of his pastoral dream.[76]

With the frontier of agricultural settlement now lying along the ninety-sixth meridian, the end of the Civil War heralded the potential advent of this agrarian Utopia. Under the stimulus of the Homestead Act of 1862, the westward surge in the postbellum years had pushed the frontier out onto the subhumid plains, so that, by the 1870s, lands were being occupied where rainfall was inadequate for traditional farming techniques.[77] The initial homesteading years were therefore bleak, but once special seeds and cultivation methods were successfully developed, the desert concept finally evaporated, and agriculture became feasible on a large scale beyond the hundredth meridian. Soon the term "Great Plains" had found its way into the vocabulary of men like John Wesley Powell, Ferdinand V. Hayden, Henry Gannett, and William Morris Davis.[78]

The United States Census Report of 1870 may be taken as the birthplace of the theoretical frontier. Its particular significance lies in the fact that the census compilers, without interpretative comment, inserted a map depicting the very first representation of a frontier line.[79] Francis A. Walker, superintendent of the census and president of MIT, became the frontier's major theoretician during the 1870s.[80] Not only did he emphasize the railroad's importance in accelerating population expansion westward but, like Henry Gannett (chief geographer of the United States Geological Survey under Powell), he recognized the dynamic, processual nature of the frontier by stressing the multidirectional and irregular character of settlement penetration.[81]

Yet, intellectual America became most conscious of its own frontier experience, through the work of Frederick Jackson Turner. That Turner repeatedly repudiated simple one-factor historical causation is, of course, beyond dispute. Nevertheless, it was the frontier as a many-sided process, described in 1893 as "this perennial rebirth, this fluidity of American life, this expansion westward with its new opportunities, its continuous touch with the simplicity of primitive society," that Turner believed had furnished "the forces dominating American character."[82] Largely reversing the emphasis of the Teutonic school's "germ theory" of American history,[83] Turner used the metaphor of the social organism's intrinsic plasticity as the central theme of his frontier hypothesis. With similar biological analogy in mind, he perceived in American frontier history a society recapitulating the stages of evolutionary advance from hunting through settlement to urban manufacture.[84] In his analysis, therefore, biosocial inheritance was envisaged as subservient to the influence of the physical environment in the shaping of the American nation.

Shaler and the West

In his evaluation of the West, Shaler exhibited that same inconstancy of attitude displayed by both his predecessors and his contemporaries. His first expressed opinion is to be found in a short 1883 article for the new journal *Science* on the native pasturelands of the Far West. Sharing with his readers the "well known fact, that the greater part of the United States west of the meridian of Omaha [the ninety-sixth] is unfit for tillage,"[85] he made it clear that those few isolated areas that could conceivably be cultivated were exceptional and that the only use he could envisage for the region as a whole was as pasture. For that purpose, however, it had untold potential because the inadequacy of water supply could be remedied by wells and storage reservoirs and any scarcity of herbage could be overcome by introducing forage plants from climatologically similar regions. Shaler's proposals were thus broadly in keeping with Powell's call, some five years earlier in 1878, for the revision of the near-sacred 160-acre limit of the Homestead Act in favor of smaller irrigated farms and very much larger grazing farms beyond the hundredth meridian.[86] Nevertheless, as Malin notes,[87] Shaler displayed his inadequate knowledge of the ecology of low rainfall regions when he suggested that other forage plants might be used among the poorest grasses to fill the "generally wide interspaces between the tussocks of high-growing species."[88]

This area was again to come under Shaler's scrutiny in his 1884 essay for Winsor's *History*. Here he divided the continent into six "fairly distinct regions," examining each in turn and bluntly dismissing the "Cordilleran region" beyond the Mississippi Valley as a curse to the nation.[89] For despite its undisputed mineral resources, the region's low temperatures and arid conditions had served to reduce one-third of the whole continent to sterility. In Shaler's eyes—the eyes of a forest lover—the region's treelessness made matters even worse; this was prairie disease and needed urgent remedial action. "In estimating the value of North America to man," he mused, "the limitation of good forests to the region east of the Mississippi must be regarded as a disadvantage which is likely to become more serious with the advance of time. . . . It seems possible that these regions [in the West] may yet be made to bear extensive woods."[90]

By 1887, however, just the hint of a changing appraisal can be detected. Now his lament over raging forest depletion in some parts of the West—particularly regrettable due to the absence of a suitable climate for spontaneous replacement—was itself recognition of some subregional diversity in the trans-Mississippi West.[91] Within three years, indeed, Shaler had allowed this hairline crack to widen into two distinct regions. Loosening the chains of his forest ideology, Shaler's newest proposal focused on economic potential. An eastern zone, incorporating eastern Nebraska, Iowa, Missouri, eastern Kansas, Arkansas, eastern Texas, all of which offered excellent soils for agriculture, he now felt needed to be severed from the states to the west, where low, unreliable rainfall and a lack of irrigation made them suitable only for livestock.[92] More elaborate still was Shaler's treatment of American regional geography in the 1894 *United States of America*. Here the arid West—that area between the hundredth meridian and the narrow Pacific coastlands—was examined in more detail. Those states stretching south from North Dakota to Texas represented a zone of transition comprising a belt of scantily watered lands about two hundred miles in width, beyond which lay the "desert region" of the upper Missouri basin and the Cordilleran plateau. This plateau environment, Shaler believed, would bring about the social evolution of a new, nomadic human type whose mind and body would be adapted to the necessities of life that "the herdsman's primitive occupation" entailed. In contrast, the Cordilleran region (as distinct from the plateau)—that area between the eastern face of the Rocky Mountain front and the chain of the Sierra Nevada—was on many accounts the most interesting portion of the United States. Thus, whereas some ten years earlier Shaler had spurned the whole region as a national scourge, he now issued an appeal for a detailed study of its potentially great mineral wealth. "It seems likely," he concluded his inventory, "that in twenty years from the present time the commercial values which will thus be won from this 'Great American Desert' will be as large as that obtained in any equal area of the continent."[93]

Following this move away from a cavalier dismissal of the West to what Malin described as "a revision which might be called a forest man's apology for his former misunderstanding,"[94] Shaler, in his turn, accounted for the emergence of the American Desert myth as the perceptual response of pioneers and explorers nurtured in a forest environment:

The American desert of our older geographies was pictured as the most inhospitable realm, fit to be compared in sterility with the arid wilderness of Asia and Africa. In part, the impression it made upon the early explorers was due to the fact that they went forth into its fields from the densely forested and superabundantly watered district of the eastern part of the continent. . . . In a word, the name of desert, which was applied to the district, is to a great extent a misnomer; it might be better termed "the arid region," or, better still "the country of scanty rain."[95]

Forest and Prairie Frontiers

Fundamental to Shaler's shifting sentiments about the West and its potential was the radical dissimilarity he saw between forest and prairie frontiers. He had hinted at this difference as early as 1884, when he attributed the timberless character of the western prairies to Indian firing[96] and at this stage had further insisted that, although the open prairies had the temporary advantage that they could be more easily brought under cultivation than wooded regions, their doom was ultimately sealed because a forest environment was so crucial to the well-being of civilization and its social structures. And yet, contrarily, in this same paper he remarked on the international significance of rapid prairie settlement and its runaway agricultural triumphs. "We are now in the midst of the great revolution that these easily won and very fertile lands are making in the affairs of the world," he wrote. "For the first time in human history, a highly skilled people have suddenly come into possession of a vast and fertile area which stands ready for tillage without the labor that is necessary to prepare forest lands for the plough."[97]

Whatever the infelicities in Shaler's portrait—and Carl Sauer was later to catalog many of the problems of prairie homesteading[98]—it served to lodge firmly in his mind the forest-prairie breach. In his 1892 monograph on soils, he made the distinction explicit when he contrasted the "dense forests of the lowland region" with the "open nature and easy subjugability" of the prairie grasslands. "It is clearly one thing to push forward the frontiers of a civilization where each acre has to be slowly and laboriously stripped of its timber and, if it be a glaciated district, of its bowlders also," he urged, "and it is quite another undertaking to extend cultivation over a prairie district where a plowman may turn a straight furrow for miles away from his starting point."[99] The impact on continental penetration was understandably marked, for the open character of the prairies and their fast-increasing accessibility by railroad and steamboat meant that the rate at which the line of frontier settlement moved to the West became vastly greater than before. So while it had taken nearly two centuries for the English colonists to break through the rough country of the Alleghenies and into the eastern parts of the Ohio Valley on the prairie margins, widespread occupancy of the lightly wooded regions of the Far West took less than fifty years.

These embryonic reflections grew to full maturity in 1894 in Shaler's rather more systematic account of the westward movement. Only those who had actively participated in clearing away a primeval forest, he began, could conceive how onerous was the job of preparing it "for the uses of civilization." Indeed, in the wooded eastern states, it could take a decade's hard labor to clear enough ground for the beginnings of agriculture and a whole lifetime of sweat and toil to transform twenty or thirty acres into good farming land. The prairies, by contrast, were mere "holiday work" for those who had battled for generations with hoary forests of the Atlantic seaboard, which forced Shaler to conclude that "after 1830 the frontier ceased to have anything like a definite line, for the immigrants swarmed to the westward all along the ways of communication."[100] Whereas Turner viewed the frontier in rather idealized terms as one continuous process through to its official closure in 1890, Shaler saw it as a twofold advance. For him, the orderly line of wilderness penetration up to 1830 was a forest phenomenon—slow, dogged, persistent—while the open grassland encouraged much more rapid and irregular settlement advance.

Besides this, Shaler, like Walker, understood that the railroad and the steamboat could not be ignored in interpreting the American frontier experience. As he put it, "The key to the swift conquest of the continent by the immigrants from northern Europe is to be found in the rapid development of the economic arts, and mainly in the inventions which pertain to transportation." While the early pioneers had only the simplest of tools and "no intellectual resources save those of tradition," the transportation revolution of the nineteenth century had allowed frontier societies to benefit from direct support "from the older seats of culture."[101] And herein lay a major difference between Shaler's frontier and its Turnerian counterpart. For just as Turner overlooked the new technology of communications and thus failed to realize how profound was the influence of accessibility on the pace of frontier settlement, Shaler pinpointed this very transformation as quite decisive in the American frontier story. Thus he summarized the "peculiarly rapid advance" of settlement in the northwestern states as

undoubtedly due to the singularly perfect organization of the ways by which the people were brought to the country. The waters of the Great Lakes, of the upper Mississippi and of its navigable tributaries, with the aid of cheap steamboats, at once laid open a large portion of the area to settlement, under conditions which permitted the incoming people to bring an abundant store of household comforts with them. The rapid construction of railways, which under our system of Government land grants had their termini generally on the very frontier, afforded a yet more convenient method of access to the land. The prairie character of the surface made it possible for the tillage swiftly to occupy so large a part of the area that the people at once won their way to wealth through their exports of grain and cattle.[102]

Given that this account appeared in 1894, it might seem that Shaler's sketch of the frontier process was merely a gloss on the Turnerian thesis made available the previous year. This conclusion, however, would overlook the appearance in 1892 of one of Shaler's rather less well known works, *The Story of Our Continent*. For in this "Reader in the Geography and Geology of North America," Shaler had already provided a coherent framework for interpreting frontier history, a framework indeed lending support to James C. Malin's claim that Shaler and Turner arrived at their frontier hypotheses quite independently.[103] The fundamental forest-prairie distinction, the centrality of transportation, and the use of the term "frontier" were all part of Shaler's historical consciousness by the time that book appeared prior to the publication of Turner's essay. To demonstrate just how far advanced his project was by that date, I append the following rather lengthy extract.

Although the numerous wagon-ways which were built between the years 1815 and 1840 from the old settlements of the Atlantic coast to the Mississippi Valley enabled a large number of emigrants from the older states to gain access to this country, it was not until railways were pushed through these mountains that the rapid occupation of the wide and fertile land began. It was a great advantage for the railways, as well as for the earlier wagon roads, that as soon as they had passed the relatively narrow region of mountains they entered upon broad plains which were generally but partly forest clad, so that the cost of constructing these roads was not great. From 1815 to 1840 the settlements in the Mississippi Valley were mainly limited to the forest country which occupies the eastern portion of that valley. It was necessary for every farmer to hew down the forest from all the fields which he won to tillage, and after this serious task was accomplished it was generally from ten to fifteen years before the stumps disappeared from the ground to such an extent that tillage was easy.

After 1840 the settlements broke through into the prairie district, and rapidly dispersed the Indians from those great plains. In this region the task of winning farms did not demand more than the tenth part of the labor required in the forest-clad portions of the country. The construction of railways proceeded with amazing celerity. The great rivers of the Mississippi Valley afforded ready access to a large portion of the surface, and so the white population swept over the plains with great rapidity, winning, in fifty years, a larger area of the continent to the uses of civilization than had been won in the two centuries of previous growth.

If the whole of this country had been clad with forests, as is much of the eastern portion of the Ohio Valley to this day, the movement of the population could not have been anything like as rapid. It is probable that if the country had been densely wooded, the frontier could not, at the present time, have advanced much beyond the line of the Mississippi. We thus see how important to man are the simpler physical features of the land. The Appalachians restrained the colonial population to the Atlantic coast. The rich forest-lands of the Ohio Valley gave a new field for a portion of this people, but it required more than fifty years to push the farming country up to the edge of the prairies of Indiana and Illinois, while it has taken a third of the century to extend the culture from the margin of the front district to the base of the Rocky Mountains. . . .

We may thus divide the westward progress of the English-speaking peoples in North America into three periods; the first extending from the time of settlement in the early years of the seventeenth century to near the end of the eighteenth, a period of 150 years, in which this people were confined almost altogether to the narrow strip of fertile lands to the eastward of the Appalachians. Next, a period from about 1790 to the middle of the nineteenth century, in which they obtained, substantially, control of the Mississippi Valley. Third, the last forty years of the present century, in which they have overcome the difficulties which beset the occupation of the Cordilleran district.[104]

This 1892 exposition of the American frontier process clearly demonstrates that Shaler did not derive his version from Turner's paper. Indeed, contrariwise, we might well wonder whether Turner was aware of Shaler's work on the theme. No conclusive answer can be given, but there are suggestions that Turner could well have read some of Shaler's contributions on the geographical context of American history. At the same time, it has to be remembered that, in a reply to Merle Curti, Turner insisted that he had not been influenced by Shaler.[105] Still, he is known to have used for his student dissertation an article published in the same volume of Winsor's *History* as Shaler's "Physiography of North America"; moreover, E. D. Neill, author of this article, had referred his readers to Shaler's introductory essay.[106] Besides, *Nature and Man in America* was familiar both to Van Hise and Chamberlin, intellectual colleagues of Turner at Wisconsin,[107] and of course Turner did later turn to Shaler's work on regional and soil geography in his series of introductory lectures on American regional physiography.[108] It indeed is possible that Shaler may well have had some influence on Turner's thinking, perhaps before the 1893 paper on the frontier.

Yet none of this, it must be understood, is intended to undervalue the brilliance or originality of Turner's historical imagination, particularly the way in which he conceptualized the causal links between America's unfolding history and its frontier experience. But it is equally to insist on the vitality and independence of Shaler's own historical skills. By a piece of historical introspection on America's domestic scene, Shaler had come to recognize the dual significance of grassland and transportation in the changing phases of continental subjugation—factors so vital that, in Malin's opinion, Turner's failure to perceive them meant that he never really got his frontier out of the woods.[109] Indeed, notwithstanding the flourishing industry of historical writing in the Turner mold, Shaler did have his own band of disciples who continued to spread his message. His discussion of the barrier effect of the Appalachians, for example, was echoed in the later work of Semple and Brigham,[110] while his more general frontier narrative so impressed Malin that he felt compelled to write that, "in his use of the frontier concept as applied to American history, he [Shaler] appears to have been original, and in some respects his application was more significant and enduring than Turner's."[111]

The Cultural Heritage and the Frontier Experience

While the themes of transportation and grassland figured prominently in Shaler's reconstruction of the westward movement, they did not mark the limits of his frontier commentary. From time to time, he called attention to other cultural factors, notably when he described the influence of Old World inheritance. Like Herbert Baxter Adams, Woodrow Wilson, and John Fiske, Shaler—as we have already seen—adopted the "germ theory," which posited a Teutonic origin of American democratic institutions in the English shire and ultimately in the primitive folk meetings deep in the heart of the old Germanic forests.[112] Certainly Shaler's was a qualified Teutonism. For one thing, he took seriously the direct effects of environment on human society. Besides, he restricted the influence of the Teutonic town meeting to New England and attributed the difficulties of local government in other states to the absence of this "precious seed of the Folkmote."[113] Still, the thrust of the Teutonic vision greatly attracted him, especially when wedded to ideas about immigration restriction and Aryan supremacy. In many ways American history could be seen as the offshoot of those ancient racial-national seeds grown in northern Europe. Pursuing this argument Shaler believed that the success of grassland pioneering in the Mississippi Valley was due to the cultural strengths of English colonialism. If the occupation of Ohio had been attempted by "ordinary immigrants from Europe," for example, it was doubtful whether those early settlements would have survived. For whereas the political history of the Spanish and French in North America was a tale of "internecine difficulty, distrust, and treachery," English frontiersmen and women, benefiting from the experience of outpost settlers to the west of the Alleghenies, had found the resources to consolidate sophisticated defense systems.[114]

In recounting the frontier's role in the evolution of American society, then, Shaler allowed neither environment nor inheritance to occupy a senior authorship. The story was, rather, a genuinely cooperative venture. On the one hand, America simply had the best environment to house the institutions of an already-advanced civilization; on the other, only Aryan society could truly respond to the stimulus of the New World's landscape. Indeed it was the very unsuitability of North American geography to cradle civilization that eventually provided it with unparalleled opportunities for disseminating the racial qualities of the Teutons. In a word, history had at last arranged for "the delivery to the English-speaking people of the one continent which seems eminently fitted for their use."[115]

Nowhere, in Shaler's mind, did the reciprocal interaction between environment and inheritance come more sharply into focus than on the American frontier itself. Here the environment had induced in the British pioneers new survival skills—both political and social—that, when acquired, were passed on to succeeding generations. The long educational experience of subjugating a new country had fostered not only a high degree of independence and self-reliance but also the ability to construct newer appropriate systems of government. In a social Lamarckian vein, Shaler suggested that the forest frontier had given birth to a new kind of social being, the frontiersman. A thoroughly pragmatic creature, this slayer of woods believed he served the "god of progress by the sacrifice of the forest."[116] Indeed "Forest Man" played the leading role in Shaler's evolutionary account of human development.

According to this schema, the first human beings, originating in the trees and then dwelling on the forest floor, eventually emerged from their ancestral woods as forest-destroying farmers who, only now, were beginning to sense the need for developing a conservation policy. While other commentators denounced the system of pioneer farming as wasteful, myopic, or immoral, Shaler instead conceived of it as an inevitable stage in society's historical development. In the case of the American frontier, the colonists' ready adoption of the strong-rooted Indian corn, or maize, had contributed to rapid settlement advance.[117] Thus Shaler saw that the crop species and tillage traditions that were developed in northern Europe, however indispensable for the establishment of colonial settlement on the Atlantic seaboard, were utterly inadequate with interior penetration. For this reason the initial experience of breaking new ground was never easy. Only patience, resourcefulness, and tenacity had overcome the pessimism of those first years of harvest failure and desperate frustration. Like Turner, who observed that "at the frontier the environment is at first too strong for the man,"[118] Shaler also concluded "that when a new portion of our frontier is occupied there is apt to be a period of rude and painful experience in which there is a frequent loss of crops, leading to the impression that the area is more or less unsuited to the needs of agriculture. After a time the culture becomes more or less reconciled to the environment, and the people become more hopeful of their future."[119]

The impact of frontier geography, however, extended beyond the modification of traditional Aryan agriculture. Political institutions and social fashions were no less subject to the frontier's transforming power. Conditions of life on the margins of the nation had always forced settlers to devote their energies to practicalities rather than to abstract, theoretical issues and had produced that "inventiveness" so distinctive of the American spirit. Of course every member of the Aryan race had innovative capacities, Shaler reminded his readers, but "their efficiency in this work depends upon the stimulus that comes to them from their surroundings."[120] In the American case their ingenuity was manifest not just in technological artistry but more especially in the institutions specially cultivated to meet social and political needs. In particular, the frontier had become "the nursery of freedom" by offering a rare opportunity for emancipation from the tyranny of age-old customs and ways of life. The wilderness had provided the widest scope for enterprise and self-expression, and the settlers willingly threw off the shackles of the "cramping influences of their ancient life" and "became accustomed to think and act freely, guided only by the ideals which their religious and social traditions and their written law enjoined upon them."[121] Indeed it was this newfound freedom from the time-honored traditions of the Old World that sparked the American Revolution; it was nothing less than a rebellion against the tightening reins of authority to show that the nation had come of age. Forged on the anvil of frontier experience, the American Declaration of Independence was therefore a clear and masterful statement of social and political liberation. Furthermore, the frontier was also a place of religious opportunism. But while Turner saw the proliferation of religious sects as the result of a struggle for power, Shaler felt that such church rivalry was a genuine expression of ecclesiastical antiauthoritarianism. After all, the achievement of individualism was, for Shaler, the final test of any social system; it was the goal toward which civilization pressed, the point where a people were "in the largest sense free."[122] On the American frontier, therefore, the long process of social evolution had reached the apex of development, and America was poised to become the home of the greatest race on the face of the earth.

Shaler's sustained exposition of the relations between American history and geography certainly merits his identification as a founding father of historical geography in America. Yet, if this designation is accepted,

Shaler in 1900. Courtesy of the Harvard University Archives.

it is noteworthy that the first concerted attempt to write American historical geography was by a geologist inspired by a biological theory. For the precepts of Lamarckian evolution, I want to suggest, particularly sensitized him to the importance of environment and inheritance, whether in biological or in social theory. By reference to social Lamarckism, Shaler found he could construct a framework for the interpretation of society and nature in general and American history and geography in particular, in which both the determining impact of environment and the formative influence of biocultural heritage were held in balance. I do not mean to imply here that Shaler's writings on history and geography were a conscious effort to reconcile some supposed tension between nature and culture but rather that they were the natural outgrowth of a more general Lamarckian vision of the world. The fact that a similar evolutionary social philosophy underlay the metaphors with which so much of Turner's rhetoric was suffused might suggest that it provided some overarching canopy under which a history informed in very specific ways by geography most naturally had its place. For when Shaler confirmed that the nation's "political and mechanical innovations have, as the naturalist would express it, been closely related to their environment," it was the continued vitality of Lamarckian theory that allowed him to bind social history and natural history so closely together.[123]

7   Toward an Environmental Ethic

Another American, Nathaniel S. Shaler, was an outstanding
figure in the last quarter of the nineteenth century because, like
Marsh, he looked at conservation from a world point of view
and because he was interested in so many phases of the subject.
— Glacken, "Origins of the Conservation Philosophy"

Les études ultérieures ont montré qu'il fallait joindre au nom de
Marsh celui d'un autre précurseur, Nathaniel Southgate Shaler:
ce géologue est surtout connu des géographes parce qu'il avait
su remarquer les qualités de William Morris Davis, dont il fit
son assistant à Harvard. On oublie son ouvrage: *Man and the
Earth* . . . dans lequel il développe le thème de la conservation
des ressources naturelles.
— Claval, *Evolution de la géographie humaine*

The years between the American Civil War and the First World War
marked the shift from the natural world of St. John de Crèvecoeur,
who saw in America an unflawed moral order, to the world of mass
mechanization, wholesale materialism, and urban sprawl. It was a
period during which the vision of a pastoral ideal began to lose
credibility, an era when the "image of the land as holding inexhaustible
economic opportunity gave way to the vision of technological abun-
dance,"[1] and a time when a new environmental ethic was sought to
accommodate the needs and the wants of an industrial society. The
movement for the conservation of natural resources, as it is thought

of today, began to take shape during these years.[2] The historical configurations of this mood, its social and economic context, and the biographical profiles of some of its leading political crusaders are, by now, well known.[3] But despite the plethora of historical reviews of the growth of the conservation spirit in late nineteenth-century America, the significance of Shaler's contribution has been largely ignored.[4] While a few, isolated comments punctuating these general histories represent him as a precursor of the later twentieth-century movement, most fail to do justice to the comprehensiveness and eclecticism of his work on the conservation theme.

Although J. Russel Smith paid tribute in 1924 to Shaler as "a pioneer appreciator of the relationship between earth and man"[5] and Lewis Mumford in 1931 praised his *United States of America* as the first work to adequately take account of the American environment,[6] not until the late 1940s and 1950s did sporadic references to his conservationism begin to reappear. Malin, as we have seen, interpreting Shaler's conservation policy in terms of a "closed frontier," rated his contribution more highly than Turner's,[7] while the interdisciplinary symposium *Man's Role in Changing the Face of the Earth* was liberally sprinkled with Shaler references.[8] Even more recently, as the epigraphs above reveal, Glacken and Claval have underlined the importance of Shaler's ecological studies, while Worster describes him as "a recognized resource expert" and Koelsch predicts a rediscovery of his insights in the light of the renewed interest in the aesthetics of the humanized landscape.[9] Still, none of these has provided any comprehensive account of the structure of Shaler's environmental ethic. Certainly Shaler's ecological musings exhibit their fair share of conceptual untidiness, and this seeming incoherence may account for their neglect. But these very tensions are themselves, in many ways, reflections of the historical ambivalence the American nation has displayed towards its own landscape.

The profound sense of ecological responsibility that foreshadowed the emergence of an articulate conservation movement was indeed predominantly of nineteenth-century origin. Yet the environmental images that even then continued to dominate America's national consciousness had their roots in the colonial experience.[10] To the first European settlers the wilderness presented a formidable threat to survival itself. Of course the forest furnished the Puritan pioneers with fuel and building materials, but it was fundamentally an enemy occupying land needed for crops and providing cover for "marauding savages." As Roderick Nash reminds us, "Successive waves of frontiersmen had to contend with wilderness as uncontrolled and terrifying as that which primitive man confronted. Safety and comfort, even necessities like food and shelter, depended on overcoming the wild environment."[11] Small wonder that the sinister symbols that the wilderness evoked did not begin to disperse until the romanticism of the eighteenth century portrayed solitude, mystery, and even chaos as coveted entities.[12] Beginning in the cities, the idea that the sublimity of the divine was best revealed in untamed nature heralded a "back-to-nature" movement that most stylishly manifested itself in New England transcendentalism.[13] Disenchantment with the complacency of rationalism and religion alike and a distrust of the new science with its materialist resonance paved the way for the transcendentalists' mystic celebration of the wilderness as the place par excellence of spiritual renewal, divine revelation, and personal regeneration.[14]

The preservationist ethic implicit in this new faith was, in turn, challenged in the 1870s by the conservationist, "wise-use" approach to nature championed first by George Perkins Marsh and later by Gifford Pinchot.[15] In 1912 the previously episodic exchanges between the two groups were crystallized in their public squabble over the proposed reservoir site for San Francisco in Hetch Hetchy, a valley of spectacular scenic grandeur in the Sierra Nevada. Whereas John Muir, decrying the "temple destroyers' contempt for nature," elevated Hetch Hetchy into a symbol of purity and beauty and used it to castigate the vulgar commercialism of American life, Gifford Pinchot steadfastly maintained that "the whole conservation policy is that of use, to take every part of the land and its resources and put it to that use in which it will serve the most people."[16]

These conflicting attitudes to nature inherent in the American ideology provide, I think, a clue to understanding the distinctive and, in some ways, prophetic environmental ethic that Shaler developed. Belonging to a generation caught between two sets of values, one old and one new, Shaler was neither willing nor able to wholly embrace or repudiate either. Indeed an elucidation of his writings on conservation leaves the reader with a feeling of déjà vu, for the ambiguities exhibited in his work continue to feature in the contemporary ecological literature. His feelings about wilderness preservation for transcendental-spiritual ends, for example, are echoed in modern romantic calls for environmental sensitivity and the rejection of mechanized society,[17] while the countervailing "scientific" response to the demands of a conservation economy finds parallel in the sophistication of modern environmental systems analysis.[18] Those diverse traditions of Western thought[19] that have produced these two conflicting standpoints—mystic naturism[20] and quantitative ecological energetics[21]— were held in balance by Shaler, perhaps the very reason why his work has been lost in a no-man's-land between the landscape aesthetics of the transcendentalists and the empirical resource studies of the National Conservation Commission. If Walt Whitman was correct in identifying the basic American dilemma as, to use Herbert London's vocabulary, "the irreconcilable opposition between Nature and civilization and all its ramifications: rural versus urban; spontaneity versus calculation; unconscious versus self-conscious; heart versus head," then Shaler strove to reconcile the irreconcilable.[22]

The Sources of the Synthesis

Shaler's synthesis of preservationist and conservationist thinking fitted comfortably into a conceptual framework derived from a variety of sources. Of fundamental importance, of course, was the work of George Perkins Marsh.[23] So impressed was Shaler with Marsh's *The Earth as Modified by Human Action* (the revised 1874 edition of *Man and Nature*) that he inserted in his annual report of the Kentucky Geological Survey for 1875 a forty-one-page appendix comprising extensive extracts from this "great masterpiece." It was a work that, he announced, should be read by "all who desire to understand the effects of man's action on our earth's history."[24] Clearly Shaler was both familiar, and in basic agreement, with Marsh's environmental policy, so that, while others were entering the camps of extreme optimism or pessimism, he was "among the few who maintained a more balanced outlook."[25] Shaler's contribution, however, was far from a slavish rehearsal of Marsh's doctrines, for his conservation proposals were more broadly based: pioneer studies of the potential agricultural value of reclaiming the coastal and freshwater swamps and marshes of the Atlantic shore, preliminary surveys of unexploited energy resources and mineral depletion, and a review of the dangers of soil erosion and its effect in altering forest ecosystems were just some of the additional themes to come under Shaler's scrutiny.

Doubtless the implicit utilitarianism exhibited in Shaler's agreement with Marsh was further advanced by his close friendship with some of the leaders of pragmatism, the then new, distinctively American school of philosophy. American pragmatism was officially launched in 1877 by Charles Peirce, whose attachment to concreteness and observability in philosophical analysis was all of a piece with his emphasis on the public nature of scientific knowledge.[26] After all, this pragmatist stance must be seen against the background of Peirce's zoological studies under Agassiz and the fact that the only permanent job he ever had was with the United States Coast and Geodetic Survey. Shaler's friendship with Peirce (and indeed with his father, Benjamin Peirce) dates from as early as 1869,[27] and was so strong that Berg suggests it as the reason for Shaler's recruitment to the survey.[28] Besides this, any early exposure to pragmatism, even in its most nascent form, would have been further reinforced by Shaler's equally firm friendships with Chauncey Wright, Josiah Royce, and William James, all Harvard colleagues closely associated with the developing school of pragmatism. Moreover, the United States Geological Survey's policy of keeping scientific work subordinate to economic concerns would have struck a sympathetic cord with Shaler's utilitarianism and would have reinforced his own refusal to divorce pure from applied science.[29]

Alongside these influences, Shaler's education under Agassiz had introduced him to two fundamental principles that were to dominate his conservationist outlook. First, there was the strongly empirical emphasis inherited from Alexander von Humboldt that had prompted Agassiz to undertake a *Forschungsreise* to Lake Superior. The resulting publication, a volume of major ecological importance,[30] incorporated two chapters on vegetation in which Agassiz provided a detailed botanical inventory as well as discussing those environmental influences that seemed to affect the distribution of species. The painstaking observation and recording essential for such cataloging composed the very fabric of Agassiz's scientific method and was perpetuated in Shaler's own educational philosophy and scientific practice. And second, the inspiration Agassiz had drawn from the romantic metaphysics of Germany's nature philosophers had led him to appeal to the purposes of some "Supreme Intelligence" as the ultimate explanation of each and every datum of scientific investigation. For Shaler, this faith in the overriding control of a beneficent Providence, expressed, as we have seen, in a reconstituted natural theology, engendered an emotional attachment to the world of nature. As his wife was later to observe, Shaler's sentimental affection for the land sprang from a conviction that the spiritual and the material were closely intertwined. "In his opinion," she wrote, "the tiller of the soil... is preëminently a naturalist, and more than any other kind of man is in a position to gain the spiritual profit which may come from an intimate relation with the forces which control the development of the world."[31]

Shaler's concern for environmental quality and the preservation of natural beauty was thus another link in the chain by which he hoped to reunite nature, humanity, and God. Indeed, by the turn of the century, he was so committed to the importance of these transcendental values that he confessed that, "without spiritual reconciliation with nature," his work would be altogether sterile.[32] The imperious scientism of the evolutionary reductionists, with their unruffled faith in the all-sufficiency of natural law, left him intellectually restive and spiritually dry. How much better it was to see the realm of nature as a place of incessant re-creation, a dynamic world in which a sense of "the infinities of the latent, of the unmanifested" expunged "all the brutal suggestions of the mechanical view of nature."[33] Only by approaching so close to nature that the throb of life could be felt, the forces producing intelligence observed, and the aesthetic sensibilities aroused could modern society be liberated from the tyranny of the Machine.

In mapping out the borderlines of Shaler's environmental ethic, the utilitarianism of Marsh and the pragmatists and the romanticism of the German Naturphilosophen as mediated through Agassiz stand out as major landmarks. Diverse indeed as these environmental stances are, it is worth pausing to remind ourselves that the selfsame polarities still exist today. The belief that the deterioration of the environment and the loss of human dignity are the net outcome of a technocratic society,[34] a theme stretching back to a Thoreauvian indictment of the machine age, stands in stark contrast to the optimism of the economic-engineering approach to resource problems that has its roots in the utilitarian philosophy of Marsh and Pinchot and that gave impetus to the New Deal policies of the 1930s.[35] Shaler's historical significance, therefore, lies in his concern to straddle these two perspectives—the aesthetic and the pragmatic, the primitive and the progressive—by accommodating both amenity and commodity views of the world of nature.[36]

The Problem: Culture versus Nature

For Shaler, the exploitation of nature was part of the very process of human evolution and could be traced to two fundamental causes, one practical, the other philosophical: the activity of the frontiersman, particularly conspicuous in North American history, and a growing spiritual alienation from the material world.

The arboreal model of human evolution that Shaler advanced taught that, although the first humans were derived from a long line of tree-dwelling primates, the agricultural expertise typical of their subsequent development had brought a steadily more subversive exploitation of forest and soil. Shaler's own reverence for the forest was thus so profound that he regarded it as the wellspring of civilization, not least because it had furnished the fertile earth in which society's roots continued to be deeply embedded. How bitterly ironic, then, that the human species should so inevitably desolate the very environment that had created it! And yet Shaler could find grounds for confidence that the final chapter in the drama of evolution would ultimately portray the human species as a race of forest preservers, when men and women had at last come to realize that "this tree-covering of the lands is necessary for the maintenance of those conditions of climate and timber-supply on which the utility of the earth . . . in good part depends."[37]

Initially in North America, the frontier experience was synonymous with similar forest clearance, for without it agriculture, even of subsistence scale, would have been impossible. The practice indeed was inevitable and therefore understandable. It was nothing less than a recapitulation of early civilization's emergence from barbarism. Shaler found it objectionable, however, when the environmental attitude bred under conditions of virgin land was perpetuated after the passing of the frontier. Thus the wanton devastation that continued to be wrought by fire and ax was evidence of a social atavism originating in humanity's early evolution and replicated in America's more recent frontier past. It prompted an emotional appeal for greater care in the use of the forest: "The forests are a precious heritage of man—they provided him a cradle; they furnished him the soil, and they still offer him their help in some of his greatest needs. No man has the right to destroy them when their destruction means calamity to his fellows or his successors."[38]

Certainly for Shaler, the coming ecological crisis was as much the product of forest frontiering along the Atlantic seaboard and of grassland pioneering on the interior prairies as it was the legacy of humanity's evolutionary history. This belief that America's resource problem was the offspring of the colonial spirit was soon to become so conventional a tenet of conservationist thought that over twenty years later Charles Van Hise urged that unrestrained forest exploitation was a direct consequence of the country's rapid colonization. As he put it, "It was due to this fact, combined with the feeling that the resources were illimitable, more than enough for all, there has come this recklessness, this wastefulness, which is unmatched in the history of the world."[39] Indeed this interpretation has continued to be invoked, as in the case of Lewis Moncrief, who connects it with the picture of American society as recapitulating the steps in the evolutionary cycle from primitivism to civilization.[40]

Shaler's reflections on humanity's spiritual estrangement from nature, supplementing this evolutionary account, were most extensively presented in his series of lectures at Andover Theological Seminary, which, as we have already seen, had reverted by the 1890s to the doctrinal liberalism of the Harvard Divinity School.[41] Here, Shaler proposed that the alienation of human culture from the natural world was the consequence of those philosophical speculations and theological dogmas that stressed the discontinuity of humanity and nature. While subhuman species enjoyed a very substantial adjustment to their environment, the emergence of human consciousness had brought both a detachment from and an opposition to nature through the effort to elicit life's necessities from an unyielding world. These antipathies soon led primitive men and women to conceive a realm beyond the natural that they populated with other-worldly beings, propitiable if not friendly, and powerful enough to render them assistance. Now, however, society was coming to acknowledge that its ancient enmity towards the material world was the greatest error of its intellectual infancy. So while it might appear to the supernaturalist that the naturalist was doing a cruel work, Shaler could assure his readers in a thoroughly pragmatic vein that "the cause of truth will not suffer by the change, and that the best of the old view will be preserved in the new. There is room in the actual universe for all the good which the ideals of man have ever contained. There is on earth a firmer foundation for heaven than it has had before."[42]

Clearly, a restoration of the old sense of empathy with the external world, lost since the demise of polytheistic thinking, was now needed. Monotheism, for all its other virtues, had failed to nourish this close relation by neglecting the "phenomenal world" in favor of the "infinite." Even the design arguments of the theologians, with which Shaler was much in sympathy, had done little to lead the mass of humanity to a love of nature. The only way by which a real sense of kinship could be aroused was thus by the awakening of a sense of beauty and order in nature and by a recognition of its ultimately "spiritual" character.[43]

Shaler's Resource Review

Having thus accounted for the origin of society's profligate dealings with earth resources, Shaler's conservationism displayed its thoroughly utilitarian thrust in his proposals for tackling the practical problems of forest depletion and soil wastage. His wise-use approach to forests, for example, is effectively illustrated in his discussion of the Kentucky white oak stands, "by far the most valuable of all the timber trees" in the state, but in danger of extinction in many districts owing to the complete absence of young trees.[44] The program he outlined, while demonstrating a willingness to intervene in natural processes for economic ends, also indicates an awareness of uncontrolled interference. His plan to girdle worthless trees and to plant valuable species on extensive areas of impoverished land would, he believed, check nature's course and also humanity's indiscriminate destruction.

In 1877 Shaler directed the attention of the Kentucky Geological Survey to the indirect significance of forest cover. Although his allusions to their influence on rainfall, temperature, and water storage may merely reflect the popular, if erroneous, tradition that supposed that forest removal had modified the seasons,[45] his subsequent analysis was rather more penetrating. Not only did the spongy character of the forest bed tend to restrict direct surface runoff, thereby increasing the effectiveness of precipitation, but it also affected water quality by limiting the quantity of detrital material borne away. Both these detrimental repercussions were subsequently attested in the detailed studies of Zon, Wilm, and Kittredge during the 1940s.[46]

If Shaler's insights on this subject had been tucked away only in the dusty folios of the Kentucky Geological Survey reports, then it would be understandable how twenty years could elapse before the Inland Waterways Commission, appointed in March 1907, emphasized that forest preservation was a prerequisite for water control. But even though Shaler had discussed these forest-hydrological relationships in such prominent publications as the first volume of *Scribner's Magazine* (1887) and again in the ninth chapter of *The United States of America* (1894), Gifford Pinchot and W J McGee continue to be credited with the first systematic analysis of the relation of forests to stream control.[47]

Closely allied, and similarly conservationist, was Shaler's work on soils. His 1891 monograph "Origin and Nature of Soils," described by Glacken as a "landmark in the history of soil concepts,"[48] demonstrated that he appreciated, unlike so many other scientists, the value of folk custom and farming tradition.[49] Rather like Sauer, he envisaged agricultural knowledge as a combination of local experience and scientific experiment. Indeed Shaler surmised that many of the nation's soil problems stemmed from its relative lack of accumulated earthlore, due to recent colonization, especially as compared with the Old World.[50] His monograph was therefore designed as a practical guide for the farmer, and he began with a forthright declaration of mankind's profound dependence on the earth, thereby hoping to dispel the popular image of the soil as a trivial commonplace. Instead, he insisted, this "slight, superficial and inconstant covering of the earth" was of such importance to society's well-being that it was a fundamental moral responsibility of science to help supplant this common misconception by disseminating a sense of the wonder and beauty of the soil.[51]

Whereas the Russians had already formed a school of pedology in St. Petersburg under the leadership of the renowned Vasilii V. Dokuchaev,[52] soil science as an independent discipline had still to be established in North America. Certainly Shaler's work lacked the systematic development of the Dokuchaev school, but his original thinking on pedogenic processes makes him one of the pioneers of contemporary soil geography.[53] So modern was his dynamic, organic view of soil processes that Nikiforoff in 1959 used the following extract from Shaler's monograph to portray the soil as "a sort of bridge between the organic and mineral worlds."[54]

[The soil] is the realm of mediation between the inorganic and the organic kingdom, [but] it is by the variety of its functions more nearly akin to the vital than to the lifeless part of the earth. It is not unreasonable to compare its operations to those of the plants which it sustains, for in both these are the harmonious functions which lead matter from its primitive condition to the higher state of organic existence.[55]

In the United States, Shaler's student Curtis F. Marbut subsequently became the founder of the American school of soil science. Through his reading of Hilgard's work and the contributions in German translation of Dimitriev Glinka, Marbut eventually was able to incorporate Russian terms and concepts into the work of the United States Soil Survey.[56]

Of greater importance from the standpoint of conservation history are Shaler's comments on soil erosion. As we have seen, he wholeheartedly agreed with Marsh on the role of the human species as a geological agent, but now he disputed the contention that the great enemy of nature was the savage, not civilization. While Marsh depicted primitive society as "a warfare of extermination, a series of hostilities against nature,"[57] Shaler by contrast insisted that the soil problem had its origins in the evolution of agricultural technology. For if, at the hunting, fishing, and gathering stage of social evolution, human culture had little or no impact on the soil, with the emergence of even the simplest agricultural techniques, difficulties of an altogether new order arose. The removal of natural vegetation and the overturning of the soil were exceedingly subversive processes. And nowhere was this more conspicuous than in the case of soil erosion. As he later estimated, "When the field is stripped of the native plant covering and stirred by the plough . . . [a] single gale may strip off from the dry surface and bear away more of the soil matter than would have been blown away in a geological period, and an hour's torrential rain may wash off to the sea more than would pass off in a thousand years in the slow process of erosion which the natural state of the earth permits."[58] Agricultural practices therefore introduced both quantitative and qualitative changes into the history of environmental disruption. Summarizing the economic consequences of unrestrained agricultural exploitation in 1896, he thus wrote:

The primitive man disturbed the conditions of the soil no more than did the lower animals. He made avail of the natural products of the forest, field, and stream, never stirring the earth except, it might be, to bury his dead; but in the first step upward began his manly career as a devastator. He became a soil-tiller, and with the invention of this art began the greatest revolution in the economics of the earth that has ever been instituted by a living being. Each extension of civilization has widened the field of destruction, until nearly one-half of all the land is subject to its ravages. It is now a question whether human culture, which rests upon the use of the soil, can devise and enforce ways of dealing with the earth which will preserve this source of life so that it may support the men of the ages to come. If this cannot be done, we must look forward to the time—remote it may be, yet clearly discernible—when our kind, having wasted its great inheritance, will fade from the earth because of the ruin it has accomplished. It should be the province of science to point the way to the remedy for this ill.[59]

Clearly, as Glacken has quite rightly said, Shaler had warned his fellow countrymen "in vigorous language that their soils would wash away if they neglected to care for them."[60] But Shaler went further than merely crying "danger!" He specified a number of measures that would counteract soil erosion. Allowing the need for farming and thereby distancing himself from a too facile primitivism, he believed that much soil wastage was due to a neglect of simple and inexpensive precautions that would prevent a progressive decline in productivity and tillage value. It seemed logical, for example, to attempt to replace those substances that under natural conditions were returned to the soil by decaying vegetation. This replacement might be accomplished, he suggested, by subsoiling—a technique whereby the lower compact layers of the soil and even the decayed portions of bedrock were disturbed so as to make available to plants new supplies of mineral nutrients. Thus, while the surface might still wear down at a rate much more rapid than when wooded, the lowering of the base of the soil would be kept in pace with it. In addition, the development of the hard pan produced by repeated shallow plowing could be avoided by the use of some subsoil-breaking method that would permit water and roots to penetrate the lower portions of the detrital layer. Among the means he suggested was the familiar device of sowing crops like red clover, whose taproots infiltrate to a greater depth than that to which the plow is ordinarily driven, thus providing waterways and paths for the roots of weaker species into the subsoil. Furthermore, since soil erosion was primarily due to "the reckless effort to win for plow-tillage land which is fit only for other and less unnatural forms of culture," he judged that, wherever slope inclination exceeded five degrees of declivity, the land should be retained in grass or orchard or used as timber nurseries.[61]

While Shaler's neglect by the National Conservation Commission, organized by President Roosevelt in 1909, is a historical enigma (none of the contributors referred to Shaler's writings), America's neglect of them both was a national tragedy.[62] The commission's total disregard of Shaler is particularly arresting in view of Theodore Roosevelt's laudatory epitaph,[63] Gifford Pinchot's acclamation of Shaler's unsurpassed authority on soil erosion,[64] and the latter's close friendship with T. C. Chamberlin, whose essay "Soil Wastage" pointed out that, before the protective covering of the soil was removed for agriculture, a normal ecological balance was maintained.[65] Still, this omission seems inconsequential when set against the Dust Bowl horror of the 1930s, the heavy price America paid for its failure to heed the commission's recommendations. Writing in 1938, Sauer could remind his readers that soil devastation was "the most widespread and most serious debit to be entered against colonial commercial exploitation,"[66] and bringing ethics into the discussion of land use under profit economies, he fashioned his concept of destructive exploitation.[67] While the creators of the New Deal policies and of the Soil Conservation Act of 1935 looked back with respect to the utilitarian, wise-use advocates of the turn of the century, Sauer's particular praise was reserved for the work of Shaler.

The utilitarian ethic undergirding Shaler's conservation philosophy is also clearly exposed in his commitment to the need for reclamation through irrigation. Consider, for example, his firm conviction that the desert would indeed bloom. Those areas of the world's deserts where irrigation was feasible he believed had far greater agricultural potential than the semiarid regions where great natural fertility was contingent on an unpredictable rainfall. In North America the great valleys of the Rio Grande, Colorado, Arkansas, and upper Mississippi were eminently suited to irrigation, but the Upper Missouri, despite its inadequate seasonal rainfall, surpassed them all by virtue of its unrivaled storage potential.[68] Nonetheless, only since the implementation of the Pick-Sloan Plan since 1944, with its 150 reservoirs and some five million acres of newly irrigated land, has the capacity that Shaler envisioned moved steadily toward its realization.[69]

Since the reservoir question provided one arena for the rancorous feud between the conservationists and the preservationists in the early 1900s,[70] Shaler's proposals for reservoir construction appearing in 1905 leave no doubt about the basic orientation of his stance on natural resources. Sharing the convictions of John Wesley Powell,[71] whose political crusading occasioned the congressional act withdrawing reservoir sites from public entry,[72] Shaler displayed his truly conservationist colors when he cast the resolution of the irrigation question as a microcosm of the wider resource issue. "The probable gain from the subjection of the land-waters to the needs of man," he reflected, "is not only to be found in the hundreds of millions of people who may thereby be fed, but in the better order of the earth for the uses of man, and in his bettered adjustment to its conditions."[73]

The activist thrust of these latter proposals was merely the culmination of a sustained attempt to inventory the economically untapped resources of the country. His earlier work in 1884 for the United States Geological Survey, for instance, had brought him to an awareness of the very substantial agricultural value of the seacoast swamps of the eastern United States. And their already rich economic potential, he noted, was further enhanced by the "great advantage [that] the more northern marsh areas . . . are generally near the larger centres of population of the country, where they will have a high value as market-garden soils or fields for the raising of hay."[74] In the region of Plum Island alone, he estimated that reclaimed land could maintain an agricultural population of around five thousand. Again, in 1890, his account of the fresh water morasses of the nation in general, and of the Dismal Swamp region in particular, was suffused with economic interjections.[75] So when he came to recap on his lifelong resource experience, he prophesied that reclamation would become an increasingly important practice in a world of diminishing natural resources. The mud flats, marshes, and mangrove swamps along the shorelines alone, he estimated, amounted "to an aggregate area of not less than 200,000 square miles of land which with a fully peopled earth will be brought into tillage. As this land is of rare fertility and enduring to the tax of cropping beyond that of any upland fields, it has a prospective value as a human asset far beyond an equal area of ordinary ground."

With a vision for reclamation encompassing both the land and its seashore margins, it was only to be expected that Shaler's hopes would drift more and more seaward. Why should the resources of the ocean not be used to their full potential? Such usage would of course necessitate an international study of marine life in addition to the excellent government commissions on local fisheries already in existence. But the benefits could be incalculable, for the inevitable increase in food supply would be merely a prelude to the "greater task of domesticating so much of marine life as would lend itself to the service of man."[76]

As one final illustration of Shaler's belief in the maximum utilization of natural resources for the benefit of society, mention must be made of his writings on the future supply of metals, minerals, and energy. On this topic, mirroring the confidence of the age, he pinned high hopes on the technological wizardry of modern science. Not that he displayed unbridled optimism. There was no room for complacency. While his review of the two principal gold- and silver-producing regions of the country (in the Appalachian and the Cordilleran ranges) led him to the conclusion that the American precious-metal mining industry "is certain to make a very rapid growth in what is left of this century,"[77] the condition of the other metal resources was much less healthy. Drawing on his experience with the geological surveys and various railway commissions,[78] Shaler concluded his 1905 review of iron ore, copper, mercury, lead, zinc, tin, and platinum with the warning that many of these metal reserves were approaching exhaustion. Iron ore, to take just one example, was becoming scarce in Britain and the Mediterranean, and a conservative reading of the American situation showed that "at anything like the present output of ores the existing production cannot be maintained for fifty years."[79] No less was this the case with natural oils and gases: the former, he judged, would not outlast the nineteenth century, while the latter were "destined to speedy exhaustion."[80]

Nevertheless, despite this rather gloomy forecast, examples drawn from both sides of the Atlantic persuaded Shaler that a great permanent energy source was provided by the world's rivers and would increasingly be harnessed through existing and projected hydroelectric power schemes. With the panache of a practiced optimist, he predicted that "the crafty mind of man will be certain to win his needs for all his time" from the wind, the sun, and the sea.[81] But such buoyant histories of the future are, as often as not, the triumph of hope over experience. In this case the variability of winds and the irregularity of tidal heads have continued to present such difficulties that they still have not been successfully exploited. No doubt Shaler would have justly incurred the judgment of Van Hise, who urged discrimination between the practicable and the wholly visionary.[82] Still, in acknowledging the need to develop alternative sources of energy, Shaler does emerge as a figure of perceptive analysis and prophetic insight.

Solutions and Synthesis

Not content to specify particular solutions to problems of soil and forest management, hydrological regulation, and energy supply, Shaler also wanted to instill in the American consciousness a sense of ecological accountability; for only by overhauling the social conscience could the nation ever seriously come to grips with its resource problem. In such an enterprise, education was of fundamental importance. Shaler's conception of education, typically, drew both its strength and its optimism from the belief that Lamarckian use-inheritance operated not only in the biological but also in the intellectual realm.[83] This belief implied that, if society could be taught what the soil meant to the human species and its long-term future, then the major step in solving the resource problem would have been achieved. Shaler sought to effect this understanding by popularizing scientific knowledge in articles and books for the general reader and, institutionally, by establishing in 1902–3, in consultation with Gifford Pinchot, a forestry program at the Lawrence Scientific School.[84] Indeed this initiative was to become the precursor of Harvard's Forestry School founded in 1904.[85] The driving force of Shaler's logistic enthusiasm may be gauged from the letter he addressed to the President and Fellows of Harvard College in March 1903:

On the part of the Administrative Board, as well as on my own part, I beg you to take steps to establish a programme in Forestry in this School. . . .

In order that I might see the situation more clearly, I have taken a nominal position in the Department of Forestry of the Bureau of Agriculture; and I have thereby gained a clear insight into the opportunities of employment. It is now evident that we can for a number of years to come find places for well-trained men in forestry in the government work, and there are many demands for men to take charge of corporation and private estates. Mr. Gifford Pinchot, who is in charge of the government forestry work, strongly advises the establishment of the proposed programme.[86]

The utilitarian thrust of this strategem, with its implications for technical education at the Lawrence Scientific School, it may be noted in passing, aroused the wrath of Alexander Agassiz, who regarded the university purely as a research institution.

As if to balance the pragmatic complexion of these proposals, Shaler also tried to educate his compatriots in the ways of contemplating nature's beauty. Despite his desire for reconciliation with nature and the longing for that mental isolation that he termed "the calm, affectionate forthgoing to the environment," Shaler admitted that pushing against raw winds in open country and peering wet-eyed from a bleak mountain top did little to induce spiritual contact with nature. His article "The Landscape as a Means of Culture" (1898) was an attempt to bring scientific principles to bear on this aesthetic problem. Appreciating the difficulty that viewers face in focusing on a single phenomenon when surrounded by multifaceted scenic beauty, he made the simple suggestion that, after a preliminary reconnaissance, spectators should position themselves with precision so that they could best see into "the heart of things." The reason was purely psychological. Human perception was such that the eye could not receive stimuli from several different directions at once, which caused the observers frustration when they were "forced to attend to a circular view."[87] Considerations of angle, range, geological structure, and geographical knowledge all contributed to an appreciation of the landscape.

But Shaler knew that scientific principle was not enough, and so, in language reminiscent of John Muir,[88] he reminded his readers that serious landscape contemplation demanded the reverence of a "worshipper, and . . . the spirit of devotion which befits a temple."[89] Without this attitude there could be no flowering of the contemplative mood, especially in days when the meditative spirit ran against the grain of the times. Still, where this instinctual response to nature still lingered, science could be of service. Some lovers of nature, of course, might resent the suggestion that their spontaneous pleasure could be the subject of cold rational inquiry, but Shaler felt that, just as the appreciation of music and drama had been enhanced by study, so training could enlarge the aesthetic experience of the primitive wilderness. If, as he put it, "the student has a true love of nature, and will train his eye and mind in the knowledge and understanding of the earth which is just about him, he will everywhere gain that sense of the majesty and beauty of his dwelling-place which is the best fruit of all the naturalist's labor."[90] Indeed, instruction in the art of scenic appraisal was particularly needed in North America because the landscape was much less picturesque than the Old World and because the public needed to be reeducated in their understanding of what constituted natural beauty rather than remaining enslaved to the traditional impressions derived from European literature.

As time passed, however, Shaler came more and more to concede that his counsel was incomplete, for education of itself did not bring with it a sense of responsibility. Information, it was now clear, could not create obligation. Realizing that knowledge was different from wisdom, Shaler had come to recognize, as many have done since, that answers to the environmental dilemma could not be formulated purely in terms of finding scientific solutions to ecological puzzles.[91] Nor indeed was a revolution in personal morality the solution.[92] Indeed, with the passion of a cri de coeur, he declared at the end of his life that society's duty to the earth ought to be taught as gospel: "Unless the world of men should become philosophers, we must look in the future as in the past for the leading spirits, the rare men, to be guides to the new dispensation, the masses following in the ancient dumb way—taking their light not directly from nature, but in the good old way, mediately through their prophets."[93]

Convinced now of the need for collective action, Shaler pressed for governmental control of the environmental emergency. The Appalachian forest, for example, presented a problem covering such a large area that it was unrealistic to expect effective remedial action from the dozen or so separate states. Efficient management demanded federal intervention at the highest level. As to ethical justification for this move, Shaler insisted that the wealth of individuals was theirs to use but not theirs to waste, since such improvidence inflicted suffering on their natural heirs and upon the state. The primary function of government was thus to ensure that the material foundations of society were not heedlessly dissipated. Already the American people had caused so much irreparable environmental damage that, at best, they could only transmit their natural resources "to the future in a very impaired state."[94] Measures had to be taken immediately to arrest the spread of this national disease. As with Pinchot and Van Hise after him, Shaler saw this action as a patriotic venture, for true statesmen planned, not for the next election, but for the next generation. They did not think primarily of themselves, but rather of the race and its future well-being. The nation's later citizens would judge the current generation by how they had used their environment and would "date the end of barbarism from the time when the generations began to feel that they rightfully had no more than a life estate in this sphere, with no right to squander the inheritance of their kind."[95]

No less enthusiastic about resource management than about beauty in nature, Shaler did his best to steer a middle course between the spiritualizing of the romantic preservationists and the utilitarianism of the economic conservationists. For this project his conception of the humanized landscape was a vital ingredient. In *Man and the Earth* Shaler sought to show that an aesthetic concern was not incompatible with environmental engineering. Some feared, he began, that the modern world would soon become so artificial, so planned, that its primitive beauty would be utterly lost. Careful reflection, however, would show that, while the character of the beautiful would doubtless be greatly altered, there was reason to believe that a humanized land-scape would actually enhance artistic value and thereby bring even greater spiritual enrichment. Not that the untamed wilderness would be totally lost. The work of those societies whose purpose was "the preservation and enhancement of the natural beauty of the country"[96] was to be highly commended, as was the government's protection of Yellowstone Park "from the depradations of hunters or the deface-ments of business."[97]

And yet Shaler felt that in the future the main source of national beauty would lie in the works of material culture. His evolutionary account of the early emergence of artistic sensitivity lent support to this claim. With an ancestry essentially lacking in aesthetic sensibility, the appearance of a creature with a sense of creative discrimination was one of the greatest marvels of our evolutionary history. Since this aspect of human development was originally part of the more general process of mental evolution, it was, in principle, thoroughly compat-ible with technological progress. The absolute wilderness, whatever its nobility, had scenic grandeur for really rather few, and even for these it lacked the charm of a landscape bearing the marks of human culture. Either in sight or in imagination, a region had to be populated in order that the viewer could experience a sense of empathy with it.[98] "Hence it is only the more expanded souls can rejoice in the untrodden deserts, the pathless woods, or the mountains that had no trace of culture," he confirmed. "Such people these places with their fancy— at least they feel the Lord is there; and so they have their bond with what else would be utterly savage to them."[99]

For the future, Shaler believed the human animal would greatly alter the face of the earth in its progressive subjugation of the planet. People would certainly create further ecological disturbances, for cer-tain species could only maintain themselves in such large numbers and with such a measure of freedom that Shaler deemed incompatible with social advancement. Some indeed would have to be eliminated, and to that extent the humanization of the earth would inevitably involve suffering. But these changes, far from being detrimental to the human spirit, would bring both the benefits of a modern technological way of life and the opportunity for greater artistic imagination.

In order to grasp how Shaler used his concept of the humanized landscape to synthesize beauty and utility, his treatment of highways is worthy of consideration. His book on the subject included a chapter entitled "The Relation of Public Ways to the Ornamentation of a Country," in which, drawing on the expertise of the professor of architecture at the Lawrence Scientific School,[100] he announced:

It is a good feature of our day that the people of this country, so long neglectful of all considerations of beauty in the landscape about them, have not only become interested in that element of culture, but are willing to make considerable sacrifices in order to adorn the land about their dwelling places. Inasmuch, therefore, as roads are important elements in a landscape, serving greatly to elevate or to debase the view, it seems fit to give some attention to the aesthetic quality of roads.[101]

Shaler's overriding concern was to ensure that roads would be specifically planned to blend harmoniously with the physical geography of an area, rather than being "forced across the field of view, climbing the hills abruptly and in other ways disobeying the injunctions of nature." One of the best ways of maintaining a balance between the needs of transport and the preservation of natural beauty, he suggested, was by planting roadside trees, not in stilted regimented lines, but by "systematic plantations of groups of trees on either side of a traveled way, the species being varied and the outline of the plantations toward the road broken so as to promote pleasing vistas." Furthermore, roadside parks should be provided as well as stone bridges, of which he believed there were "no other architectural features attainable in our American landscapes so well calculated to enhance their beauty as the sight of well-shaped masonry arches over the streams."[102]

In his concept of the humanized landscape, Shaler thus found a means of integrating, to some degree, the artistic with the pragmatic, the aesthetic with the economic. Moreover, it allowed him to find new ways of conceptualizing the relationship between culture and nature.[103] After all, human beings were, through and through, part of the material world and at the same time distanced from it by their ability to transform it for their own purposes. For Shaler, a domesticated world, "an intensely humanized earth, so arranged as to afford a living to the largest possible number of men," was entirely consistent with the belief that society's alterations of the landscape "will enhance its aesthetic value, making its features far more contributive to spiritual enlargement than they were in their primal wilderness state."[104]

It is plain, then, that the rejection of civilization and urban life, commonplace in the then current environmental literature and stylishly articulated in a long literary tradition of intellectual antiurbanism, was not echoed in Shaler's writings.[105] Like his friend and colleague William James, Shaler loved nature, but his love of it was tempered by a fondness for the sociability of civilized society that made him unable to subscribe wholeheartedly either to Thoreau's primitivism or to Spencer's materialism. In the domesticated earth that he envisaged, the quality of the primal that only the wilderness can evoke was certainly lacking. But then, so too was the unrestrained exploitation of the earth by "economic man." His longing was for the dissemination of an environmental ethic in which the "nobility of the primitive fitly recognized may have its due place even in an earth subjugated to the uses of man."[106]

Paradoxically, perhaps, the neglect of Shaler's work on the environment theme would seem to stem from those very qualities that render it a genuine precursor of Harold Rose's more recent appeal for a "broader view of conservation than that which has prevailed in the past . . . considering the growing importance of spiritual values in a materialistic oriented society."[107] But these two themes, represented in the diverse schools of environmental thought that developed in the late nineteenth- and early twentieth-century America, each had distinct, if unspoken, criteria of acceptance within their respective communities.[108] The scientific conservationists seem to have viewed Shaler as a describer of the problem rather than as a rigorous exponent of the systematic resource inventory. On the other hand, Shaler's welcome of technological innovation and his enthusiasm for what he termed the humanized earth were scarcely palatable to the back-to-nature romantics, despite his enthusiasm for the transcendent spirituality manifest in the natural order.

As with many other issues he contemplated, in seeking a solution to the problems of environment, Shaler stood between two traditions: an earlier, specifically religious perspective and a later, secular scientific outlook. At the confluence of these two intellectual streams, Shaler's religious recasting of evolutionary science provided the vantage point from which he interpreted the duty of society to nature.

Coda

It seems singularly appropriate that the prophet of the New Conservation consciousness, Lewis Mumford, should constantly draw inspiration from the afterglow of America's Golden Day in the 1870s—for him the last era in history when life had been sanely poised between old values and new, and when "Americans in general had succeeded for a tantalizing moment in casting off the shackles of the Old World but not the civilizing virtues."[109] As Mumford observed, "A genuine culture was beginning to struggle upward again in the seventies: A Peirce, a Shaler, a Marsh . . . were men that any age might proudly exhibit and make use of. But the procession of American civilization divided and walked around these men."[110] And yet persistence of these diverse beliefs in the contemporary environmental literature seems to suggest, as Erisman has put it, that they "lurk in a cultural memory exerting their influence in devious ways and sporadically surfacing in new incarnations."[111]

Aspects of Earth History 8

It is perhaps as one of the great teachers of geology that he will be best remembered by us, and may the inspiration he gave long continue to bear fruit.

—Wolff, "Memoir of Shaler"

When a certain Father Blumhardt took Shaler for a student of theology rather than of geology, the misunderstanding was certainly due to an initial mishearing at their first introduction.[1] But the confusion could well have arisen for quite other reasons, for during the nineteenth century, geological inquiry played a key role in the theological dispute about God's providential care of the history of the earth.[2] At least since the time of Hutton, whose geology brooked "no vestige of a beginning—no prospect of an end" and thereby banished the supernatural from the kingdom of earth science, the Newtonian Heavenly Clockmaker became progressively unsatisfactory as a means of integrating theology with geology.[3] The latent threat to natural theology in its traditionally mechanistic garb from geology's boldest theories, however, has often obscured the ways in which the new science could be accommodated to, or rather more precisely, synthesized with, the teleological spirit. In an age when the "reign of law" enjoyed a hitherto unprecedented intellectual vogue, advocates of the rejuvenated natural theology that began to emerge in the 1840s found that, by imbuing the processes of nature with theistic impulse, they could give substance

215

to their talk of a grander vision of the Creator.[4] For them, natural law was not merely the codification of precedent; it was nothing less than the intrinsic telos of a material world impregnated with spiritual reality.

This understanding, of course, does not mean that geology was invariably, or even explicitly, stage-cast as the handmaiden to the reconstituted physicotheology or that teleological threads were woven into the fabric of every geological publication. Rather, it implies that natural theology in one form or another continued to provide the context within which much of the nineteenth-century debate about earth history was conducted and, more particularly, that this context changed in response to the new assumptions of the uniformity of nature and the even newer principles of evolutionary natural history. By the end of the century, it must be added, many who retained that taste for things teleological satisfied their appetite in purely personal reflection rather than in scientific practice. And for many of them, privatizing natural theology was only the first step toward naturalizing it, particularly as the professional courts of science outlawed metaphysical interference in the test-tube world of scientific theory.

Shaler's efforts at recognizing the constituent elements of the Ray-Butler-Paley cosmology in the light of the theory of evolution and his fusing of natural law with divine immanence have already been explored. His contributions involved subtle accommodations to evolution theory that at once diminished active divine intervention in nature and yet made divine rule both more grand and more intimate. Thus in his introduction to *Outlines of the Earth's History,* he made it clear that the principle of natural law undergirded all geological research and that through the study of physiography the student would come to appreciate "the larger truths which may help him to understand the beauty and grandeur of the sphere in which he dwells." So crucial was this methodological precept, both as an organizing principle and as the touchstone of genuine science, that Shaler began with a thumbnail résumé of how it had emerged from primitive animistic accounts of the world, of how "the conception of natural law replaced the earlier idea as to the intervention of a spirit," and of how "science departed from other forms of lore and came to possess a field to itself." Yet for all that, as we have seen, Shaler's "theisized" version of evolution allowed him to affirm that the historical "conflict between the religious authority and the men of science has practically ceased. . . . Men have come to see that all truth is accordant, and that religion has nothing to fear from the faithful and devoted study of Nature."[5]

As might have been expected, the evolutionary gradualism of which Shaler often spoke[6] permitted him to draw parallels between progress in the organic world and progress in the geological history of the earth.[7] More precisely, the principle of uniformity oiled the joints between bio- and geohistory, and Shaler found he could use it against critics to affirm that "all the facts with which the geologist deals are decidedly against the assumption that terrestrial changes in the organic or the inorganic world ever proceed in a spasmodic manner. . . the earth has gone forward in its changes much as it is now advancing."[8] On the face of it, Shaler's proposals would seem to justify Huxley's ultimately misleading dictum that "consistent uniformitarianism postulates evolution as much in the organic as in the inorganic world."[9] But this conclusion would underestimate the significance equally attached to those "critical points" of which he was so fond and that were thoroughly compatible with some versions of catastrophism. As he outlined that theory in 1890:

Let us consider the element of the unexpected which arises from the variations in the application of force to the elemental combinations, variations which are independent of the material associations that exist in these various substances. It is my purpose to call attention to the well-known but much disregarded fact that with variations in the application of force the behavior of matter may alter in a manner which is calculated to bring about the most unforeseeable consequences. I desire also to show that these variations may take place with extreme suddenness, indeed with revolutionary rapidity, and that through this action there may come about in the physical world very great modifications in conditions made, so to speak, in the twinkling of an eye. The circumstances under which these revolutions occur I shall term critical points. By a critical point I mean a station in the series of changing conditions at which a new method of action is suddenly introduced. . . . we may see how by successive gradual changes, each infinitely small, we may pass in the end a critical point, leading to consequences which are infinitely great.[10]

Thus, while rigidly adhering to the classical uniformitarian principle that all geological history could be explained by reference to forces now in operation, he incorporated into this reign of law sudden, unpredictable natural events. Moreover, on this catastrophist foundation he grounded "the analogy . . . between the interaction of the inorganic elements one with another and a similar interaction between the separated but ever-combining motives which guide the animal body." For just as "in the elemental world the combination of two substances commonly gives us a third substance different in quality from either of the original ingredients, so, when the motives or impulses of inheritance combine in the organic body, the results may have very great complexity."[11]

At first blush these proposals seem conflicting, even self-contradictory. But such a judgment would ignore the crucial distinction between uniformitarianism as a method of research and as a substantive system. A brief perusal of this subject will therefore throw some light on Shaler's geological philosophy. Diverse though its early implications were,[12] uniformitarianism found its most systematic expression in Lyell's dictum that the course of nature had been uniform from earth's earliest times and that the geological causes now operative were sufficient to account for former changes of the earth's surface. Its main import therefore, on the principle of Ockham's razor, was to ban the preternaturalism of so many contemporary catastrophists from geological discourse. But in Lyell's case it had additional implications. The thoroughly ahistorical drift of his version led him consistently to deny direction or progression in either organic or inorganic history and therefore to rule out the possibility of evolutionary change. And yet uniformitarianism could easily be given other more dynamic renderings, for, as Hooykaas puts it, "*Uniformity* might refer not so much to uniformity of the *situation itself* as to uniformity of *change of the situation*. When a small rate of progressive or directed change prevailing now, is assumed to have prevailed always, a situation that is non-uniform throughout the ages is uniformly changed."[13] Plainly this uniformitarian scheme could easily be reconciled with evolution, but no more so than the catastrophism of, say, Conybeare. His was a genuinely historicist model, postulating a time scale of almost any length short of infinity and opposing Hutton, Lyell, and later dynamic steady state theories by identifying "secular, unidirectional trends in earth history."[14] Indeed, for this reason Cannon felt confident enough to argue that certain versions of catastrophism provided a far more favorable context for the development of evolutionary theories than strict uniformitarianism.[15]

At this juncture it is important to distinguish, as do Hooykaas, Vysotskii, and Gould,[16] two logically quite separate aspects of uniformitarianism. One, sometimes termed "actualism" or "methodological uniformitarianism," is the proposition that natural laws are invariable and that those now operating are sufficient to explain the geological history of the planet; the second, referred to as "substantive uniformitarianism" or simply "uniformitarianism," is the belief that geohistory exhibits "uniformity in the intensities and rates of natural processes and in material conditions."[17] This distinction has more than theoretical significance, as may be gauged from the historical fact that, while Darwin was strongly influenced by Lyell's geological uniformitarianism as a *method* and of course by its concomitant postulate of a lengthy earth history, the paleontological basis of his theory drew on the work of progressivists in the catastrophist tradition, notably Buckland, Sedgwick, Miller, Agassiz, and Elie de Beaumont. As Hooykaas again notes, "One could say that both Lyell and Darwin used a uniformitarian *method,* but that Lyell arrived by its help at a uniformitarian *system,* whereas Darwin's theory of descent with modification is not a uniformitarian, but an evolutionist system."[18]

These nuances of interpretation are, doubtless, rather more clearcut in hindsight than they were for many geologists active in the nineteenth-century debates, but they do nevertheless help to give structure to the apparently inconsistent statements of some geologists by softening such dichotomies as the uniformitarian–catastrophist and the gradualist–saltatory polarities. Thus Elie de Beaumont, to take one example with whom Shaler was particularly familiar, was an "actualistic catastrophist" whose account of orogenesis combined an actualist method with a catastrophist system. To him, the " 'slow and continuous' phenomenon of cooling of the earth causes a slow and progressive diminution of its volume, from which ensues the rise of the mountains. This cooling, which acts as a slow and gradual cause, has as its effects violent and sudden cataclysms."[19]

To return to Shaler, his pronouncements can now be read in a new light. His method was clearly uniformitarian or actualist; his system, however, not at all inconsistently, allowed for quite sudden and dramatic changes in nature, catastrophes that were in fact fundamental to drawing the vital gradualist analogy between the history of life on earth and the history of the earth itself. From this source too may spring the divergence between Shaler's conception of earth surface processes and that of his celebrated pupil W. M. Davis. As is well known, Davis's systemic uniformitarianism—a "direct descendant" of Hutton's classical model—was thoroughly cyclical in its import. Indeed, as Higgins notes, it was to become "so necessary, central, and ingrained in American geologic and geomorphic thought that the very idea of non-cyclic development was not only incomprehensible but *taboo*": so much so that the noncyclic theories of Penck and Gilbert were long ignored.[20] At the same time, Davis, like Dawson,[21] realized that an exaggerated uniformity in geological rates was untenable. "Uniformitarianism, reasonably understood," he wrote in 1895, "is not a rigid limitation of past processes to the rates of present processes, but a rational association of observed effects with competent causes. Events may have progressed both faster and slower in the past than during the brief interval which we call the present, but the past and present events differ in degree and not in kind."[22] At the very least, Davis required that degree of uniformitarian latitude in order to squeeze into his system those fundamental, initial, tectonic uplifts.[23] But whereas he used that admission to construct an overarching cyclical earth history, Shaler marshaled sudden catastrophes in the cause of *cumulative* rather than *cyclical* interpretations in geology and biology alike. Small wonder that, as Stoddart notes, Davis remained "profoundly more impressed with the mystery of growth from egg or seed to adult [a life cycle] than he was with the cumulated effect of small-scale changes over many generations."[24]

Nowhere, perhaps, in the history of the earth sciences was the uniformitarian-catastrophist confrontation more marked than in discussions about mountain formation and ice ages, and it therefore is appropriate now to turn to Shaler's perspective on the problems of orogenesis and the glacial theory.

Orogenesis

Even if judged solely by the content of his undergraduate reading, Shaler's geological sympathies would be clearly evident. For when Benjamin Peirce asked him to review and evaluate the "Résumé pentagonal" of Elie de Beaumont's *Système des montagnes* during his final examination, he was referring to a work that Shaler had thoroughly digested.[25] Shaler had, besides, familiarized himself with William Hopkins's critical response and with his application of astronomical and chemical findings to theoretical geology, and he used this information to good effect that day. But unlike so many students, Shaler did not lose his mastery of the subject once it had served the useful purpose of earning him a Harvard degree. On the contrary, it became the springboard for a lifetime's work on the problems of mountain building.

To make any contribution to this subject, of course, required a solid grasp of the long-standing debates about the center of the earth, and Shaler made his first foray into this disputed territory. In the wake of the Neptunist-Vulcanist feuds, Elie de Beaumont had proposed that the forces that raised continents and brought about the elevation of mountain chains had operated at quite different rates in the past. Accepting the doctrine of central primitive heat (the view that "the heat now present inside the earth is the residue of the original heat possessed by the earth at its formation"), he believed that mountain building occurred during periods of intense, spasmodic earth movement.[26] These movements were the result of stresses in the solid crust created by a cooling, contracting earth nucleus, which could only be relieved by folding, local sinking, and compensatory uplift.

The model of the earth that Beaumont's theory of mountain-chain formation assumed was that of a thin, solid crust surrounding a hot liquid interior; this latter claim was subjected to the searching scrutiny first of William Hopkins and then, in more refined and amplified form, by his student William Thomson, Lord Kelvin. Yet at the same time, Hopkins shared Beaumont's catastrophist instincts. By opposing the Lyellian formula and advocating the progressivist principle inherent in the idea of a slow, cooling earth imposing an irreversible directionalism on geological history, he too bridged the old gap between catastrophism and uniformitarianism.[27] For while the *cause* of earth heat loss was uniformitarian, Hopkins noted that the *effects* were often catastrophic—a methodological stance akin to Shaler's own. On the more specific point at issue, however, Hopkins argued that the current state of knowledge regarding the solidification of a molten sphere, the effects of pressure on the melting temperature, the differential conductivity of certain rocks, and the astronomical rotation of the earth all pointed to an earth crust of at least one thousand miles.[28] These arguments, as I have said, were pursued with greater rigor by Thomson, who went even further by proposing that the earth was entirely solid. His theory was that solidification must have started at the earth's center and was now essentially complete; in his 1862 paper on the rigidity of the earth, he argued the case for solidity based on the mathematics of precision and nutation.[29]

Beaumont's theory had been widely accepted since the 1840s, when Dana had introduced it to American audiences.[30] But during the 1860s and 1870s, some geologists became impressed with the critical riposte of Hopkins and Thomson, and so several attempts were made to reconstruct geological theory dispensing with the assumption of a liquid interior. As Brush indicates, Shaler's 1866 paper "On the Formation of Mountain Chains," first printed in the Boston Society's *Proceedings* and reprinted in the *Geological Magazine* for 1868, was one of the first.[31] This paper, however, was not his first word on the subject, for the previous year he had made a preliminary effort to account for the elevation of continental masses, on the assumption of a thick outer crust "of solid material gained from the viscidly fluid nucleus as the downward cooling progresses."[32] Charles Babbage and Sir John Herschel, he reported, had explained that the changing position of isogeothermal lines in the earth's mantle was determined by the thickness and conducting power of its crustal materials. This relationship implied that an increment of nonconducting matter, by virtue of its blanket effect, would bring the isogeothemal lines closer to the original surface. Now what, Shaler pondered, would be the horizontal effects of such a deposition? The response of a compound metal bar to heating gave a clue to the answer. Warping, he observed, was always in the direction of the material having the greatest rate of expansion, which suggested that areas over which deposition was occurring would tend to bend downward while there would be a compensatory upward movement of those surfaces where the processes of denudation were active. Sea bottoms, therefore, would have a tendency to subside; those portions of the earth's crust above water level would be elevated. Besides this mechanism, such movements could also result simply from transfers in weight following erosion, and these would be accompanied by the crustal fractures observable along shorelines in the form of volcanic fissures. Readjustments in the wake of deposition, whether geothermal or isostatic (to use Dutton's later neologism), were thus intimately linked with the elevation of continental masses.[33]

Shaler's 1866 contribution focused more sharply on the question of mountain formation. Already he had considerably modified his model of the globe, although it was still rooted directly in the mathematics of Hopkins and Thomson. Now, however, he departed from the former's thick earth shell surrounding a liquid interior (his own earlier view) and the latter's totally solid globe in favor of the idea that the process of solidification from the center outward was matched by a parallel process of hardening in the other direction. In this view, the earth was almost entirely solid but had a very thin liquid layer under the outer shell. As Brush points out, this version was to become the "most popular post-Thomson model."[34]

For Shaler, the new scheme had major implications for understanding mountain building. It implied, for example, a fundamental distinction between the nature of the processes at work in the solid core and those in the outer shell, and this distinction in turn had further implications for the genesis of mountains and continents: "Without any particular examination of the facts, it seems to have been assumed by most geologists that all the phenomena of corrugation, whether exhibited in mountain ranges, or in continents, are to be regarded as effects of one and the same cause differing only in magnitude. It is manifest that it is a matter of first importance in seeking an explanation of the origin of these phenomena, to determine whether this assumed identity of cause is true or not."[35] If, Shaler began, both continental and mountain elevations were merely the differential effects of the *same* cause, there would merely be a difference in the magnitude of the two phenomena—a difference proportional to the size of the area disturbed and a graduated series from continental folds "to the most inconsiderable flextures." Manifestly, such features simply did not exist. So Shaler was forced to conclude that the thermal contraction theory operated in two quite different spheres. The *continental* folds, he suggested, were corrugations of the whole thickness of the crustal shell, brought about by the cooling of the interior and the accommodation of the hardened outer crust to a diminished nucleus. *Mountain chains*, by contrast, were another effect of contraction but could not "be referred to the shrinking of the whole mass." Their formation originated from "changes going on within the crust itself, and in no way connected with the regions below."[36] As to the mechanics of this theory, Shaler argued that atmospheric temperature would eventually prevent further heat loss at the earth's surface and therefore arrest shrinkage action; the layers below, however, would continue to contract—even though their temperature was higher than at the surface—and the resulting lateral pressure would lead to the upfolding of the superficial strata on the outer shell. Continental elevation and mountain-chain formation were thus of different origin.

By the 1870s, the idea of a solid earth had gripped the minds of many geologists. Joseph Le Conte, for example, announced to the readers of the *American Journal of Science* in 1872 that "the whole theory of igneous agencies which formed the foundation of theoretic geology, should be reconstructed on the basis of a solid earth";[37] Osmond Fisher more closely approached Shaler's model when, in 1873, he abandoned the solid earth in favor of the fluid substratum.[38] Such conversions, however, were not universal, as is witnessed by the vacillations of James D. Dana. During the 1870s he seemed convinced by the Hopkins-Thomson thesis.[39] But by 1880, when the third edition of his *Manual of Geology* appeared, he had revived his earlier fluidist theory. Even his temporary acceptance of a solid earth was altogether diffident, for he began his 1873 series of papers on the subject by reaffirming his 1846 statement that the definition of continental and ocean areas had begun with "the earth's solidification at surface," and therefore that the "principal mountain chains are portions of the earth's crust which have been pushed up, crumpled, and plicated by the lateral pressure resulting from the earth's contraction." Indeed in a footnote to these remarks, he argued that Le Conte's principle did not differ from his own 1846 contribution, "except that it is connected with the idea of a solid globe." This qualification is important too because it helps explain Dana's apparent misreading of Shaler's 1866 paper. Ever sensitive to questions of scholarly priority, Dana had dismissively complained that "Professor N. S. Shaler in 1866 . . . presented, *as original*, the idea that 'mountain chains are only folds of the outer portion of the crust caused by the contraction of the lower regions of the outer shell'; . . . which is essentially the view that Le Conte attributes to me." Dana, however, had made reference neither to Shaler's idea of a *double* contraction—of the earth nucleus and of the outer shell—nor indeed to the thin fluid substratum, although he now accepted it. In fact, he went on to say that "this restriction of the interior liquidity of the earth to an undercrust layer, does not require in itself any modification of the view I presented more than twenty-five years since, on the results of the earth's contraction, since there is still a flexible crust and mobile rock beneath it."[40]

Although Shaler subsequently added some embellishments to this basic theory, it was to remain his considered judgment for over twenty years. Indeed he often summarized it when he needed to treat orogenic questions, whether for popular audiences, as in the case of the *Atlantic Monthly*, or for scholarly peers at the Boston Society of Natural History.[41] In the latter context, he supplemented his account by arguing that the different elements of a particular mountain system might be the cumulative result of crustal contractions at different periods. The Appalachian complex, for example, showed signs that its separate components were of different ages, as did the mountain axes in the Cumberland Gap. Throughout, the thrust of this augmented argument was that empirical studies had vitiated Beaumont's monogenetic explanation. As he put it, "It would be nearer the truth to say that mountain systems are more likely to be the product of parallel upheavals occurring in successive geological periods than of single epochs of elevation." Yet at the same time, he did make it plain that his post-Beaumont version was not excessively uniformitarian; while he announced that "the greater part of these dislocations have been made slowly," he immediately conceded that "some of them have been formed with a great suddenness, and attended by movements of extreme violence."[42]

Shaler was now sure that mountain building was not a monogenetic process. Neither was it monocausal. Consider, for instance, his 1888 paper for *Science,* "The Crenitic Hypothesis and Mountain Building." This theory was originally put forward, according to Le Conte, by Osmond Fisher in 1875 to account for volcanic activity.[43] Shaler himself had already resorted to it for that purpose and continued to refer to it.[44] As he saw it, materials deposited on the seafloor would retain water in the interstices of the accumulation. When heated by successive series of deposits, the steam encased in these subterranean rocks at a depth of between ten and twenty miles could reach a temperature of some 2000° F, and the only way in which these imprisoned gases could escape was by lateral migration within the level of now softened rocks toward any point where fissures would lead to the surface. And this process, Shaler argued, was "the sole *essential* phenomenon of volcanic eruptions."[45] For all his confidence in its theoretical virility, however, the crenitic hypothesis was thoroughly contested by Israel C. Russell in 1904. Quoting extensively from Shaler's work, he attacked his premise of vast thicknesses of accumulated stratified rocks, pointing out that marine sedimentation continued only as long as subsidence brought the deposited material below sea level. Sedimentary accumulation, he conceded, would certainly cause an increase in temperature, but even where this did occur, the result would be a more general expansion in mass volume and an elevation of the surface above water level. While he found Shaler's "a very interesting and suggestive" theory, he was thus forced to conclude that it was ultimately flawed.[46]

The kernel of Shaler's current proposal to link the crenitic theory with mountain building lay in his conviction that the retention of internal heat and crustal downwarping were not the only results of sediment accumulation. This action would also "induce an upward migration of the imprisoned waters, and consequently, in time, a transfer of material to higher levels in the rocks." The subsequent expansion of the lower rocks would eventually cause the upper strata to buckle and produce fold mountains. The "intensification of deposition" could therefore "in time lead to a reversal of the down-sinking movement and the construction of a mountain system in what was previously a basin of sedimentation."[47] To Shaler, then, it was now clear that several different processes were involved in mountain formation, namely, the secular refrigeration of the earth (the thermal contraction theory), the transfer of weight by erosion and deposition (isostatic readjustment), and the subterranean migration of rocks (the crenitic hypothesis).

The specific question of mountain growth did not engage Shaler's interests again until the December 1893 meeting of the recently formed Geological Society of America. Here the significance of isostatic processes received renewed emphasis. In his account of the Pleistocene distortions of the Atlantic coast, he pointed out that the development of a "synclinal" by the import of extensive detrital sediments preceded orogenic movements.[48] Moreover, he now felt that these readjustments provided some grounds for reestablishing links between continental elevation and mountain formation, or, as Gilbert put it in his pioneering *Lake Bonneville* study, between epeirogeny and orogeny. Shaler's work on the Italian peninsula, for example, provided supportive testimony, for "the folding and other compressive phenomena which here produced the mountain-axes of this district" were accompanied by "a progressive uprising, in a massive way, of the deposits which form the outer part of the earth."[49] The subterranean movement of rocks toward the mountain-built zones as a result of the parallel actions of erosion and deposition seemed a viable explanation, and corroborative evidence was forthcoming from the fact that crustal folds that did not rise above sea level never attained continental development, whereas those on the land, subject to denudation, were continuously uplifted.

Shaler's talk of a synclinal factor in orogenesis was made more explicit by 1899 in reports on the geology of the Richmond and Narragansett basins. Doubtless this reflects some infiltration of James Hall's synclinal theory of 1859, which proposed that the accumulation of sufficient coastal sediments to cause crustal depression was a prerequisite for mountain building. As James Geikie explained the full-fledged geosynclinal theory (a term coined incidentally by Dana): "Towards the bottom of the basin the down-bent strata would be rent or fractured, while the diminished width of surface above, caused by the sagging, would result in the strata being wrinkled."[50] Hall's own version, however, had serious flaws, for by ignoring the role played by folding, faulting, and nappe formation in the orogenic story, he had presented, as Dana caustically quipped, an explanation of the origin of mountains with the origin left out.[51] Shaler was no less

sensitive to these grave weaknesses, and so his reports were designed
to confirm those theories that, like those of Fontaine and Russell, tied
synclinal depression to the processes of uplift, folding, and faulting.
In these 1899 contributions, he therefore rehearsed the now standard
geosynclinal portrait, emphasized its particular relevance to the Appalachian system as sculptured out of a former erosion basin, and
concluded with a synopsis of where he and Hall parted geological
company.[52]

The shifting focal point of Shaler's orogenic thought over more
than thirty years thus mirrors the evolution of mainstream geological
opinion during the second half of the nineteenth century. No less was
this the case with another subject interwoven with his reflections on
tectogenesis, namely, the glacial geomorphology of the Quaternary
era.

The Glacial Theory

As a student of Louis Agassiz, Shaler was in direct, daily contact with
the leading glacial theorist of his generation. In the 1830s, De Charpentier's speculations about the possible transportation of erratics by
ice became the launching pad for the young Agassiz's own thinking
about a former ice age, and he daringly proposed that the advance and
retreat of Alpine glaciers was not a local incident but part of a more
widespread climatic change that had affected the whole of Europe.
Synthesizing and yet superseding earlier conjectures, Agassiz's *Etudes
sur les glaciers* (1840) received such critical acclaim, especially in Britain,
that after its publication "the principle that glaciers accounted for
morainic ridges and much of the scattered deposits was never seriously
disputed," although his enthusiasm for vast continental ice sheets was
not so readily shared. Agassiz's ideas preceded him to the United
States, and by the time he settled into his new post at the Lawrence
Scientific School, the import of his glacial theory was already widely
appreciated. As Hitchcock confessed, "Agassiz's ice mechanism afforded the first really satisfactory explanation of such apparently dis-
similar features as morainic accumulations, furrowing of rocks, and
erratics and perched blocks on top of gravel deposits."[53]

The earth-shaping power of ice was thus understandably never far

from Shaler's mind. And yet, while he had come to a realization of
the mechanism of isostatic readjustment by erosion and deposition by
1865, he did not link this up to glacial matters until the mid-1870s.
Certainly in 1866 he was sensing some association between the
"depression of the shore and the laying down of the boulder drift,"[54]
while in 1868 he had hit upon the influence accumulated ice would
exert in crustal downwarping because of the changing position of the
isogeothermal lines. Indeed by this date he felt sure that "the phenomena of glaciation and subsidence stand to each other in the relation of
cause and effect, and that the ice sheet operates. . . directly through
the means of forces brought into action by, and proportionate to, the
thickness of the glacial sheet."[55] But except for the agency of geothermal action, he was at a loss to conceive of any other mechanism.

At this stage Shaler temporarily abandoned the problem and restricted his glacial concerns to discussions of such topics as the apparent
lack of glacial relics in the Yukon Valley, Alaska, drift evidence for a
double submergence of the New England coast, and the characteristics
of the Charles River Valley's glacial moraines.[56] By 1870, taking stock
of his findings, he could issue his personal progress report on the
attempt to reconstruct the history of the last glacial period: "first, the
accumulation of an ice sheet, so deep as to move over all the summits
in Southern New England; second, a subsidence of the land to a depth
of from one hundred to one hundred and twenty feet below its present
position, accompanied, or succeeded, by a rapid melting of the glacial
envelope . . .; third, an elevation of the land which restored it to
almost its present level, possibly to a point a little above the level it
now has."[57]

Now working for the United States Coast Survey, Shaler had the
opportunity over the next few years to fortify his glacial armory with
further empirical stores. The geology of the island of Aquidneck and
the Narragansett Bay region, for example, had convinced him of the
error of those who continued to interpret the U-shaped valleys as the
topographical remnants of former marine action. The amplification
of original stream excavations by the erosive action of a moving ice
sheet seemed a much better explanation, not least because almost
every rock face he examined exhibited the rounded, smoothed, scored
surface so typical of glaciated landscapes, while much of the island

was "buried beneath a coating of detrital material from two to forty feet in thickness."[58] These erratics further reinforced his glacialist intuitions, and he marshaled the findings against those opponents of continental ice sheets who still tried to explain them by the action of floating ice.[59] In Shaler's mind, the marks of boulder abrasion simply could not be "explained without supposing that they were derived from the floor of the ice mass, and passed through the inevitable jostling which must occur while they were at the bottom of the pack."[60] More glacialist still, Shaler was now persuaded of the truth of James Croll's theory that glaciation was a recurring phenomenon in the history of the earth. To recognize this, moreover, was to completely redraw the map of physical and organic evolution, because periodic ice ages provided ideal conditions for exposing organic life to ever-changing environments. So for Shaler, the simplicity and fertility of the glacial theory in both the geological and biological sciences gave it appeal above all rivals.

This unmatched experience of the geology of the New England coast eventually enabled Shaler to piece together the principle of glacial isostasy. Indeed his 1874 statement of the theory, along with T. F. Jamieson's later 1882 contribution, has come to be regarded as the initial successful step toward grappling scientifically with the problem.[61] The remains of marine animals in stratified deposits above high-water mark, Shaler reported, had first alerted him to a former subsidence of the Maine coast, but now, instead of invoking the old geothermal formula, he advanced the simpler proposal that the depression was due purely to the weight of the accumulated ice on the crust. The crucial paragraph reads as follows:

It is very important to notice, however, that in the case of a rigid mass, such as we suppose the crust to be, supported on material having sufficient mobility to give way under strain, much as a fluid would do, then the imposition of any weight upon one extremity of a given section. . . would necessarily produce a change in the position of the pivotal point. Now in the ice accumulation of the glacial period we have just such a change of weight as would be likely to bring about considerable effects of this kind. A great mass of water is taken from the sea and heaped to the depth of a mile or more upon the land. A mile in depth of ice weighs about as much as half a mile of ordinary rock, so that by covering the continent of North America with a deposit of this kind we more than double its altitude above the sea. Now if the weight of the mass uplifted be an element in determining the height to which the continents are raised, then we must allow this ice mass a decided influence in depressing the continental areas.[62]

The confidence Shaler felt in his newfound theory may be gauged from his reiteration of it at both the December 1874 and January 1875 meetings of the Boston Society of Natural History. And he made good his initial formulation by integrating his findings with the parallel principle of eustacy. "In endeavouring to account for the changes of level of the continents," he began his December discourse, "it is necessary to consider not only the changes of the land, but those of the sea as well."[63] But whereas Adhémar accounted for sea level change by the gravitational attraction an ice cap alternately occupying the poles would exert on the sea, Shaler felt that the earth's thermal contraction, the changing height of the sea floor due to deposition, and the transfer of sea water to the land in the form of ice sheets were far more likely causes.

Despite the weight of evidence that Shaler had adduced and the long-lasting significance of these contributions, his theory of glacial isostasy, departing as it did from geological convention, was not immediately welcomed. James Dana's letter to Asa Gray in August 1878, for example, effectively illustrates the reserve with which Shaler's work was received. (Brackets enclose marginal notes added as postscripts to Dana's letter.)

Shaler I do not understand. From his Memoir in Mem. Boston N. H. S. printed in 1874, you would think him heels over head in the Glacial theory; for he makes the polar ice cap thick enough and extended enough *to cause a sinking of the earth's crust* and a submergence in part of the land all the way from the Pole to New Haven. The glacier must have been long growing to attain such thickness and dimensions; so that the era of the growing ice was long, before any submergence began. (The era of submergence I call the Champlain period.) I never was so much as a Glacierist as Shaler's theory makes of him. In another article in the Proceedings, presented at the meeting in January 1875 he endorses again this theory with regard to accumulating ice over the northern regions causing a subsidence; but there he puts his ice in before the Miocene period & makes it cause a subsidence of the Alaskan region such as would let in the warm Japan current of the Pacific. & Thus he accounts for the warm miocene climate, just as is done in my Geology (pp.

754–756). [To produce the cold of the Glacial climate, the geology supposes the Behrings (*sic*) Strait to be closed as now, and also the Gulf Stream to be excluded from the Arctic by elevation of the sea bottom between Europe & Greenland.] I know of no facts to warrant the idea of ice-accumulation before the miocene, and I am sure he has none; but he shows himself to be a strong Glacierist stronger than is reasonable.

You will ask whether he gives credit to the Geology with regard to that idea as to miocene climate—brought out by him 6 months after the book was published. He does in a partial way. He evidently did not derive his theory from the Geology; for he says this, in the remark that while his paper was going thro' the press he found in my Geology a reference to Mr J. H. Bradley's view on the subject advocating such a theory. My geology does not give a mere *reference* to Mr Bradley's views, [Mr Bradley aided me in proofreading. . .] but presents the theory in the chapter on changes in climate "as the most probable both for the warmth of the Miocene & the cold of the Glacial era," and shows its efficiency by reference to Croll; calculations with regard to the heat carried N by the Gulf Stream, just as Shaler does.

Shaler advocates his theory that an ice cap may cause submergence also in an art. read before the B. N. H. S. in Dec. 1874; (Proc. p. 288) and here the subsidence spoken of is not a general continental submergence, such as icebergs would require (2000 feet & upwards for New England) but only the submergence of a border region, & such a one as I and others admit to have been submerged, seashells proving it.

Are you not wrong in saying that Shaler supposes the Glacial era one of submergence & Icebergs, a la Dawson?[64]

Whatever Dana's misgivings, the substance of Shaler's statements proved to be well founded and do provide one of the first systematic treatments of the theory of glacial isostasy. His criteria for delimiting former sea levels—stratified sands, silts and clays, and marine abrasion surfaces—are those still employed for mapping raised marine deposits.[65]

I do not deny that earlier inklings of the isostatic principle are entirely lacking. W. B. Wright, for example, could observe that a "theory of Isostasy was first put forward by Jamieson in the year 1865 to account for the presence of elevated marine sediments of late-glacial age round about the Scottish centre of glaciation."[66] Indeed it is true that Jamieson already by that date felt that "the enormous weight of ice thrown upon the land may have had something to do with this depression. . . and then the melting of the ice would account for the rising of the land, which seems to have followed upon the decrease of the glaciers."[67] But for all that, it was not until 1882 that he presented his full-fledged version, thus leaving Shaler's 1874 treatment as one of the earliest coherent statements of the theory.

Aside from questions of scholarly priority, Shaler's presentation certainly displayed its share of individual subtlety. Consider, for example, the point described as "ingenious" by Baron Gerard de Geer, the distinguished Swedish geologist.[68] In accounting for the processes of isostatic readjustment, Shaler had speculated on the position of the pivotal point between elevation and depression. Take the situation where the continent was rising while the ocean bed was sinking. Suppose, he pondered, the fulcrum lay somewhere along the inner stretch of the shoreline; then the coastal fringe would participate in the depression of the seafloor. Evidence from this source about general continental movements would thus be profoundly misleading, for the interior landmass would in fact be rising, not falling. De Geer was greatly impressed with this theoretical possibility and suspected that it might apply to the oscillation of the Scandinavian lands.

For some years after 1875, the administration of the Kentucky Geological Survey deflected Shaler from glacial studies, and he did not return to the topic until the 1881 publication, with W. M. Davis, of *Illustrations of the Earth's Surface: Glaciers.* The text was written by Shaler, and the diagrams, bibliography, and plates were selected and annotated by Davis, the junior author. The book's attractive production and comprehensive treatment ensured a good reception. Although designed for student use, not surprisingly it was, as one reviewer put it, "far more than a mere text-book. It abounds with original suggestions resulting from a wide survey of facts."[69] More leisurely in pace, the book form perfectly suited Shaler's current needs. It allowed him space to recapitulate his findings to date, to provide a genuinely synthetic review of glacial geomorphology as a subfield of geology, and to draw on fieldwork experience in the Swiss Alps. Thus he could supplement an exhaustive inventory of European, Asian, American, and Antipodean glaciers with an overview of the contributions of Agassiz, Lyell, Croll, Reclus, Hooker, Haast, Berghaus, King, Hayden, and Whitney.

At the same time, Shaler now pushed on into new territory. First, he charted his own course through a mushrooming literature on the origin of glacial periods. Evidence for the variability of past climates was copious, but explanations were more common than convincing. Thus Shaler found Lyell's suggestion of the geographical redistribution of all land in the polar regions no more plausible than Croll's speculations about the influence of ocean currents, while the idea of a progressively cooling globe was inconsistent with the episodic nature of glacial periods. For Shaler, determining the *nature* of past climatic conditions was the first step toward constructing any viable theory. Turning to his dossier of excavation findings at Big Bone Lick, he noted that the record of interglacial animal life, which presupposed an abundant vegetation, attested to only a marginally lower temperature, while evidence from Salt Lake revealed that rainfall had been substantially heavier during the last glacial period.[70] These findings seemed to confirm Croll's portrait of former glacial environments, and Shaler was therefore impressed with his theory of periodic modifications in the earth's elliptical orbit,[71] although he conceded that Henri Le Coq's idea of changes in solar heat might also play a part. Undoubtedly, as Shaler recognized, a synthetic solution was called for.

Second, Shaler now focused more sharply on the problem of glacial motion. He had already advanced his own theory of pressure melting in 1875, which, drawing out the implications of James Thomson's research on water refrigeration,[72] proposed that the sheer pressure of the glacier would melt a portion of the ice base. Any upward movement of melt water through ice crevices would alleviate pressure and therefore be refrozen. So this theory provided both an explanation for the movement of the ice sheet "in the direction of least resistance" and an account of "the water-worn look which is so prominent a feature in the drift pebbles of the greater part of North America."[73] Shaler was even more confident about it now in 1881 because of the continued absence of any compelling alternative. The dilatation theories of de Charpentier and Agassiz, for example, suffered the drawback that the expansion of freezing water in ice fissures would not be extensive enough to cause glacial motion; Forbes's viscosity theory, he suggested, was rather inconsistent with the rigidity of ice and the lack of acceleration on steeper slopes, although he did confess that it was the most "captivating" hypothesis; Tyndall's theory, based on Faraday's research, of a succession of fractures and regelations he deemed true but inadequate; and Croll's theory of diathermancy, which posited the melting of each particle in succession, faced the objection that thin ice sheets should, but in fact did not, move faster than thick ones.[74] Overall, Shaler's criticisms were fundamentally sound, and most of his objections were sustained by Charlesworth as recently as 1957, who also confirmed the accuracy of Shaler's own pressure melting theory.[75]

Finally, Shaler did not miss the opportunity this lavish production afforded him to peruse the implications of the glacial theory for evolutionary history. The ice age, for example, had compelled a southward migration of both plant and animal species, and in true Lamarckian spirit, the new environment induced immediate modifications. The "mere act of migration" was thus itself an evolutionary mechanism, and "the enforced migrations of glacial times" in particular had combined to "preserve the beneficent changes and extinguish the others." There were, besides, other evolutionary by-products. Ice ages intensified the struggle for existence by forcing populations into confined territories. To this cause alone, the richness of European flora and fauna would well be traced. So Shaler pressed home the evolutionary implications of episodic glaciation in no uncertain terms:

Considered as an action often repeated in our geological history, glaciation cannot but strike us as a most effective helper in the great struggle for a higher and more perfect life. Through the extensive wanderings to which it compels created things, it constantly proves their fitness to endure the stress of life. Through its action the weak species are the sooner brought into the scales of the stern justice that finds all weakness fit to be punished by death. When we remember that probably very many such periods have done their work we must grant that the effects of glaciation on life have been more extensive than its physical action upon the surface of the earth.[76]

Still, if the ice age's evolutionary effects were extensive, they were not universal. The human species had remained immune to its selective power. Shaler was certain not only that human history predated the close of the last glacial epoch but also, accepting the evidence of the disputed Calaveras skull,[77] that preglacial humans were neither intellectually nor culturally inferior to their postglacial counterparts. Indeed this conclusion explained the ice age's evolutionary impotence, because, with fully developed intellectual skills, the human race had already marched beyond the borderlands of natural selection.

This project now completed, Shaler devoted his remaining glacial labors to elucidating the drift deposits with which New England was so liberally endowed. Medial and lateral moraines had already come under his scrutiny in the 1881 volume, and over the next few years he sought explanations for drumlins, kame and terrace deposits, frontal or "shoved" moraines, and boulder trains, in a series of reports for the geological survey on Martha's Vineyard (1881), Nantucket (1889), Mount Desert Island (1889), Cape Ann (1889), Cape Cod (1898), Richmond Basin (1899), and Narragansett Basin (1899).

Shaler's definitive account of drumlins appeared in 1889 and was the culmination of some twenty years' work on the subject. Indeed, when W. Upham affirmed in 1896 that "we owe the earliest observations and descriptions of drumlins to Kinahan and Close, in Ireland, and to Shaler in Massachusetts," he was referring to a paper Shaler had composed in 1870 on what he called the "parallel ridges of glacial drift."[78] At this early stage, Shaler had suggested that these ridges were the remnants of once-continuous drift sheets now dissected by tidal currents or drainage streams—a two-stage model of drumlin formation involving postglacial fluvial or marine sculpturing. Since these processes were invoked as agents of the reexcavation of former erosion channels, his account seemed plausible both because the drift deposits were of similar height[79] and because he long maintained that glacial erosion would not obliterate, but only modify, the preglacial topography.[80] By the late 1880s, however, Shaler had supplemented this account by introducing another chapter into the story—further glacial action. This carving agent he saw as decisive, for, as he put it in 1881, drumlins were "formed during a period of glacial retreat and during a subsequent extension of the ice have been eroded by the readvancing glacier in the manner in which any other rock is eroded by glacial action."[81] As he summed up their three-part history the following year:

During the first stage of the Glacial Period the ice sheet had a very much greater extension than during the second portion of the Glacial Period. When at the close of the first stage the ice retreated, it left upon the surface irregular but in places extremely thick deposits of till. During the interval between the first and second stages of the period the southern part of New England remained for a long time free from ice. During this time the till deposit left upon the surface was much eroded, perhaps by the action of the sea, in part doubtless by river action. When the second advance of the ice came to repossess the surface, it wore away a large part of the till formed during the first period, leaving the remains of it carved in the characteristic form of our drumlins.[82]

Shaler's erosion theory of drumlin formation, or, to use Tarr's designation, "the destructional theory," was subsequently adopted by Barton in his 1892 and 1894 drumlin studies.[83] H. C. Lewis and R. S. Tarr also defended it, as did G. F. Wright in his earlier days. These geologists opposed the constructional or accretion theory, upheld by men like Agassiz, Dana, Davis, Salisbury, and Russell, according to which drumlins were "accreted on gentle slopes by successive additions from englacial or subglacial debris under a thin or weak ice-border."[84] Indeed drumlins continue to remain an intriguing and enigmatic feature of many glaciated landscapes, and it is still felt that both theories have some validity.[85]

During the 1880s, Shaler also sought explanations for the glacial deposits known as "kames" and "eskers." He himself did not distinguish, terminologically, between these fluvioglacial ice contact features, but no significance can be attached to this because the terms "esker," "ose," and "kame" have not enjoyed systematic usage. Shaler acknowledged that kame formations had a more complex structure than any other glacial deposit and that their absence from many of the world's glaciated regions increased the difficulty of formulating an adequate account. The fact that the deposits were stratified in nature suggested to him that kames had been formed under water "at the points where sub-glacial rivers discharged their streams into the sea or into the temporary lakes that abounded over the land surface during the glacial period."[86] And what of those even more delicate deposits that he christened "Indian ridges" or "continuous kames"—features that would now be termed eskers and that run at right angles to the aligned kame deposits parallel to the ice front? These deposits, Shaler pointed out, were to be found where a proglacial lake was formed during the last stage of ice retreat. The waste material brought into this lake by subglacial streams remained in the valley floor as the ice receded, leaving behind a ridgelike morainic deposit. Where these forms were deposited, moreover, glacial retreat must have been regular; otherwise they would be fretted and disjointed. On both counts, then, Shaler had advanced broadly compelling explanations, although his resort to marine action to account for the accompanying kettle formation was discarded (along with capricious eddies in extraglacial lakes, rotating icebergs, and the subsidence of underlying strata) in favor of the theory that they arise from the irregular settling of fluvioglacial deposits over or about melting masses of dead or stagnant ice at glacier dissolution.[87]

The Iron Hill boulder train in Rhode Island, so frequently associated with Shaler's name, was the object of his investigation in 1893. Four years earlier, for a rather more popular audience, he had claimed that boulders from Iron Hill had been transported by ice some sixty miles south to the western end of Martha's Vineyard.[88] The pattern of these erratics, it was plain, could be used to retrace the direction of ice movement and, at the same time, to throw light on the conditions of erosion beneath deep glaciers. The highly distinctive mineralogical composition of the Iron Hill formation was a great boon because it enabled the origin of the erratics to be located with absolute precision. As to the boulder train itself, its most striking feature was the way it fanned out from less than ninety feet at Iron Hill to over twenty thousand feet near Providence, and twice that at the sea front. Toward the south, the greater distance between the pebbles was matched by a proportional diminution in fragment size due, Shaler believed, to the processes of attrition. To account for this fan-shaped pattern, he rejected the idea of successive forward movements of the glacial sheet in slightly different directions and proposed instead that subglacial streams—formed by pressure melting, the effects of rock crushing, and heat loss from the earth's interior—transported quantities of debris toward the margin of the glacier. Besides, the presence of such subglacial waters would account for the relatively small amount of ice erosion in regions near to the center of glaciated districts, for Shaler had come to believe that the erosive power of glaciers was less than previously thought: "The more accurate our knowledge as to the genesis of the topography within the ice-worn region becomes, the more clearly is it proved that the essential features of the surface are not due, as was formerly supposed, to the erosion effected during the Glacial Period, but are to be ascribed to the ordinary agents of erosion which operated on this district during the pre-glacial ages."[89]

By 1899, however, Shaler had revised this judgment, modifying his earlier estimates of the amount of material eroded at Iron Hill and urging that preglacial weathering had greatly facilitated subsequent glacial erosion.[90] At various times and in different contexts, it is clear, Shaler attributed relatively more or less erosive potential to glacial, fluvial, and marine action. Still, whatever his sense of theoretical irresolution, Shaler's Iron Hill was a particularly fine case of the boulder train formation, and revised versions of his accompanying map were used by Hobbs in 1926, Flint in 1947, and Charlesworth in 1957.[91]

Shaler's commentary on New England's glacial geomorphology was plainly painstaking, not to say prolix. Discursive though his inventory was, however, he did not lose sight of the broader historical and theoretical implications. Ever since his two-stage glacial interpretation of drumlin formation during the 1880s, he had doubted the long-held assumption that the North American drifts were the product of a single ice advance and retreat. And yet, while he had also discussed the complexities of the glacial succession in his survey of the geology of Nantucket in 1889,[92] it was not until his 1896 monograph—published under the collected names of Shaler, Woodworth, and Marbut—that the first claim was made for a plurality of glacial periods along the Atlantic coast plain, based on a discrimination of the stratified marine clays of eastern Massachusetts. But now, with the added reinforcement of Woodworth's research, the implications for Pleistocene geology could no longer be ignored. "From a purely scientific point of view," Shaler concluded, "the most important point which it has been the intention to develop in this paper concerns the division of the so-called Glacial period into three great epochs of ice action separated from one another by very long intervals."[93]

The recognition of glacial periodicity, of course, had other consequences too. Systemic uniformitarianism was ruled out. Research on the value of saliferous deposits as evidence of former climatic conditions, for example, led Shaler to conclude that the "facts clearly indicate sudden climatal revolutions at various stages in the past."[94] The same was true of fluctuating marine levels; indeed the most recent isostatic readjustment, he believed, "was not accomplished continuously but by successive rapid movements with long continued intervening periods of repose."[95] So with a lifetime's exposure to the geomorphology of the Atlantic coastlands, Shaler would have found much in their glacial topography to justify his efforts to transcend the conventional limits of uniformitarianism and catastrophism alike.

Sea and Land

If Shaler found the idea of glacial isostasy captivating, he was not blind to the corresponding principle of eustacy. Since the late 1860s, he had been convinced of pronounced, periodic changes in sea level and had affirmed its importance for reconstructing the geological history of Cape Hatteras and the phosphate beds of South Carolina.[96] But not until 1874 did he issue his first full statement of the principle, explicitly linking it with the processes of erosion and deposition and thus anticipating Suess's later mechanism that explained eustacy by the slow sedimentary infilling of the ocean basins, interspersed with rapid subsidences of the seafloor leading to much greater ocean capacity.[97] Not surprisingly, Suess himself referred to this early communication by Shaler and, in his magisterial three-volume *Das Antlitz der Erde* (1883–1908), noted that in "1875 [*sic*] N. S. Shaler . . . expressed his conviction that it was the sea and not the land which is subject to movement."[98] As we now know, of course, this was not quite the whole story. Still Shaler had announced in this 1874 paper:

Sudden and great changes in the contour of any continent may occur without necessarily affecting the sea level, but all changes of the sea floor, be they elevation or depression, unless the alterations of each movement compensate each other, must be attended by modifications of the average shore level upon the whole earth. One of these changes is in a determined direction. The average erosion of the land at present is probably not far from one foot in five thousand years. With the present areas of land and sea this would raise the sea level one foot in about fifteen thousand years. . . . It is to be noticed that this change is constant and in one direction, so that the other vacillating changes will not overcome it. It will cause a steady accumulation of height through all accidents of elevation and subsidence.[99]

Marine accumulation of sediments, however, was not the sole cause of positive eustatic changes, and Shaler went on to link fluctuating sea levels with the glacial theory. The incorporation of seawater in great ice sheets, for example, would have a marked influence, and he estimated that, if a quarter of the earth's surface was ice covered to a depth of one mile, there would be a corresponding fall in sea level of some twelve hundred feet. Because the sheer weight of ice sheets would offset, by isostatic adjustment, some of these changes, clearly glacio-isostasy and glacio-eustacy were mutually dependent mechanisms.

These clear, if embryonic, statements might well have heralded a new diastrophic model in geomorphology. But over the next years Shaler tended to focus on the glacial side of the equation, explaining marine erosion features by continental elevation and subsidence. This position is clear, for example, in the 1881 volume *Glaciers*, where only passing reference was made to sea level variation. Again in his 1890 study of the freshwater morasses of the United States, he was compelled to confess that he had "for convenience spoken of all these movements as if they were due to the elevation and depression of the continent, [although] it is by no means impossible that some of the oscillations were due to actual changes in the level of the sea and not to movements of the land."[100]

Even with these intermittent observations, some twenty years were to elapse before Shaler returned in any serious way to the eustatic theme. Now in 1894 he urged that such factors as submarine folding and the gravitational attraction of continental landmasses had also to be taken into account. But the main thrust of the paper was to examine the evidence for changing sea levels because with the exception of Gilbert's *Lake Bonneville* study, little serious work had been done on the discrimination of the various criteria. Chief among the problems of delimiting the position of higher shorelines, Shaler reported, was the fact that onetime marine features were subsequently subject to the ordinary effects of denudation and could therefore be all too easily confused with landforms sculptured solely by subaerial erosion. Former marine cliffs, the submarine shelf, and barrier beaches were, for this reason, rather untrustworthy; the characteristic "sub-ovate" form of beach pebbles, however, convinced Shaler that here was reliable evidence. For lower sea levels there were even fewer trustworthy records. Indeed he was forced to conclude that flooded valleys were "the only proof of value."[101] And yet, whatever the complexities in identifying marine relics, Shaler had not the slightest doubt that changes in shoreline position were "due either to the rise or fall of the sea or to a positive movement of elevation or subsidence of the land" and that its precise location at any time was "likely to be determined by an equation between these two independent swayings."[102]

Shaler's diligent efforts to keep the principles of isostasy and eustacy in balance might well be seen as a symbolic attempt to straddle the European and American traditions of geology. For with general American opposition to the eustatic theory emanating from practitioners like Leverett, Alden, Johnson, and Davis, particularly on the question of peneplanation, Shaler made it clear that the baseleveling formula could not be contemplated in isolation from eustatic events. Thus, almost a decade before "the coming of age of the eustatic theory," marked according to Chorley by Chamberlin's 1908 paper on diastrophism,[103] Shaler had urged that one of the reasons "for the disappearance of the topographical relief of the East Appalachians can be found in the marine erosion to which they have been subjected."[104]

Besides these effects, changes in the relative heights of sea and land had other implications too, particularly for deciphering the history of shoreline evolution. Indeed the quality of Shaler's work on this theme is nowhere so glowingly celebrated than in Douglas Johnson's introduction to *The New England-Acadian Shoreline* published in 1925, the research for which had been assisted by a two-thousand-dollar grant from the Shaler Memorial Fund:

The reader will understand from what has just been written how eminently appropriate it is that the present volume should be included in the Shaler Memorial Series of Harvard University. But there exists an even stronger reason for connecting the name of Nathaniel Southgate Shaler with any treatise which aspires to discuss the origin and evolution of the Atlantic Shoreline. Every student of this coast will quickly discover in his investigations no name so often confronts him as does that of the distinguished geologist of Cambridge, who for many years devoted no small share of his energies to the elucidation of our coastal phenomena. His contributions to different aspects of this subject provide a wealth of valuable information; and if the student finds that many theories entertained in earlier days of our science must be revised in the light of fuller knowledge now at our command, he will at the same time be constrained to acknowledge the heavy debt we owe to the fertile and suggestive imagination of one whose name is forever associated with man's attempt to decipher the physical history of the Atlantic seaboard.[105]

From as early as 1871, Shaler had been seeking to identify shorelines of subsidence and elevation and indeed had the temerity even then to challenge the eminent Charles Lyell's interpretation of the recent geological history of the South Carolina coast.[106] But in his later work, particularly his 1895 monograph on the beaches and tidal marshes of the Atlantic coast, later students like Johnson found the building blocks from which a commanding theoretical edifice could be constructed. Here, for instance, he depicted offshore bars as features of emergent shorelines by assuming an uplift of the continental shelf as the initial step in bar formation. After the shallowing of the offshore zone by the processes of deposition, the breaking of storm waves at some distance from the shore would result in the accumulation of a ridge parallel to the coast.[107] Johnson was later to adopt this chronicle of events when identifying the hallmarks of young shorelines of emergence, as he likewise adopted Shaler's reconstruction of the normal history of lagoon plant colonization after the creation of the offshore bar.[108] As to the telltale signs of submergent coasts, Johnson shared Shaler's misgivings about the significance of fjords. For Shaler had contested the view of Dana, Upham, and Fairchild, who insisted that land subsidence was the essential cause of fjord topography. Having long regarded the Maine coast as "the product of the peculiar form of erosion brought about by one or more glacial periods," he affirmed— as did Gilbert in 1904—that fjords were formed by glacial channel overdeepening without any reference to continental sinking.[109] Certainly his designation of the Maine embayments as "fjords" rather than "rias" was subsequently rejected, although Johnson did note that the Mount Desert region, where Shaler had done substantial fieldwork, was a possible exception. Still, Shaler did believe that where they existed, flooded valleys—rias—were the unmistakable marks of the submergent coast.

With such theoretical and empirical riches to be gleaned from the study of marine action, it was only to be expected that Shaler would find other forms of hydrological activity attracting his curiosity. And so a brief survey of his thinking on fluvial processes will serve to complete the contours of his geological contribution.

Fluvial Action

Compared with the hefty file of manuscripts on other geological matters, Shaler's folio of published work on fluvial action might seem rather thin. Few in number though they were, these papers are important both because they demonstrate the vitality of non-Davisian geomorphology and because some of his proposals have stood up to the scrutiny of the latest earth science technology.[110]

As initially expressed, Shaler's interest in fluvial processes was the outgrowth of his survey of the Carboniferous limestone caverns of Kentucky for the state survey, for here the results of subterranean stream erosion were dramatically exposed. The distinctive karst topography of the region, with its typical pattern of surface drainage, "sink hole" features, and natural bridges, he found captivating and periodically returned to the subject in popular writings for over twenty years.[111] The processes of chemical and mechanical erosion were particularly fascinating, and Shaler estimated that as much as 100,000 miles of cavern tunnels in the state had been excavated by these processes, a figure later repeated by both Dana and Tarr.[112] As to the mechanisms, Shaler observed that "excavation is altogether done by the dissolving action of the water" along the limestone's joint-planes until a channel large enough to accommodate a stream had been excavated.[113] From that point, the processes of mechanical erosion would move into action, and both erosion methods would continue in operation until a more resisting bed retained the stream in one level for long enough for a whole tier of caverns to be sculptured.[114] Such a process could recur many times at different levels until an entire suite of caverns had been manufactured.

This explanation of the genesis of cavern gallery systems later found its way into I. C. Russell's effort to correlate their creation with the local baselevel provided by the surface river into which subterranean drainage discharged. Thus he proposed that the highest of the Mammoth Cave galleries had been "formed at a time when Green River flowed at the level of their place of discharge, and each lower series dissolved out during subsequent stages in the deepening of master rivers."[115] In this case, Shaler's idea of less permeable layers was relegated to secondary status, but he did concede, as Tarr later put it, that the level of underground drainage was "determined either by the presence of an impervious layer, or by the influence of the surface rivers or other water to which the percolating water is tributary."[116]

Shaler's first substantive discussion of surface drainage appeared in the August 1888 issue of *Scribner's Magazine,* and while it was addressed to a popular audience, it reveals his approval of the newest geomorphological concepts of baselevel and antecedent drainage as developed in the writings of Powell and Gilbert. A generously illustrated portrait of the course of a normal river valley with its waterfalls and rapids, meanders and oxbows, paved the way for a few concluding observations of a more theoretical nature. Since continental areas were subject to periodic oscillations, Shaler pointed out that the energy of rivers was determined by the relative position of sea level, the "base level of erosion." In the unlikely case of a peculiarly lengthy period of stable baselevel, Shaler speculated on the formation of what Davis was to term a "peneplain": "If the continent should continue for some geological periods without any change in the level of the sea, the mountain brooks would gradually carve down the hills in which they lie, the table-lands would slowly disappear, and the surface would return to its primitive state of a great swamp. The rocks beneath this swamp would be subjected only to interstitial or corrosive decay, for the reason that the streams would not have fall enough to work upon their beds by mechanical erosion."[117] If Shaler's concept of baselevel was, like Powell's, a theoretical one, he clearly had gone beyond that to a recognition of the nearly featureless plain as the speculative last stage of subaerial erosion. Not that Shaler believed he had observed the latter in reality, although that same year, 1888, W J McGee claimed to have identified such denudation surfaces along the Atlantic slope, and Davis the following year made his celebrated claim for the existence of an uplifted peneplain of Cretaceous origin in the uplands of southern New England.[118]

Indeed in his paper "The Rivers and Valleys of Pennsylvania," Davis first put forward his systematic exposition of the cycle of erosion. While he did acknowledge something of an intellectual debt to Powell,

Jukes, Dutton, and Gilbert, he later maintained that the idea struck him "while working on the Northern Pacific Railroad Survey in Montana in 1883, as rather like the blinding flash of understanding experienced by a prophet in the wilderness."[119] Davis's ideal cycle involved several quite distinct phases: initial tectonic uplift so rapid that denudation processes had a minimal effect on the landscape; destructive agencies carving a whole series of sequential forms; and finally the reduction of the land surface to its ultimate form, a low plain of imperceptible relief—the peneplain. In the case of the Appalachian drainage, with its complex geological history, Davis argued that it had passed through a series of complete erosion cycles and concluded that present-day ridges were merely the remnants of former baselevels.

So simple and elegant was the new Davisian model that it was to mesmerize generations of geomorphologists. Indeed, compelling alternatives then advanced were long ignored and have only recently been resurrected. Among them was Shaler's 1898 presentation to the Geological Society of America entitled "Spacing of Rivers with Reference to Hypothesis of Baseleveling." It constitutes Shaler's reaction to his pupil's proposal and its elaboration in the writings of Keith, Hayes, and Campbell. Shaler began by reporting observations, gathered some twenty years earlier, that the rivers of Kentucky displayed a high degree of regularity in spacing. This pattern was not unusual, moreover, for Shaler recalled that, according to Albrecht Penck, whom he had heard lecture during a visit to North America in 1897 as a member of the British Association excursion, it was replicated both in the Alps and in the Cordilleras. Processes of stream abstraction and river capture were portrayed by Shaler as preliminaries to the emergence of an integrated stream network pattern, whose regularly spaced components would show a tendency to be separated by divides with accordant summit levels. As he put it, regularly spaced streams could well explain "the origin of the coincidences in mountain crests, which is so generally held to indicate the existence of ancient baselevels of erosion which have been lifted to a hight above the level of the sea and then dissected by rivers."[120] The broadly accordant summit levels of the Appalachians and indeed of the Alps might therefore have been formed without reference to Davis's uplifted peneplain but simply by the normal erosion processes of a mature river system.

Shaler's resistance to Davis's account seemed all the more plausible for a number of independent reasons. First, it could more easily account for the recorded observation that "the measure of accord in [Appalachian] summit levels is not what it seems at first to be . . . [for] at a distance of, say, 40 miles, differences of hight of 500 or even of 1,000 feet are not conspicuous." Second, given the rates of erosion since the Mesozoic, when the uplift of the peneplain supposedly took place, the original planation surface must have been some 5,000 feet above the existing summits; even if this estimate was halved, it seemed unreasonable "to assume its sometime existence to account for the slight measure of uniformity which exists in the hights of the existing crests." And third, the peneplain theory did not square with the periodic and extensive changes in land and sea level, whether by isostatic or eustatic processes.[121]

Responding to this and other criticisms of his theory in 1901, Davis acknowledged Shaler's as an "interesting" contribution to the subject of land sculpture but urged that the forms produced by stream spacing could be "initiated by dissection after peneplanation."[122] Some method of discriminating between the different origins of accordant summits was now necessary, he said, rather than an argument for excluding one at the expense of another. More significant still, in 1919, some twenty years after the publication of Shaler's paper, Albrecht Penck, apparently independently, advanced a similar theory to Shaler's to account for the Alpine Gipfelflur. Again in opposition to Davis, whose theory he had abandoned the previous year, he suggested that "the development of equally spaced valleys would lead to an accordance of summit heights between them."[123] Subsequent research too has tended to confirm this interpretation of the Gipfelflur.[124]

The roots of Shaler's discomfort with the erosion cycle sprang from many different sources. Appreciative of catastrophism though he was, Shaler felt decidedly unhappy with the exaggerated rapidity of tectonic uplift that Davis's theory presupposed. Then, long-held suspicions that planation surfaces were often the result of marine action (another indication, incidentally, of European influence) also fitted in with his belief in periodic oscillations of sea and land, which, in turn, implied a fluctuating baselevel. The different climatic conditions of the Pleistocene introduced a further complication into the idealized geographical cycle, besides which his own alternative noncyclical explanation of those accordant summits so beloved by Davisian disciples had its own attractions. Nor is it fanciful, I believe, to suggest that cyclical conceptions of earth history fitted much less comfortably than linear versions into a progressivist vision of evolutionary biology. But most of all, Shaler did not miss the significance of the fact that "the advocates of the hypothesis in question have not yet shown us a region which has been and remains effectively baseleveled."[125]

Throughout his career as an academic and practicing geologist, Shaler clearly made substantial contributions not only to elucidating the geological history of North America's Atlantic seaboard but also to more general questions of geological theory. And yet his reports on any of these topics were rarely confined to matters of pure geology. No less sensitive to the evolutionary implications of his glacial inter-pretations than to the economic value of a ria coast naturally suited to harbors, he was ever concerned to integrate the physical and organic history of the earth with the social development of its people. And this penchant, as much as anything else, confirms that, while Shaler was a geologist by occupation, he was a geographer by inclination.

Culture and the Curriculum   **9**

Nathaniel Southgate Shaler was primarily a teacher and it was his happy boast that for nearly forty years few Harvard men took their degrees without coming under his instruction.

—Bacon, "Shaler as a Teacher"

Whatever contemporaries felt about other aspects of Shaler's work, critics and disciples alike were agreed about one thing—his outstanding gifts as a teacher. Almost every death notice, memorial, and biographical entry paused to stress his popularity as a university professor, to recall the magnetic personality that attracted students from every discipline into his classes, and to wonder at his ability to infect hearers with the contagion of his own enthusiasm.[1] As one writer observed, everyone "entering Harvard for the first time would be urged by his colleagues to take at least one course with Shaler."[2] The amount of $30,500 contributed by some seven hundred people to the "Shaler Memorial Fund" amply testifies to his standing as a teacher who was "stimulating, inventive, and adventurous," "sympathetic, vehement, generous, and just."[3] In 1925, nearly twenty years after his death, a circular from W. M. Davis "To Harvard Men who remember Professor Shaler" called on his former students to contribute to "a Shaler Fund for the establishment of four Scientific Professorships at Berea [College as] a most admirable memorial to the man whom thousands of Harvard students admired and loved."[4] Even those dis-

249

approving of what they saw as Davis's mercenary motives in seeking these and other research funds to commemorate his old teacher still wanted some appropriate memorial. R. A. F. Penrose, Jr., for instance, suspicious that Davis only wanted "to commercialize his memory instead of making it an inspiration to others," offered to, and later did, commission Robert Aiken, a well-known New York sculptor, to produce a life-size bronze bust of Shaler. To Penrose, Shaler was both "a philosopher and a far-sighted prophet," and he .herefore regarded the bust as "an inspiration not only to geologists who may look at it, but to those who realize the broad humanitarian instincts for which Shaler was given the degree of LL.D. at Harvard."[5] In a similar vein, Raphael Pumpelly, recounting his experience in Germany during the mid-1850s, recalled that, except for "Professor Shaler of Harvard," American teachers lacked that warm, personal contact with the student body so typical of Cotta and his Freiberg colleagues.[6]

Shaler, of course, had served his educational apprenticeship under Louis Agassiz and from him had caught the vision of teaching by discovery. The memory of that early fish episode, when he was abandoned with it for days on end, never left him, and his own teaching practices therefore betray the infiltration of Agassiz's romantic methods. Agassiz's mode of instruction, in fact, displayed all the trademarks of what was later to be called "the heuristic principle." H. E. Armstrong's celebrated paper "The Heuristic Method of Teaching; or, The Art of Making Children Discover Things for Themselves" did not appear until 1898, but both the term and the technique were commonplace by the mid-nineteenth century. William Ross, for example, had used the term in his 1858 volume *The Teacher's Manual of Method,* and the technique had long attracted the support of men like Bacon, Rousseau, Pestalozzi, and, later, Herbert Spencer, who insisted that children "should be *told* as little as possible, and induced to *discover* as much as possible."[7]

This was the intellectual tradition to which Agassiz belonged. Early fascinated by the study of ancient and modern geography, he was always drawn to "the study of the thing-in-itself, the raw material of nature."[8] Geography as a subject, it may be noted, occupied a crucial strategic niche in this whole educational enterprise. Rousseau felt it was central to Emile's education;[9] Pestalozzi and Saltzmann used it as the focal point of early teaching in science and technology; and Carl Ritter, urged by Pestalozzi and Saltzmann alike to succeed them at their respective experimental schools, produced his *General Geography* as the fulfillment of a promise to Pestalozzi that he would write a work embodying his educational philosophy.[10] To Agassiz, therefore, books were certainly important, but "he thought it more educational to study living things in their natural habitat"—a conviction that found ample expression in his teaching of natural science at the Lawrence Scientific School.[11] Springing from contemporary European thought, this love of nature and this hatred of "verbolatry" drew further support from the midcentury American transcendentalists, who, enamored of unfettered Nature, railed at the spiritual shallowness of the new commercial order.

Given the romantic tastes of New England's cultural life during the middle decades of the nineteenth century and given Agassiz's personal influence, Shaler could scarcely have remained unaffected. Indeed both in philosophy and in practice, these sentiments were the driving force behind Shaler's various educational initiatives, chief among which was the American summer school—an institutional achievement that ultimately stands as a finer monument to his educational statesmanship than memorial funds, bronze busts, or scientific professorships.

The Method: Naturalistic

According to his student and colleague J. B. Woodworth, "Professor Shaler was opposed to formality in natural science. The literary quality of his mind made technical terminology unattractive to him." His constant aim, Woodworth went on, was less to compel the student to follow "a carefully cut and dried system of didactic academic thought than to get the man to think and form an opinion of his own."[12] The naturalistic spirit of Agassiz that Woodworth saw incarnated in Shaler's classroom was not the figment of an enthusiast's imagination. For Shaler had insisted in *Outlines of the Earth's History* that "all study of Nature should begin not in laboratories, nor with the things which are remote from us, but in the field of Nature which is immediately about us." And, "From printed pages alone, however well they be written, he [the student] can never hope to catch the spirit that animates the real inquirer, the true lover of Nature."[13] He told the Geological Society of America as much in his 1895 presidential address. Textbook training in geology, he urged, did more harm than good, and the old didactic style, particularly in field teaching, did nothing to stimulate that "precious relic of the savage life. . . curiosity."[14]

For Shaler, then, fieldwork was always high on his educational agenda. It was Shaler indeed who first taught the pampered sons of some of Harvard's patrons how to get their boots dirty. The reason was simple: personal familiarity with nature was the truest "inspiration and test of all knowledge."[15] So while he would permit some reading, the few volumes he did suggest all bore the stamp of their authors' extensive fieldwork and powers of scrupulous observation. Lubbock, Darwin, and Gray were all advised, but especially the writings of the Scottish geologist-theologian Hugh Miller. Like these, Shaler was concerned to move from the known to the unknown. A "clue to a very large part of the earth's machinery," for example, could be gleaned by following the valley of an ordinary stream from its source to the sea; a sensitivity to the pulsations of the human heart or to the force of the wind could awaken in the student an awareness of the fundamental concept of energy.[16]

In purpose and procedure, Shaler's methods were in harmony with the tone of Huxley's *Physiography,* first published in 1877.[17] Attacking the classical curriculum of the English schools, Huxley called for the teaching of physical geography as the only real means of instructing every child "in those general views of the phenomena of Nature."[18] His proposals were revolutionary. Geography classes, where they existed, usually began with the announcement that the earth was an oblate spheroid revolving around the sun in an elliptical orbit. Nothing was less calculated to interest or instruct. The whole tenor of such traditional tuition was "in direct antagonism to the fundamental principles of scientific education."[19] How much better it was to *show* pupils the results of glacial or marine erosion, to allow them to observe directly the effects of rivers and rainfall. Such experiences were the best introductions to the basic organizing principles of causality and interconnectedness.

Shaler, we recall, had met Huxley during his 1872–73 travels in England, and he was doubtless well acquainted with his writings. But whether or not the influence was direct, Shaler's *First Book in Geology,* published in 1884 some seven years after the *Physiography,* displays a markedly similar outlook. Its purpose was to provide the beginner with "some general ideas concerning the action of those forces that have shaped the earth" and to instill the image of "the world as a great workshop." He was conscious, of course, that it was ironic for an avowed textbook hater like himself to produce a textbook. But he diligently disqualifed his own efforts by reminding his readers that, however "carefully a textbook may be prepared, and however well it may be used, it cannot itself alone give much insight into nature. This must come from the use of the student's eyes and mind." Still, if books must be written, they should be designed to help students follow their own instincts, and so, like Huxley, Shaler began with the most familiar of substances—water. An examination of river pebbles followed, for he insisted that an appreciation of how they were formed would constitute substantial knowledge of the history of the physical world. Succeeding chapters dealt in a similar way with sand, soil, rocks, coal, wind, valleys, lakes, and, ultimately, organic life itself.[20]

Meanwhile local study, or *Heimatskunde,* had become popular in Germany since the time of Pestalozzi. While it was particularly suited to the teaching of earth science, however, the physical geographies of neither Somerville (1848) nor Anstead (1867) registered the force of its revisionist currents. And even though Guyot's "geography of the school yard" became something of "a popular fad," only in the American summer school field excursions did the idea really come to full fruition.[21] Indeed it took the stimulus of the American and French summer experiments to force Patrick Geddes in Britain to call for the resurrection of "nature study" in the early 1900s.[22] So even the first formal attempts in Britain to establish the school field trip as a method of geography teaching and the formation of the Le Play Society postdate the American summer school by several decades;[23] Shaler, as we shall now see, was at once its architect and engineer.

Interlude: The American Summer School

Shaler's bibliophobia expressed in his advice to Geology 4 students that they must "read no geology lest they imbibe wrong notions"[24] unquestionably reflects Agassiz's hope that he would "live long enough to make textbooks useless and hateful."[25] The origin of the American summer school, by contrast, typically traced to Agassiz's short-lived Anderson School of Natural History on Penikese Island,[26] undoubtedly is open to dispute. Shaler himself made this clear when he put together his personal reminiscences in a "Notice concerning the Summer School." In 1868, some five years prior to the Penikese project, he himself had begun "Summer School instruction . . . with the unadvertised offering of excursions and lectures on Geology in the Connecticut Valley and the Berkshire Hills, ranging as far as the Hudson Valley." Nor was this excursion a mere flash in the pan. The success of the trip encouraged him to push for it in the public school arena and to repeat the exercise over the next three summers.[27]

Other precursors of the scheme can, of course, be enumerated. During the summer of 1869, for example, J. D. Whitney took a group from the School of Mining on a scientific expedition to Colorado;[28] O. C. Marsh and a number of other Yale professors conducted parties of students to the Rockies during the early 1870s; and Professor Orton of Vassar College was accustomed to spending some of the summer vacation with pupils engaging in geological fieldwork. But if Willoughby's official report of 1891–92 is reliable, these initiatives were rather limited and not to be compared with Shaler's systematic ventures. It was, he reported, "Prof. N. S. Shaler, who first suggested to his colleague, Louis Agassiz, the establishment and maintenance during the summer of a seaside laboratory at Nantucket for the benefit both of university students and of teachers of science in secondary schools. The outcome of this suggestion was the establishment of the Anderson School on Penikese Island."[29]

Shaler's role in the summer school movement, however, is rarely so conspicuously recorded. Indeed when the Penikese project was first launched, some of Shaler's circle were concerned that certain Harvard celebrities would step in and take the credit for the original idea. While traveling in England in 1873, for example, he received this message from an anonymous friend:

And now about this summer school of natural history. . . . Your friends think you ought to be on the ground when the thing is started even if you go back to Europe as soon as the first term is over. . . . It looks very much as if the "big wigs" and their satellites will rush in and bear away all the credit of the idea and as none of them know anything about "outdoor" teaching it will end in failure. It is true that at the tail of one of his letters ———— [Agassiz?] did admit that his young friend Shaler had originated the idea. . . . It makes me indignant when I see other people stealing your thunder.[30]

Shaler's experience of field teaching, as his unknown friend implied, was quite unsurpassed. Besides these early excursions, for example, Shaler had also experimented with Saturday field classes during October and November in 1869 and with weekend courses in the Museum of Comparative Zoology. The latter practice he conceived as a civic demonstration of "the eminent value of natural history as a branch of general education, and the extent to which our Museum, by the organization and its resources, is capable of effecting the dissemination of sound knowledge of this science."[31] Penned in 1868, these sentiments testify to Shaler's early belief that museum facilities should be used for the community at large—no doubt a reflection of Charles Peirce's emphasis on the public nature of scientific knowledge.

The weekend classes, however, failed "hopelessly" to meet their worthy objectives, and Shaler was soon forced to abandon them.[32] Still, he did persist with the experimental summer scheme and in 1871, despite a heavy teaching load due to Agassiz's illness, took a group of Harvard students to the James River in Virginia.[33] Little is known of the educational content of the excursion, but Shaler's personal impressions were charmingly recorded in three numbers of the *Atlantic Monthly* for 1873.[34]

The summer of 1872 brought Agassiz's departure for Cape Horn. On the eve of sailing, Agassiz called Shaler aboard the new United States Coast Survey steamer, *Hassler.* He was, he told Shaler, "chagrined" that the museum had done so little to arouse the interest of public school teachers in its facilities, and he "begged" Shaler to find some remedy.[35] Shaler gave the matter some thought and, drawing on his previous experience, quickly hit on the idea of establishing a summer school on Nantucket. The new scheme soon engaged his administrative energies, and he pressed ahead with all due haste. The site was eminently suitable: buildings could be hired for a mere five hundred dollars, student board was available at a reasonable cost, and the good harbor with several boats would facilitate the collection of marine specimens. But before plans could be finalized, Shaler's health deteriorated, and he sailed in late November for England on the *Siberia,* leaving the most crucial problem still unresolved: how to finance the undertaking.

On his return from South America, Agassiz inherited the draft of Shaler's master plan . . . and the cash burden. By the following March the dilemma had still not resolved itself, and Agassiz appealed for aid to the Massachusetts legislature. In response to the published plea, John Anderson, a New York tobacco merchant, offered Penikese Island as a site and, later, a grant of fifty thousand dollars to fund the venture. To Agassiz it was a godsend. He promptly accepted the offer, shelving Shaler's Nantucket site in the process. Shaler was plainly disgruntled and dashed off a note to Alexander Agassiz complaining that "life will be a good deal harder at Penikese than at Nantucket unless we can manage to get a steamer to travel there several times a week; even then it will be difficult to manage about food etc."[36] Despite the conditions the Penikese school went ahead, with Louis Agassiz in the driver's seat and holding forth on everything from glaciology to embryology.[37]

Widely acclaimed though the success of the Anderson school was, its further development was retarded by Agassiz's untimely death in December 1873. Another school was convened the next year under the leadership of Alexander Agassiz, but financial embarrassment, the poor site on Penikese, and the younger Agassiz's lack of teaching skills all contributed to its decline, and the project was not repeated. To Shaler this outcome was all too predictable, and he remained convinced that, if "the establishment had been placed at Nantucket, as it was originally designed, and not on a little harborless isle in a waste of sand flats, it would doubtless have remained successful, and hastened the development of the system of summer instruction. As it was," he went on, "the discontinuance of this School gave a blow to the project, the effects of which were long felt. Many persons were led to believe that a failure which really was brought about by geographical conditions indicated some defect in the general scheme."[38]

Its misfortune notwithstanding, the Anderson school has generally been seen as the prototype of the American summer school. Several later experiments independently conceived or in direct succession to Penikese, however, also helped to shape its subsequent biography.

Alexander Agassiz's private laboratory at Newport in 1877, the Peabody Academy of Science's summer school of biology at Salem between 1876 and 1881, and the Chesapeake Zoological Laboratory established in 1878 by the trustees of Johns Hopkins University must be numbered among the chief candidates.[39] But the other most influential precursors were the Marine Biological Laboratory at Woods Hole and the seaside laboratory established at Annisquan in the summer of 1882 by Hyatt, who shared Shaler's vision of the museum as "an instrument of public culture."[40]

In contrast to the research thrust of these latter undertakings, Shaler's summer school designs were primarily educational in intent. He freely confessed, for instance, that his "interest in the Summer School has been mainly due to the fact that it affords us an opportunity of showing our methods and resources to teachers."[41] Still, even if it was an exercise in university public relations, Shaler knew that the benefits for schoolteachers were considerable, which was always an added spur to his endeavors. At the same time, because he so relished research, Shaler seized the opportunity to use the scheme for more specialized purposes too. In 1875, for example, his summer program at Cumberland Gap in Kentucky (attended, incidentally, by W. M. Davis) was restricted to a heavy dose of stratigraphic, topographical, and dynamic geology.[42] Dana's revised 1874 edition of *Manual of Geology* and Lyell's *Principles* were the sacred texts, and the class was understandably limited to graduates, teachers, and young geologists in training.[43] With a professional clientele and a disciplinary bias, this venture might well be seen as a precursor of the periodic field conferences that later emerged in the 1920s under the auspices of the Association of American Geographers.[44]

Despite Shaler's dedication and determination, however, the summer school project gradually languished. Numbers progressively declined as Shaler's preoccupation with the Kentucky Geological Survey, the Tenth Census, and a host of publications usurped more and more of his energies. Besides, under the Eliot administration, the formal organization of summer teaching at Harvard was actively discouraged; not until the late 1880s did the lull come to an end, when, under Shaler's chairmanship, a committee was appointed to take charge of summer instruction.[45] From that point the scheme began to flourish, and by the turn of the century it had developed beyond the scope that even Shaler had envisioned.

As well as serving the needs of the school-teaching fraternity, some felt there were other markets to cultivate. Could the summer school not serve, for example, as a means of fostering international cooperation and of disseminating American culture abroad?[46] Two Harvard alumni, Ernest L. Conant, a lawyer practicing in Havana, and Alexis E. Frye, superintendent of schools for Cuba by military appointment, certainly thought so. They told President Eliot and General Leonard Wood, military governor of Cuba, that a thousand or more Cuban schoolteachers could benefit from a summer course at Harvard in 1900, and they persuaded the Harvard administration and the Fellows of the College of its value as good foreign policy.

Shaler, who had misgivings from the outset, would have nothing to do with the scheme. For a start, he believed that accommodating the visitors at Harvard would displace the normal summer school students. Besides, their different culture, as he did not hesitate to tell Eliot in 1900, would create its own problems. "It is to be presumed," he complained, "that the greater number of these Cubans will be men of African blood and of somewhat peculiar manners. Even five hundred of them will make a throng in Cambridge."[47] There were other reasons for concern, too. The change of climate would increase the chances of illness; other patrons would be discouraged from enrolling because of the large Cuban presence; and many of the visitors would doubtless be unable to meet rental and other fees. In fact the university forestalled the latter eventuality quite easily: it simply provided housing, food, entertainment, instruction, and a month's salary for the 1,181 Cuban teachers who did come for six and a half weeks. The total cost of the project—seventy thousand dollars—was met by public subscription.[48] Shaler, however, saw even this impressive philanthropic expression of international brotherhood in a poor light. "Two years of experience with the Cuban folk in contact with the Summer School," he told Eliot some eighteen months later, "has undoubtedly served to damage the temper of our schoolteachers. They feel we have petted and favored a rather unworthy lot of foreigners while our own folk have been compelled to pay for everything they have received. I hope therefore that we are at the end of the experiment."[49]

For all the publicity the Cuban arrangement received, it did little to promote the cause of the summer school. Much of the teaching was carried out by young graduates or even by undergraduates from Harvard and Radcliffe and had to be accommodated to the Cubans' rather poor English. At the same time, it did nothing to harm the normal summer school, for during and after the experiment it progressed apace. By now too, under Shaler's direction, it had become a fixed item on the university calendar. A distinctively American contribution to higher learning, the summer school satisfied a real need in the nation's educational system.

Whatever the sources of the summer school's changing fortunes, for Shaler, at least, its inspiration was thoroughly naturalistic. He always felt that it provided the best forum for displaying Agassiz's teaching methods. The unnatural compartmentalization of conventional college life, for example, simply had no place in the vacation schools, where the sustained, uninterrupted investigation of a particular problem, often in the field, was standard practice. Most college graduates, in fact, had "never done a single piece of thoroughly consecutive work" and had therefore never been exposed to what for the scientist was a way of academic life. "The elder Agassiz," he mused, "was used to say that the student of natural science must take time to 'let the facts soak into him,' and he considered a month a short time for even a small body of facts to penetrate in this manner into the student's mind. . . . The only chance for this consolidated work which our school system affords to the new education is found in the vacation periods."[50]

Echoing these sentiments in 1902, Patrick Geddes in Britain called for a return to nature study in the spirit of Rousseau and Pestalozzi. The "problem of education is thus not to increase the present overcrowding of new 'subjects,' each to be treated by its teachers and examiners in its separate hour and compartment," he urged, "but to unify and vitalise the whole by help of this literal 'Return to nature.' " For Geddes, as for Shaler, the solution to the educational problems of the day, posed by the disparate claims of naturalist and humanist, utilitarian and idealist, would be found in the methods advocated by the "geographer, as the exponent not of any one special science of nature, nor of any one tradition of culture, but as the concrete synthetician of all these."[51]

If direct contact with nature was, for Shaler, the only real method of scientific training, its objective through and through was individualistic. In the summer schools, for instance, an opportunity was provided for a different *kind* of thinking altogether, an independent, creative, indeed speculative, approach to knowledge that went against the grain of "the ordinary school-room."[52] Nor was Shaler content to allow individual expression free reign only during those fleeting summer weeks. He wanted it unleashed throughout the entire curriculum; to attain that vision, of course, the whole educational system needed overhauling.

The Objective: Individualistic

Like his pragmatist colleagues, Shaler was certainly not guilty of neglecting the social aspects of education, but like them, he would have agreed that a better society could best be achieved by "the cultivation of a dynamic, adaptable mind . . . resourceful and enterprising in all situations."[53] The new cult of self-realization in education, to which Shaler subscribed, had its roots in the wider structural changes of an increasingly professional society whose highest ideal was expressed in the "self-governing individual exercising his trained judgment in an open society."[54] For Shaler, this feature distinguished genuine education from a mere military training. The object of the latter was "to develop the will power of the individual, but at the same time to subjugate this volition to the command of the superior." The former, by contrast, should be designed "to enfranchise the man, to put him in the fullest possession of his natural powers, to quicken and elevate him in every way, and finally to leave him absolutely self-centred and free."[55]

The experimental method was clearly central to any individualist curriculum, and the advantages it offered students by liberating them from scholastic authority were so crucial that short-term gaps in understanding were but a small price to pay. Individual creativity and independent research, in fact, had more and more come to dominate Shaler's own teaching. Whereas in his earlier days he used to encourage his students to compare their "work of observation and delineation with that done by trained men on the same ground," experience had taught him that this was a big mistake. "So great is the need of developing independent motive," he now affirmed, "that it is better at the outset to make many blunders than to secure accuracy by trust in a leader. The skillful teacher can give fitting words of caution which may help a student to find the true way, but any reference of his undertakings to masterpieces is sure to breed a servile habit."[56]

There were, of course, other implications for the teacher courageous enough to strike out along the individualistic path. If, as Shaler believed, the art of education lay in the effort "to bring out of that curious body of latencies, the human mind, the good therein contained," it was clear that much time would have to be spent in trying to understand the pupil at the desk. Once again, it is not difficult to catch the strains of Rousseau-type philosophy. For as Rousseau, who gave impetus to the psychological movement in education, emphasized, education had as much to do with understanding the child's nature so that teaching could progress in pace with it as it did with the actual content of communicated information. So, like the naturalists for whom the best science teachers were those who studied student nature while the student studied nature, Shaler felt that in "the new education, the school will have to be a psychological observatory, where men who conceive the nature of human beings acquire and practice the most difficult art of discovering the capacities of each pupil, and of fitting the culture to his needs."[57] The full implications of the psychological approach later found their most rigorous expression first in the writings of McDougall and Thorndike and then in the paedocentricity of G. Stanley Hall. But this movement, which set the child, not the teacher or the book or the subject, in the forefront of the educational picture, had its roots firmly lodged in the naturalistic tradition.

The spirit of individualism with which Shaler's temperament was charged found outlet in three areas of current educational controversy, namely, student discipline, the elective system, and college examinations. "Detesting anything like mechanical treatment of a human soul," his wife observed, "he refused to be hemmed in by rules or to raise authority to a system of oppression."[58] Sympathetic by disposition, Shaler was constitutionally hostile to the routine discipline that suppressed the individual temperament. With these sentiments he was in glittering philosophical company. Locke, for example, had affirmed that "*alteris paribus* those children who have been most chastised, seldom make the best men," while Spencer had depicted education as the means of providing students with the opportunity of developing independent responsibility.[59] So too for Shaler, "The worst feature of any routine discipline is that it fails to take account of the vast differences which exist between individual pupils, and treats a whole class of students as if they were all cast in one mould."[60]

Perhaps because of his liberal reputation, Shaler was invited to participate in a symposium on student discipline printed in the *North American Review* in 1889. The implementation of the new elective system was now gathering momentum, and the air of student freedom it stirred up had again raised the question of college discipline. Apart from Shaler, who at that stage had not taken up the position of dean at the scientific school, the other contributors were all college or university presidents.[61] One of the more reformist, Shaler was unique among the participants in bringing more general educational aims into the arena of debate. Some discipline, he conceded, was necessary to preserve moral standards and to develop responsibility if, as he felt, education's ultimate objective was to fit the student "for the higher walks of life." The manner of exercising moral judgment, however, was a quite different matter. The old resort to fear was indisputably a potent agent of control, but it was psychologically repressive, morally defective, and socially manipulative and therefore to be rejected. In its place, Shaler urged, there should be a simple, direct appeal to individual responsibility—a mature strategy fully in accord with a culture that had come of age. Indeed, those who grumbled about the erosion of the old schoolhouse discipline wanted to preserve formality at the expense of freedom and therefore failed "to comprehend the deeper currents of their time" that were carrying Americans toward the "vast democratic humanization" of their society.[62] Thus, while other contributors were more explicit in their discussion of such issues as chapel attendance and student government, they were decidedly more punitive in outlook. If Shaler's was not the official Harvard line on the subject, it nevertheless was truly representative of the currents of thought there and therefore indicative of the gulf between Harvard and other leading universities in the 1880s.

In his endorsement of the elective system, Shaler's individualism again shines through. The autonomy he himself had enjoyed under Agassiz had persuaded him that allowing students a range of academic choices helped cultivate the "varied, elastic mind" capable of coping with the multifarious demands of life in a rapidly changing world. Since the extension of freedom and the closer teacher-pupil relationships it encouraged brought both moral and intellectual gains, Shaler could see no reason for restricting the privileges of the elective system to later college life. He told the Seventh Annual Meeting of the Harvard Teachers' Association in 1898, therefore, that he saw "no difference whatsoever between the profit which is had by the choice of a Senior and that which is had by the choice of a Freshman" and went on to suggest that it could even be extended to the secondary school level.[63] The Harvard administration, it should be noted, had by now taken steps to implement some of these new procedures, and Shaler presented their program at this meeting in the hope of closing the gap between the secondary and tertiary sectors. If this reform could be implemented, there were mutual benefits to be gained: it would relieve the university of teaching elementary courses; it would enrich the school curriculum; and it would encourage pupils to begin considering their future careers at an earlier age, provided that they were shielded from the worst excesses of the professional drive.

Since Shaler's individualism thus centered on creativity and self-expression rather than on elitism and superiority, it is understandable that his most radical challenge to the educational status quo lay in his critique of the examination system. For teacher and student alike, the whole process, written or oral, was counterproductive. It encouraged a restricted curriculum, superficial assessment, and examination-oriented study. In the most scathing of tones he charged that the examination replaced the true "academic spirit. . . by motives which are as low as those prevailing among professional turfmen or the speculators in a stock exchange." In its place, he proposed a number of practices designed to do justice to each student's individual capabilities. For a start, teachers' reports gleaned from personal knowledge of the candidate would be far superior to assessments based on "reading a lot of written matter, produced in hot haste." Again, the capacity to work consistently and methodically was an important skill that might be identified by perusing day-to-day notes. And of course, the ability to acquire, organize, and apply information would come to the fore in project-based teaching. His proposals, he recognized, were far-reaching and "apt to be misunderstood." It was a plea, he hastened to add, "not for the abolition of academic tests, but for the replacement of the present system of non-educative and degrading conditions, such as are induced by the examination room." Unrestrained in rhetoric, he finally pilloried the system as "a part of the rubbish inherited from other centuries, when men put less faith in youth than it is the privilege of our time to entrust. . . . The proctored examination, with its education in trickery and shams, should now be regarded as an anachronism, and be speedily cleared away."[64]

Shaler's concern that self-development should not degenerate into egoism also surfaced in one other matter of passing educational moment—athletics. If his opinions here display the faddish passion to treat every subject under the sun "scientifically," they nonetheless reveal his intellectual continuity with the earlier naturalistic tradition. Just as Rousseau advocated both physical exercise and games on the grounds of their moral and social value,[65] so Shaler believed he could support his similar diagnosis with the insights of evolutionary science. The real benefit of sport thus lay in the cooperation, leadership, and discipline it encouraged, and these of course were qualities on which the "very successes of the race may depend." The curricular implications of this evaluation were immediate. Because they lacked "the coöperative element," purely individualistic games like lawn tennis and gymnastics had "less moral value than the associated field sports" and should be correspondingly discouraged. So Shaler could not agree at all with Eliot's outspoken criticism of competitive team sports in favor of track-and-field events, lawn tennis, cycling, and bowling. On the contrary, Shaler was persuaded that the training they gave in team spirit and personal sacrifice were "in a high degree enlarging to the youth."[66]

Strange then as it may seem, Shaler's approach to the athletic question neatly encapsulates much of his more general educational outlook. Team sport could almost be seen as a kind of social model where individual strengths were harnessed for the good of all. And this position, no less than his liberal attitudes to discipline, the elective system, and examinations typified a humanistic social theory that had other educational repercussions.

The Idealist Thrust

However progressive Shaler's methods and motives were, in respect at least to the thrust of his educational thinking, he was thoroughly traditional. For here, too, the idealism that pervaded his scientific vision also found expression. The first aim of education, he once told his readers in a piece entitled "Humanism in the Study of Nature," was to restore "the old sense of close sympathetic relation to the outer world which was lost with the death of polytheism." Conventional natural theology and catechetical religion more generally had done nothing to foster empathy with nature, and so education was the last resort. Thus, when "the teacher of natural science can create or deepen the sense of the beautiful and the ordered in nature, he has done his work as minister in this great need." For Shaler, the aesthetic and the teleological were inextricably intertwined, and so, having awakened in the student the sense of beauty, the teacher must press on to communicate nature's order. Understanding the principle of "continuity of action in the world," most effectively exemplified in the history of human evolution, would induce a "feeling of . . . kinship with nature—a sense of a kindly filial relation to the earth which will widen and deepen all the ways of thought."[67] In this way, the teaching of geology and paleontology directly contributed to his idealist purposes. Sternly practical though these sciences were, their finest legacy, he was sure, would ultimately prove to be their revelation of humanity's real place in nature. After all, the intellectual and moral revolution "concerning man's origin, duty, and destiny" had materially altered society's self-understanding.[68] Indeed in his own classes Shaler evidently managed to achieve something of that vision. "The large majority of students went forth from his classes," one student recalled, "broader in their mental vision, more in sympathy with Nature . . . and with a deep conviction that they had gained a mental attitude that gave them a more human outlook on the world."[69]

If this fostering of empathy with nature was part of Shaler's evolutionary project to reconcile nature, humanity, and God, it also reflects the persistence of the educational ideals of America's Puritan fathers. Jonathan Edwards, for instance, had found an idealist synthesis of faith, reason, and science that was partly rational and partly mystical.[70] By encouraging a fusion of the natural and spiritual worlds, Shaler was clearly a latter-day exponent of this same tradition, albeit in a context where God was more and more identified with nature's uniformity. Everything that education could do to replace mechanical models of nature by more dynamic, organic metaphors was thus a step in the right direction. Not only would it help reinstate the very teleological viewpoint now under threat, but it would bring students to a holistic conception of nature "as the place of incessant creation each of the creatures in its measure affecting the whole; each sharing in the universal life as the corpuscles in the blood of man share in the work of his body."[71] This quest indeed had been Shaler's own ever since his first days at Harvard, for, as he recorded in his personal journal for 18 September 1859, "the true philosopher has more than mere curiosity to tempt him to give his life to contemplation and research, for although at last he gains but a knowledge of his own insignificance, his studies have brought him nearer his Creator."[72]

Toward a Pragmatic Curriculum

Idealist though Shaler's ultimate educational motives undoubtedly were, his thoughts on the content of the school and university curriculum were solidly rooted in the realities of the modern world economy. His close friendships with the leaders of the American school of pragmatism had evidently a part to play in this, but just as important were the social demands of the newly professionalized America. Thus the type of curricular revisions he advanced and the utilitarian conception of the university he championed foreshadowed the thrust of Dewey's progressive educational philosophy.

The pragmatist claims in the educational sector, however, had already been staked out. The philosophical threads of Peirce's logic and of James's psychology were easily woven into a distinctive educational pattern.[73] For despite internal differences between the movement's adherents, the pragmatist agenda included an emphasis on observation, experimentation, and hypothesis formulation; the functional character of thought; and an assurance that ideas arise in and through experience. In education, it left little room for pure intellectualism because, for the pragmatists, the so-called disinterested pursuit of knowledge simply had no meaning. Knowledge in abstract is mere information. Only when "wrought out in action" and "tested in the crucible of experience" did it become real knowledge.[74] Buoyantly optimistic about "the perfectibility of man," pragmatists spurned every practice designed to keep the pupil sitting deferentially at the feet of any teacher. Instead, they encouraged students to forge for themselves the intellectual skills needed for dealing effectively with the situations of real life. With an emphasis on action rather than reflection, the pragmatist conception of education thus remained problem oriented, committed to the project method of teaching and inspired by the ideal of the self-possessed individual functioning successfully in the theater of American democracy. Indeed Dewey himself was later so to reinforce the interweaving of knowledge and practice that,[75] on the educational scene, the very terms "pragmatism" and "experimentalism" came to be used interchangeably.[76]

Experimentalism and individualism, we now know, were the warp and woof of Shaler's own teaching procedures. But beyond that, his pragmatist instincts, at least in the popularized form of the social importance of "usable intelligence," also came to the fore in his revisionist views about curricular content. After all, his own long-standing conviction that utility and progress were intimately related in Western history was buttressed by the biases of a professionalizing culture. So in a series of essays published during 1895 in the *Chautauquan*, Shaler traced the value of applied science to civilization. Chief among the intellectual achievements of the scientific method was the rejection of mysticism that first emboldened the human species to seize control of nature, and that small beginning heralded the later application of science to industry.[77] From that date, the "march of invention" had been relentless, in fields ranging from textiles and agriculture to communications and medicine.[78] The union of pure and applied science, exemplified par excellence in geology,[79] was thus a marriage of such convenience that, in the educational sphere, its divorce must be averted at all costs.

Nowhere were the needs of applied knowledge more apparent than in America's own recent past. Academic culture had long been highly esteemed by the American people, but the pioneering spirit had nurtured a growing suspicion of mere "book learning." The traditional liberal education, of course, had been instituted for the benefit of the ecclesiastical, legal, and medical professions, and so the prestige of scholarship had rapidly deteriorated as a popular ideal in the new world of cutthroat commercialism. The schoolteacher, apparently detached from the real world and immune to its pressures, made few efforts to adapt the curriculum to society's needs. Small wonder that the stereotyped image of the schoolboy, just emerged from the seclusion of college life, was subject to endless parodies in the popular press. With these Shaler could only concur. For educationalists had simply ignored the profound structural changes in American society and had made no provision for the newer engineering and business professions.[80]

In this environment of educational stagnation, it was only reasonable that many parents would direct their children to technical schools, where the curriculum was career oriented, rather than to traditional colleges. Shaler understood this choice but demurred. His passion was to keep that marriage of technical and academic culture alive, and he therefore saw the future of American education in introducing applied science to the existing university curriculum. It was a narrow tightrope to walk. On the one hand, he had to cope with purists like Alexander Agassiz at the Museum of Comparative Zoology; on the other, with technophiles like Francis Walker at MIT. Nevertheless Shaler believed that, if his proposals were followed and the "combination of professional and culture work could be in any way contrived, all the interests of education would be much better served by our universities than at present . . . we should have no strong line dividing the professional from the academic training, but men would mingle their tasks in a profitable way."[81] Apart too from its utilitarian effects, cross–disciplinary fertilization would counteract the worst dangers of specialization. So if the curriculum must be modified to meet the needs of an industrial society, it should preserve at the same time the spirit of culture traditionally associated with higher learning.

These pronouncements, it is clear, are at once typical yet distrustful of what Bledstein has termed "the culture of professionalism."[82] In an era of urbanization, industrialization, and the meteoric rise of the university, traditional American colleges, inherited from the eighteenth century, seemed too loose, too informal, to cope with the changing patterns of social life.[83] And so, as the old-time college came under the pressures of a standard-conscious society, many semiautonomous institutes sprang up within the university structure to provide more specialized facilities. At Harvard itself, Charles Eliot devoted a great deal of time and energy to increasing the endowment and to restructuring the undergraduate college around several professional schools.[84] So quickly did this scheme catch on that some have suggested that the American university as it emerged at the end of the nineteenth century, with its muscular faith in the omnipotence of "usable intelligence,"[85] was nothing less than Eliot's elective system writ large—a system that welded the pressures of specialization to the demands of society and yet preserved the ethos of American individualism.[86]

Shaler's ambivalence about the new spirit of the times calls to mind Emerson's impressions when he returned to the Harvard Yard in the midsixties on his election to the board of overseers. With high hopes that Harvard would become a university for men instead of a college for boys, he still could not dismiss the old college and the values for which it had stood. When Shaler penned his essay "Relations of Academic and Technical Instruction" almost thirty years later in 1893, he had lived through many of the changes Emerson had foreseen, and like his predecessor he was only too aware of both the strengths and the weaknesses of the professional culture now dominating the educational horizon. On the surface, this article was part of an ongoing debate about whether technical training should be offered within the university sector or at a separate establishment specializing in applied science and technology. In fact it was directed specifically at Walker of MIT, who, according to Shaler, had an intense dislike of Harvard and of the Lawrence Scientific School.[87]

The need for technical education, Shaler reported, had at last been acknowledged. The problem was that, in severing professional training from literary culture, "the advantages to be derived from associating the new arts with the old learning" had been carelessly sacrificed. Immediate steps must be taken to prevent further erosion. The Lawrence Scientific School, the Sheffield School at Yale, and the scientific establishments at the Universities of Pennsylvania, Columbia, Cornell, and Michigan had surely confirmed that "an American university was incomplete without a school of applied science."[88]

For different reasons, opponents could not agree and vigorously opposed such measures. Some had been blinded by the promises of a technological utopia. Others, Shaler felt, needlessly clung to the scholastic traditions of the Old World. In his case, he felt entirely justified in departing from the classical curriculum for the benefit of society, but to jettison the whole structure would be sheer folly. After all, Lamarckian evolution had sensitized him to the processes of organic and social inheritance, and he insisted that the university was the most efficient agent for transmitting the ever-increasing store of learning from one generation to the next.[89] It transcended the mere sum of its constituent elements because it could not be reduced to buildings, teachers, libraries, or collections; it possessed a tradition, centuries old, of passing on accumulated experience to posterity. Besides, independent technical schools were culturally shallow and intellectually lopsided because they focused exclusively on "knowledge of only one side of human culture." Harvard's Lawrence Scientific School, he boasted, provided just such a forum, where students training as engineers, chemists, and practical geologists passed their days in an atmosphere of true academic culture, where "knowledge and a capacity for inquiry are valued for their own sake."[90] Here, then, was *the* model for the newly emerging American university.

In a public response, Walker's chief complaint was about the vagueness of Shaler's article. But in a letter to Horace Scudder, he made it clear that he regarded it as a direct attack on MIT. Shaler brazened it out, insisting that he had taken pains to keep his "mind clear from the Institute of Technology."[91] But Walker saw the matter in a different light. "At neither the first nor the second reading," he told Scudder, "did I apprehend the full reach of his [Shaler's] propositions and suggestions. Their 'carrying power' is simply tremendous. The article is really an attack on all that this school stands for, and on the school itself."[92] Walker's response in the pages of the journal was therefore frank. He certainly had to admit the error of many technical schools in "not providing more so-called liberal studies."[93] But he forthrightly rejected the insinuation that independent professional schools were inferior. Moreover, he pointed to Shaler's own Lawrence Scientific School as evidence of declining numbers, unaware that at the time of writing it was actually experiencing a very rapid growth.[94]

Besides this, opposition to Shaler's proposals was forthcoming from other quarters too. At Harvard itself he had to face the truculence of Alexander Agassiz, who sent the following comments to Eliot in 1902:

I fully agree with your statement of what Shaler has done for the Sc. School and Geol. Dept. But I have not the least interest in developing at Camb. a technical university. That is not the business of a University. You measure numbers and their success in life. I care only to see the original work done at Camb. by Professors and students. There never will be a lack of Technical Schools or of Professional Schools. . . . I look upon the existence of the Tech. and of a Technical School at Camb. as a misfortune. The community is neither large enough nor rich enough to support two institutions. . . .

When Whitney's Professorship was changed to one of research he was absolved from all responsibility of teaching mining & metallurgy. Certainly while he did influence mining engineers his pupils were of a very different stamp from those by Shaler & Co.

But it is useless to discuss the Shaler question. He and I are bound to follow opposite directions. I can only regret the utilitarian tendency at Cambridge and the little encouragement research receives.[95]

This indeed was not the only occasion on which Agassiz vented his spleen in letters to Eliot. In the winter of 1901–2, for example, he maligned Shaler and Davis in a series of notes to the president. He constantly grumbled about Davis, now Sturgis Hooper Research Professor, claiming that "instruction and not research is his great interest."[96] Davis indeed felt it necessary to compose a lengthy self-defense to Eliot in 1903 and to point out that much original research went into the writing of even elementary textbooks.[97] For Shaler's part, he simply regarded Agassiz as "the most unapproachable human being on the planet," especially when it came to matters about his beloved Museum of Comparative Zoology.[98]

If Shaler's integrative proposals were balanced on a knife-edge between the policies of men like Agassiz and Walker, they were also perched between even greater powers. The Harvard and MIT administrations, we recall, joined forces against his cherished Lawrence Scientific School. As we have seen, Shaler fought their schemes tooth and nail and emerged from battle saddened and disillusioned. Still, when the Harvard board laid plans for the establishment of a bachelor of science degree in the university and for the Graduate School of Applied Science, Shaler was fully involved and was thereby able to preserve institutional links between technical and literary education.[99]

An Evolutionary Process

In the years following the publication of Darwin's *Origin of Species,* as the editor of *Galaxy* noted rather wryly, there was a "universal drenching" of literature with Darwinian ideas.[100] Educational theory was no exception, for practitioners turned to evolution themes for any insights they might give into the nature of their own task. It is certainly difficult to delineate in any precise way these evolutionary influences, but it is equally clear that *particular* versions of the evolution story were quickly drawn into the educational forum.[101] Indeed it has been suggested that the prevalence of neo-Lamarckian biology in the United States was less a consequence of its scientific appeal than of the firm support it seemed to give for the possibility of improving humanity through education. As Joseph Le Conte noted, "All our schemes of education, intellectual and moral, though certainly intended mainly for the improvement of the individual are glorified by the hope that the race also is thereby gradually elevated. . . . All our hopes of race-improvement, therefore, are strictly conditioned on the efficacy of these factors—ie. on the fact that useful changes, determined by education in each generation, are to some extent inherited and accumulated in the race."[102]

With its inherently progressivist perspective, Lamarckism seemed tailor-made for the educational aspirations of New England's cultured classes. Through education the "public immorality of the Gilded Age" could be overpowered and lasting social benefits guaranteed.[103] Once statistical information on social evils had been compiled, reform Darwinians insisted, the time would be ripe for direct intervention in the affairs of society through the medium of the schoolroom. Besides, if evolutionary principles were to be brought to bear on educational methods, the Spencerian dictum that education must conform to the natural rhythms of mental development was obviously a prime candidate.[104] In the field of education, therefore, evolutionary theory proved to be as conceptually fertile, not to say prodigal, as it already was in a host of other sciences and social sciences.

From the outset Shaler was convinced that the process of "the transmission of learning through the university" could not be contemplated in isolation from "the modern view of the origin of man." The concerns of ordinary life were no longer subject solely to the dictates of religion or social law; the "guiding truths of that science which shows us how we struggled through the wilderness of the ages from the inconceivably remote time when our being came forth out of the earth and began its long way upward" had entered the scene and now illuminated the whole sphere of education.[105] And yet, when it came to spelling out the new science's implications for education, Shaler's recommendations were just as tortuous as his underlying evolutionary convictions. Even in this single article he drew indiscriminately on both Lamarckian and hereditarian versions.

Primarily, the law of the inheritance of acquired characteristics that operated in both the organic and mental realms was ground enough for Shaler to speculate that it might equally function at the level of social institutions:

> To the naturalist, the devices which men have instinctively invented in order to accomplish the transmission of learning are most interesting, for the reason that they are framed on the same general principles as those by which the ever-increasing needs for the work of the organic body are provided for. . . . So, too, in that other and vaster organism, which we term the state, civilization, or humanity, according as we view it,—a structure which, though invisible and elusive, is still perfectly real,—the separate functions are united in their action, so that the whole has a true, and in a sense personal quality. Those who would conceive the nature of human society should carefully note that the process of evolution leads to ever more and more complicated orders of association.[106]

The analogy of the social organism was clearly captivating.[107] And so, multiplying metaphor on metaphor, he now tried his hand at identifying the social body's various hereditary mechanisms or, to use his own neologism, "transmittenda." The fundamental medium for the transmission of moral norms was the priesthood; next in the hierarchy were the various means of procuring, preserving, and passing on material resources; and at the apex stood the teacher, to whom was entrusted the supreme art of transmitting the cultural and spiritual heritage of civilization. In that task, moreover, the university was the chief instrument of cultural inheritance, and it was the responsibility of the educationalist to ensure that it was allowed to continue to perform that vital social role.[108]

As with his stance on organic evolution, Shaler's propensity for Lamarckism did not temper his enthusiasm for Galton's hereditarian science. The "direction or set of the individual capacity is, in substantially all cases, determined by the conditions of inheritance," he thus insisted; "it is implanted in the individual by events which were shaped before he came upon the earth."[109] Clearly this view implied that, if accurate eugenic records could be compiled from information about students' ancestors and made available to teachers, their mission in guiding and counseling the young would be greatly facilitated. This task, indeed, was especially important because science had confirmed that the quality of mental endowment had already displayed itself by the time of puberty. Still, whatever the inconsistencies, these observations serve to highlight Shaler's concern to bring education, no less than theology, anthropology, geology, or geography, under the rule of evolutionary law.

Shaler's educational stance, it is clear, contained diverse elements derived from many sources. His idealist prejudices, encouraged by Josiah Royce, were wedded to the naturalistic methods of Louis Agassiz. The nascent pragmatism of Peirce and James took the form of a strenuous appeal for curricular relevance. Nor was Shaler immune to the pressures of an emerging professional culture or to the attractions of an infinitely flexible evolutionary theory. All were allocated their parts on Shaler's educational stage. And yet, despite an obvious absence of any coherent educational philosophy, his overwhelming success as a university teacher remains unquestioned. And here, truly, lies his real educational legacy—not in any undergirding theoretical pro-

Shaler in 1901. Photograph by Professor Charles A. Sanger. Reprinted from *The Autobiography of Nathaniel Southgate Shaler* (Boston: Houghton Mifflin, 1909), frontispiece.

nouncements but in the enriched lives of the more than six thousand students who enrolled in at least one Shaler course at the Lawrence Scientific School and who gathered round him during those long, hot, summer days in the field. These educational achievements, surely, are what made him a figure of legendary stature at Harvard during America's Gilded Age.

Epilogue

During the early decades of the nineteenth century, natural theology provided a more or less integrative, intellectual context for the study of the relationships between nature, humanity, and God. The coming of the Darwinian theory, with its inherent challenge to the argument from design, precipitated the fragmentation of this common cultural matrix as many of the central questions were devolved to separate academic disciplines.

The period of Nathaniel Southgate Shaler's life spans these years of transition, and much of his work, therefore, reflects a concern to bridge the gap between the spiritual and the secular. I have tried to argue that Shaler, like many contemporaries, sought to reintegrate a disintegrating context by reconstructing natural theology in such a way as to make divine power more immanent within the world and by rejecting mechanistic metaphors of nature in favor of an evolutionary, organic model. Certainly, with the increasing balkanization of knowledge implicit in a growing emphasis on specialization and professionalism, much of his technical writing was presented in "scientific" language with little or no reference to overarching metaphysical questions. And this approach attests both to the implicit, rather than the explicit, nature of the theological base of his scientific undertakings and to the theodicean transfer of the foundations of the social order to the new laws of nature.

With these sentiments, it is therefore scarcely surprising that much of Shaler's work represents a synthesis of old and new perspectives: his geography juxtaposed the determining impact of physical environment with the directive influence of biocultural inheritance; his geology and geomorphology accommodated the European emphasis on marine action and the American stress on fluvial processes; his racial pronouncements were a fusion of Lamarckian evolution, hereditarian eugenics, and Christian ethics; his concern for the depletion of natural resources countenanced both romantic primitivism and pragmatic conservationism; his educational writings exhibited both utilitarian and idealist aspirations. All these differing emphases, I would suggest, represent the various ways in which Shaler strove to apply those two fundamental principles, neo-Lamarckism and natural theology, to a variety of problems that long remained central to the study of nature and culture. Indeed the evolution of geography in the early decades of the twentieth century continued to exhibit many recurring themes anticipated in Shaler's work. The Davisian model of environmental control and organic response, the centrality of culture typical of the Berkeley School, Huntington's fascination with ethnicity and eugenics, the awakening interest in resource evaluation and management—all were prefigured in Shaler's multifaceted writings.

Interesting though such an exercise in retrospective history may be, there are perhaps more important lessons, both negative and positive, to be learned from Shaler. While it would be churlish to accuse him of consciously manipulating science to suit social policy, it would equally be myopic to overlook the readiness with which scientific theory and ideological praxis went hand in hand; it cannot be gainsaid that science can be used to justify specific value systems and group interests. Nevertheless, in an age when there is a renewed awareness of the limitations that environment ultimately imposes on the human race, demanding an urgent sense of cosmic humility, Shaler's call for the integration of science and ethics, the scholar and the citizen, and culture and nature seems to have particular relevance. For without a pervading sense of the social responsibility of knowledge, "what we call Man's power over Nature," as C. S. Lewis once put it, "turns out to be a power exercised by some men over other men with Nature as its instrument."[1]

Appendix: Analysis of the Writings of Nathaniel S. Shaler

Note: Works shown for the years 1908 and 1909 were published posthumously.

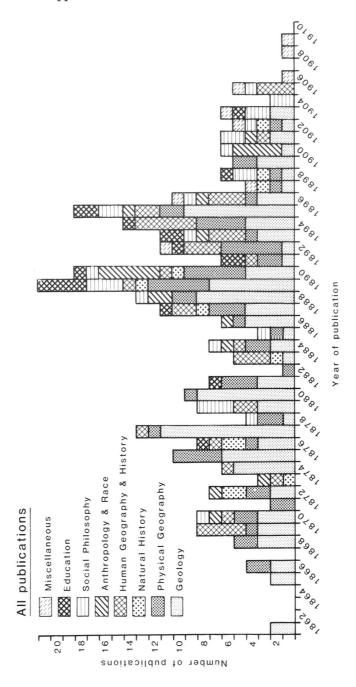

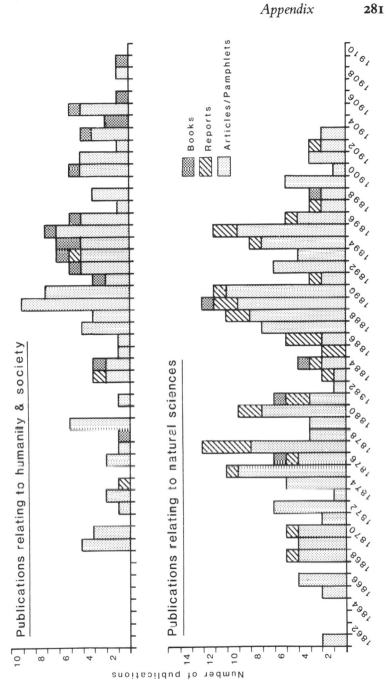

Publications relating to humanity & society

Publications relating to natural sciences

Books
Reports
Articles/Pamphlets

Number of publications

Notes

Complete bibliographic information regarding works by Nathaniel S. Shaler appears in the Bibliography.

Chapter I

1. Representatives of these approaches are Robert K. Merton, "Science, Technology, and Society in Seventeenth Century England," *Osiris* 4 (1938): 360–632; J. D. Bernal, *Science in History* (London: Watts, 1957); Joseph Needham, *Time: The Refreshing River* (New York: Macmillan, 1943); Robert M. Young, "The Historiographic and Ideological Contexts of the Nineteenth-Century Debate on Man's Place in Nature," in *Changing Perspectives in the History of Science,* ed. M. Teich and R. M. Young (London: Heinemann, 1973), pp. 344–438; J. Ben-David, "Introduction," *International Social Science Journal* 22 (1970): 7–27; Diane Crane, *Invisible Colleges: Diffusion of Knowledge in Scientific Communities* (Chicago: University of Chicago Press, 1972); Thomas S. Kuhn, *The Structure of Scientific Revolutions* (Chicago: University of Chicago Press, 1962); Michel Foucault, *L'archéologie de savoir* (Paris: Gallimard, 1969). See also Joseph Agassi, "Towards an Historiography of Science," *History and Theory* 2 (1969): 1–117; John Higham, "Intellectual History and Its Neighbors," *Journal of the History of Ideas* 15 (1954): 339–47.
2. See, for instance, D. N. Livingstone, "Some Methodological Problems in the History of Geographical Thought," *Tijdschrift voor economische en sociale geografie* 70 (1979): 226–31; idem, "The History of Science and the History of Geography: Interactions and Implications," *History of Science* 22 (1984): 271–302; D. R. Stoddart, "Ideas and Interpretation in the History of Geography," in *Geography, Ideology, and Social Concern,* ed. D. R. Stoddart (Oxford: Blackwell, 1981), pp. 1–7; Vincent Berdoulay, "The Contextual Approach," in ibid., pp. 8–16; Paul Claval, "Epistemology and the History of Geographical Thought," in ibid., pp. 227–41; Henry Aay, "Textbook Chronicles: Disciplinary History and the Growth of Geographic

Knowledge," in *The Origins of Academic Geography in the United States,* ed. Brian W. Blouet (Hamden, Conn.: Archon Books, 1981), pp. 291–301.

3. These approaches are represented in the following works: Milton E. Harvey and Brian P. Holly, "Paradigm, Philosophy, and Geographic Thought," in *Themes in Geographic Thought,* ed. Milton E. Harvey and Brian P. Holly (London: Croom Helm, 1981), pp. 11–37; R. J. Johnston, "Paradigms and Revolutions or Evolution? Observations on Human Geography since the Second World War," *Progress in Human Geography* 2 (1978): 189–206; idem, *Geography and Geographers: Anglo-American Human Geography since 1945* (London: Arnold, 1979); D. R. Stoddart, "The Paradigm Concept and the History of Geography," in *Geography, Ideology, and Social Concern,* ed. Stoddart, pp. 70–80; Anne Buttimer, "On People, Paradigms, and 'Progress' in Geography," in ibid., pp. 81–98; Vincent Berdoulay, "Professionnalisation et institutionnalisation de le géographie," *Organon* 14 (1980): 149–56; Olavi Granö, "External Influence and Internal Change in the Development of Geography," in *Geography, Ideology and Social Concern,* ed. Stoddart, pp. 17–36; Horacio Capel, "Institutionalization of Geography and Strategies for Change," in ibid., pp. 37–69; Elspeth Lochhead, "Scotland as the Cradle of Modern Academic Geography in Britain," *Scottish Geographical Magazine* 97 (1981): 98–109; Derek Gregory, *Ideology, Science, and Human Geography* (London: Hutchinson, 1978).

4. Thomas F. Glick, "History and Philosophy of Geography," *Progress in Human Geography* 8 (1984): 275.

5. Some of these pitfalls are outlined in John C. Greene, "Objectives and Methods in Intellectual History," *Mississippi Valley Historical Review* 44 (1957–58): 58–74; Maurice Mandelbaum, "Concerning Recent Trends in the Theory of Historiography," *Journal of the History of Ideas* 16 (1955): 506–17; Quentin Skinner, "Meaning and Understanding in the History of Ideas," *History and Theory* 8 (1969): 3–53; Philip P. Wiener, "Some Problems and Methods in the History of Ideas," *Journal of the History of Ideas* 22 (1961): 531–48.

6. David Hull, "In Defense of Presentism," *History and Theory* 18 (1979): 5.

7. As Clarence J. Glacken has written: "A historian of geographic ideas . . . who stays within the limits of his discipline sips a thin gruel because these ideas almost invariably are derived from broader inquiries like the origin and nature of life, the nature of man, the physical and biological characteristics of the earth. Of necessity they are spread widely over many areas of thought." *Traces on the Rhodian Shore: Nature and Culture in Western Thought from Ancient Times to the End of the Eighteenth Century* (Berkeley: University of California Press, 1967), p. xiii.

8. John Kirtland Wright, "A Plea for the History of Geography," *Isis* 8 (1925): 484.

9. Young, "Historiographic and Ideological Contexts," p. 350.

10. Robert M. Young, "The Impact of Darwin on Conventional Thought," in *The Victorian Crisis of Faith,* ed. Anthony Symondson (London: S.P.C.K., 1974), p. 21. See also Alvar Ellegård, "The Darwinian Theory and Nineteenth Century Philosophies of Science," *Journal of the History of Ideas* 18 (1957): 362–93; W. F. Cannon, "The Problem of Miracles in the 1830's," *Victorian Studies* 4 (1960): 5–32; Richard Yeo, "William Whewell, Natural Theology, and the Philosophy of Science in Mid Nineteenth Century Britain," *Annals of Science* 36 (1979): 493–516.

11. Robert M. Young, "Natural Theology, Victorian Periodicals, and the Fragmen-

tation of a Common Context," in *Darwin to Einstein: Historical Studies on Science and Belief,* ed. Colin Chant and John Fauvel (Harlow: Longman, 1980), pp. 69–107.

12. Carl O. Sauer, *Geography of the Pennyroyal: A Study of the Influence of Geology and Physiography upon the Industry, Commerce, and Life of the People,* Kentucky Geological Survey, ser. 6, vol. 25 (Frankfort, 1927), p. ix; Clarence J. Glacken, "The Origins of the Conservation Philosophy," *Journal of Soil and Water Conservation* 11 (1956): 65; letter to the author from Lewis Mumford, 22 November 1980. In his 1954 presidential address to the Association of American Geographers, J. Russell Whitaker added his testimony in the following terms: "I dare say that we geographers have never learned to make truly effective use of our great men. We have ignored them perhaps, as we certainly have, Matthew Fontaine Maury, Nathaniel Shaler, and John Wesley Powell." "The Way Lies Open," *Annals of the Association of American Geographers* 44 (1954): 238.

13. Walter Berg, "Nathaniel Southgate Shaler: A Critical Study of an Earth Scientist" (Ph.D. diss., University of Washington, 1957). Berg, moreover, makes use of only some 117 out of Shaler's 308 publications.

14. As Wiener writes: "The intellectual biographer of an unknown or neglected minor figure may properly claim to find in his subject a revealing key to the opinions of a larger cross-section of the contemporary population than the more advanced thinkers of the same era." "Some Problems and Methods," p. 540.

15. Charles T. Copeland, "Biography of Prof. Shaler," *Boston Transcript,* 18 July 1906.

Chapter 2

1. Charles T. Copeland, "Biography of Prof. Shaler."

2. William Herbert Hobbs, "Nathaniel Southgate Shaler," *Transactions of the Wisconsin Academy of Sciences, Arts, and Letters* 15 (1907): 924.

3. Newspaper clipping attached to J. B. Woodworth's personal copy of Shaler's *Autobiography,* held in the library of the geology department, Harvard University. This poem appears in Henry Ware Eliot, *Harvard Celebrities: A Book of Caricatures and Decorative Drawings* (Cambridge, Mass.: Harvard University Press, 1901). See also the chapter on Shaler entitled "Poetic Geologist" in Rollo Walter Brown, *Harvard Yard in the Golden Age* (New York: Current Books, 1948), pp. 105–18.

4. *Dictionary of American Biography,* s.v. "Shaler, William."

5. Shaler, *Autobiography* (1909), pp. 7–8.

6. Ibid., pp. 24, 25.

7. Ibid., p. 57.

8. Ibid., pp. 57, 63.

9. Shaler, "Father Blumhardt's Prayerful Hotel" (1870), p. 713.

10. Bruce Kuklick, *The Rise of American Philosophy: Cambridge, Massachusetts, 1860–1930* (New Haven: Yale University Press, 1977), p. 5.

11. Stow Persons, *Free Religion: An American Faith* (Boston: Beacon Press, 1963), pp. 99–156. Useful general accounts include Morton White, *Science and Sentiment in America: Philosophical Thought from Jonathan Edwards to John Dewey* (New York: Oxford University Press, 1972), pp. 71–110; Sydney E. Ahlstrom, *A Religious History of the American People* (New Haven: Yale University Press, 1972), pp. 597–614; Herbert W. Schneider, "Religious Enlightenment in American Thought," in

Dictionary of the History of Ideas, ed. Philip P. Wiener (New York: Scribner's, 1973), vol. 4, pp. 109–12. See also George F. Whicher, ed., *The Transcendentalist Revolt against Materialism* (Boston: Heath, 1949); Stow Persons, *American Minds: A History of Ideas* (New York: Holt, 1958).

12. Ralph Waldo Emerson, *On Nature* (Leipzig: Insel-Verlag, n.d.; first published 1835), p. 5.

13. For general surveys see John C. Greene, *The Death of Adam: Evolution and Its Impact on Western Thought* (Ames: Iowa State University Press, 1959); John W. Burrow, *Evolution and Society: A Study in Victorian Social Theory* (Cambridge: Cambridge University Press, 1966); Peter J. Bowler, "The Changing Meaning of 'Evolution,'" *Journal of the History of Ideas* 36 (1975): 95–114.

14. Neal C. Gillespie, *Charles Darwin and the Problem of Creation* (Chicago: University of Chicago Press, 1979), pp. 89–90.

15. Francis Bowen, "A Theory of Creation," *North American Review* 60 (1845): 426–78; Edward J. Pfeifer, "United States," in *The Comparative Reception of Darwinism,* ed. Thomas F. Glick (Austin: University of Texas Press, 1972), p. 171.

16. For general reviews of the Darwinian debate, see Loren Eiseley, *Darwin's Century: Evolution and the Men Who Discovered It* (New York: Anchor Books, 1961); Peter Vorzimmer, *Charles Darwin: The Years of Controversy: The Origin of Species and Its Critics, 1859–1882* (Philadelphia: Temple University Press, 1975); David L. Hull, *Darwin and His Critics: The Reception of Darwin's Theory of Evolution by the Scientific Community* (Cambridge, Mass.: Harvard University Press, 1973); Michael Ruse, *The Darwinian Revolution: Science Red in Tooth and Claw* (Chicago: University of Chicago Press, 1979); Michael Ghiselin, *The Triumph of the Darwinian Method* (Berkeley: University of California Press, 1969).

17. See James R. Moore, *The Post-Darwinian Controversies: A Study of the Protestant Struggle to Come to Terms with Darwin in Great Britain and America, 1870–1900* (Cambridge: Cambridge University Press, 1979).

18. Young, "Historiographic and Ideological Contexts"; idem, "Natural Theology, Victorian Periodicals"; idem, "Impact of Darwin."

19. J. R. Lucas, "Wilberforce and Huxley: A Legendary Encounter," *Historical Journal* 22 (1979): 313–30; Sheridan Gilley and Ann Loades, "Thomas Henry Huxley: The War between Science and Religion," *Journal of Religion* 61 (1981): 285–308; James R. Moore, "1859 and All That: Remaking the Story of Evolution-and-Religion," in *Charles Darwin, 1809–1882: A Centennial Commemorative,* ed. Roger G. Chapman and Cleveland T. Duval (Wellington, New Zealand: Nova Pacifica, 1982), pp. 167–94. The traditional story is told in William Irvine, *Apes, Angels, and Victorians: A Joint Biography of Darwin and Huxley* (London: Weidenfeld & Nicolson, 1956).

20. David N. Livingstone, *Darwin's Forgotten Defenders: The Encounter between Evangelical Theology and Evolutionary Thought* (Grand Rapids and Edinburgh: Eerdmans and Scottish Academic Press, 1987).

21. See Frank M. Turner, "The Victorian Conflict between Science and Religion: A Professional Dimension," *Isis* 69 (1978): 356–76.

22. These details are largely drawn from Edward Lurie, *Louis Agassiz: A Life in Science* (Chicago: University of Chicago Press, 1960). On Oken see William Coleman, *Biology in the Nineteenth Century: Problems of Form, Function, and Transformation* (New York: Wiley, 1971), pp. 25–26.

23. See Ian F. A. Bell, "Divine Patterns: Louis Agassiz and American Men of Letters," *Journal of American Studies* 10 (1976): 349–81.
24. Elizabeth Cary Agassiz, *Louis Agassiz: His Life, Letters, and Correspondence* (London: Macmillan, 1885), vol. 1, p. 138. The following year, writing again to Sedgwick, Agassiz declared: "I find it impossible to attribute the biological phenomena which have been and still are going on upon the surface of our globe, to the simple action of physical forces. I believe they are due, in their entirety, as well as individually, to the direct intervention of a creative power." Ibid., p. 389.
25. Ernst Mayr, "Agassiz, Darwin, and Evolution," *Harvard Library Bulletin* 13(1959): 168.
26. Asa Gray, "Louis Agassiz," *Andover Review* 9 (1886): 38.
27. [Katherine M. Lyell], *Life, Letters, and Journals of Sir Charles Lyell, Bart.*, 2 vols. (London: Murray, 1881), vol. 2, p. 331.
28. Quoted in Edward Lurie, "Louis Agassiz and the Idea of Evolution," *Victorian Studies* 3 (1959): 98.
29. The standard biography is A. Hunter Dupree, *Asa Gray* (Cambridge, Mass.: Harvard University Press, 1959).
30. Asa Gray, *Natural Science and Religion* (New York: Scribners, 1880), p. 47.
31. Darwin himself dispatched copies of Gray's short work to a long list of individuals, among them Samuel Wilberforce, Robert Chambers, and Charles Kingsley.
32. White, *Science and Sentiment*, pp. 121–22. As an opponent of the transcendentalism of his day and an ardent devotee of the scientific method, Wright was a precursor of the later Harvard pragmatists. His philosophical stance, Wiener suggests, owed much to both Gray and Jeffries Wyman. Philip P. Wiener, *Evolution and the Founders of Pragmatism* (Cambridge, Mass.: Harvard University Press, 1949), p. 253. On Wright's death, Gray wrote, "He points out clearly the essential difference between Darwinism, which is scientific, and Spencerism, which is philosophical." Dupree, *Gray*, p. 302.
33. See *Proceedings of the American Academy of Arts and Sciences* 4 (1859): 410–16, 424–41.
34. This debate is reconstructed in Pfeifer, "United States," p. 180.
35. Shaler to Mrs. E. C. Agassiz, 8 January 1873 [*sic*], Agassiz Papers, Museum of Comparative Zoology, Harvard University Archives.
36. James Lee Love, *The Lawrence Scientific School in Harvard University, 1847–1906* (Burlington, N.C.: n.p., 1944), p. 3.
37. Shaler, *Autobiography* (1909), p. 98.
38. This point is discussed in chapter 9.
39. Shaler, *Autobiography* (1909), pp. 111, 128.
40. Ibid., pp. 121, 120, 124, 126, 130, 139, 141, 155.
41. Ralph W. Dexter, "Three Young Naturalists Afield: The First Expedition of Hyatt, Shaler, and Verrill," *Scientific Monthly* 79 (1954): 46. According to Hyatt (p. 47), assignments for the collections were as follows: (1) recent animals: polyps, insects, and birds (Verrill); bivalves, crustaceans, and fishes (Shaler); gastropods, cephalopods, worms, and mammals (Hyatt); (2) fossils: corals (Verrill); bivalves and trilobites (Shaler); gastropods, cephalopods, and crinoids (Hyatt). The continued friendship of the three students is reflected in Verrill's diary entry for 21 November 1860 regarding the student natural history society: "We held a special meeting of the club and voted to make three constitute a quorum. This will

enable Hyatt, Shaler and I to keep up the Club in spite of all the rest and without any of their assistance." Verrill Diary, Harvard University Archives.

42. Verrill later wrote a description of the island entitled "Notes on the Natural History of Anticosti," *Proceedings of the Boston Society of Natural History* 9 (1862–63): 132–35.

43. Shaler, "List" (1865); Alpheus Hyatt, "Remarks on the Beatriceae, a New Division of Mollusca," *American Journal of Science and Arts* 39 (1865): 261–66.

44. Dexter, "Three Young Naturalists," p. 51.

45. Quoted in Berg, "Shaler," p. 25.

46. Shaler, *Autobiography* (1909), p. 225.

47. He wrote, for example, to his wife on 28 June 1864: "I met Professor Agassiz this afternoon and received from him a *somewhat* cordial greeting. The old fellow is in a great rage against all students, and is worn with the trouble they have lately given him. I am inclined to think that there has been much wrong done him in the way of petty spite by the rebels of the M[useum of] C[omparative] Z[oology]. The students have failed to show him the consideration his age and preeminent services entitle him to receive, and have done much to embitter his declining years." Shaler, *Autobiography* (1909), p. 225.

48. The difficulty of obtaining such reliable assistance was noted by Agassiz himself when he commented that "young men are offered tempting situations before they have gone through the last stages of their professional studies, and these temptations raise unduly the aspirations of even the least competent students. Under such circumstances a devoted student is a blessing to his teacher, as he is likely to be an honor to his country and a successful promoter of science." *Annual Report of the Trustees of the Museum of Comparative Zoology* (1864), p. 8.

49. The following outline of work for the winter of 1865 is recorded by Shaler's wife: (1) July 15: essay on the intellectual relations of the four types of animal; (2) September 1: on the formation of continents; (3) optional: on the relation of philosophical systems to scientific methods; (4) November 1: on the changes of coast line in New England. See Shaler, *Autobiography* (1909), p. 226.

50. L. Agassiz to M. Barrande, 4 November 1866. Reproduced in Shaler, *Autobiography* (1909), pp. 226–27.

51. This latter experience led to his article "Father Blumhardt's Prayerful Hotel" (1870).

52. *Report of the Museum of Comparative Zoology* (1868), pp. 41–44.

53. See John E. Wolff, "Memoir of Nathaniel Southgate Shaler," *Bulletin of the Geological Society of America* 18 (1906): 594.

54. Shaler, "On Changes in Geographical Distribution of the American Buffalo" (1869). Shaler added a brief appendix "On the Age of the Bison in the Ohio Valley" to Joel A. Allen's study *The American Bison, Living and Extinct* (Cambridge, Mass.: Harvard University Press, 1876), pp. 232–36. See also Frank Gilbert Roe, *The North American Buffalo: A Critical Study of the Species in Its Wild State* (Toronto: University of Toronto Press, 1951), p. 850. Shaler later applied this research to the more general problem of the origin of the prairies, suggesting that the eastward movement of the buffalo had resulted from western deforestation by fire. This theory was later used by Carl Sauer, who refers to Shaler's contribution in his 1927 monograph, *Geography of the Pennyroyal*.

55. Quoted in Berg, "Shaler," p. 61.
56. Shaler to C. W. Eliot, 23 July 1870, Charles W. Eliot Papers, Harvard University Archives.
57. See W. M. Davis, "Nathaniel Southgate Shaler: Estimate of the Teacher and the Man," *Harvard Bulletin* (16 May 1906), p. 1; Shaler, *Autobiography* (1909), p. 369.
58. See Charles W. Eliot, "The New Education," *Atlantic Monthly* 23 (1869): 203–20, 358–67. Also Henry James, *Charles W. Eliot, President of Harvard University, 1869–1909*, 2 vols. (Cambridge, Mass.: Constable, 1930).
59. See the discussion in chapter 9.
60. Shaler (Oxford) to W. James, 20 April 1873, William James Papers, bMS Am 1092 (1008), Houghton Library, Harvard University.
61. "I am so thoroughly convinced that it would be a good fortune for your state to secure the services of Professor N. S. Shaler as Geologist to direct your survey," wrote Agassiz to Preston H. Leslie (governor of Kentucky), "that though he is absent and I know nothing of his intentions I take the liberty of calling your attention to his eminent abilities and perfect qualifications for such work." Quoted in Shaler, *Autobiography* (1909), p. 270. When Shaler was eventually nominated as director of the survey, Benjamin Peirce wrote to Leslie expressing his "own gratification at the appointment of that able man as State Geologist." Ibid., p. 271.
62. See Berg, "Shaler," p. 69.
63. See *Report of the Superintendent of the United States Coast Survey, Showing the Progress of the Work for the Fiscal Year Ending with June, 1876* (Washington, D. C.: Government Printing Office, 1879), p. 42. Through his attendance at this summer camp, W. M. Davis began his professional career as a teacher at Harvard. See Richard J. Chorley, Robert P. Beckinsale, and A. J. Dunn, *The History of the Study of Landforms; or, The Development of Geomorphology*, vol. 2, *The Life and Work of William Morris Davis* (London: Methuen, 1973), pp. 61–62.
64. Shaler made this need clear in a speech on 15 July 1875 to the Society for the Advancement of Education in Frankfort, Kentucky. Berg, "Shaler," p. 75. In a letter to *Nature* the same year, he lamented that in American schools geology "is taught in the air, not on the earth." Shaler, "Geology in America" (1875), p. 6.
65. Shaler, *Autobiography* (1909), pp. 284–85.
66. Quoted in Berg, "Shaler," p. 78.
67. A brief review of the work of the second Geological Survey of Kentucky may be found in George P. Merrill, *The First One Hundred Years of American Geology* (New Haven: Yale University Press, 1924), pp. 477–78. Shaler's career with this survey is charted in Ivan L. Zabilka, "Nathaniel Southgate Shaler and the Kentucky Geological Survey," *Register of the Kentucky Historical Society* 80 (1982): 408–31.
68. See Thomas G. Manning, "Geological Survey, United States," in *Dictionary of American History* (New York: Scribner's Sons, 1976), vol. 3, pp. 163–64; Michele L. Aldrich, "Geological Surveys, States," in ibid., pp. 164–66.
69. The standard reference work is William H. Goetzmann, *Exploration and Empire: The Explorer and the Scientist in the Winning of the American West* (New York: Knopf, 1971). See also Henry Nash Smith, "Clarence King, John Wesley Powell, and the Establishment of the United States Geological Survey," *Mississippi Valley Historical Review* 34 (1947): 37–58; A. Hunter Dupree, *Science in the Federal Government: A*

History of Policies and Activities to 1940 (Cambridge, Mass.: Harvard University Press, 1957); Wallace Stegner, *Beyond the Hundredth Meridian: John Wesley Powell and the Second Opening of the West* (Boston: Houghton Mifflin, 1954).

70. Goetzmann, *Exploration*, p. 588.
71. See Preston Cloud, "The Improbable Bureaucracy: The United States Geological Surveys, 1879–1979," *Proceedings of the American Philosophical Society* 124 (1980): 155–67.
72. W. Stull Holt, *The Bureau of the Census: Its History, Activities, and Organization* (Washington, D. C.: Brookings Institution, 1929).
73. Shaler, *Autobiography* (1909), p. 319.
74. Shaler, *Campobello* (1881).
75. Shaler, *Autobiography* (1909), p. 325.
76. *Illustrations of the Earth's Surface* (1881) had been described by its reviewer in the *Geological Magazine* 9 (1882): 34, as a book that would find itself on "many a drawing-room table." Later, James Croll called it a magnificent work. *Discussions on Climate and Cosmology* (New York: Appleton, 1886).
77. See his reports on resources for railway companies in 1881 and 1884.
78. Berg, "Shaler," p. 95.
79. Powell to Shaler, 8 March 1884, Shaler Papers, HUG 1784.10, Harvard University Archives.
80. Berg, "Shaler," p. 98.
81. See Shaler, "Introduction" to Penrose (1888).
82. See Stegner, *Powell*, pp. 283–93: William Culp Darrah, *Powell of the Colorado* (Princeton: Princeton University Press, 1969), pp. 288–98.
83. Berg, "Shaler," p. 103.
84. Powell to Shaler, 6 October 1885, Shaler Papers, HUG 1784.10.
85. Darrah, *Powell,* p. 294.
86. Shaler to Alexander Agassiz, 27 June 1886, Agassiz Papers, bAG 766.10.1.
87. Shaler, "Swamps" (1886), p. 232.
88. *Nation* 50 (23 January 1890): 79.
89. *Nature* 53 (28 March 1895): 507–8. Nevertheless, this reviewer did conclude that some of the essays contained "not a few interesting and suggestive passages."
90. Helen R. Fairbanks and Charles P. Berkey, *Life and Letters of R. A. F. Penrose, Jr.* (New York: Geological Society of America, 1952), p. 106.
91. Shaler, *Autobiography* (1909), pp. 349, 354.
92. Shaler, *Elizabeth* (1903), p. vii. For more general discussion of this point, see John Angus Campbell, "Nature, Religion, and Emotional Response: A Reconsideration of Darwin's Affective Decline," *Victorian Studies* 18 (1974): 159–74; Robert E. Fitch, "Charles Darwin: Science and the Saintly Sentiments," *Columbia University Forum* 2 (Spring 1959): 7–12.
93. Brown, *Harvard Yard*, p. 113.
94. G. K. Gilbert to Shaler, 21 January 1890, Shaler Papers.
95. Shaler to O. C. Marsh, 24 May 1892, Marsh Papers, Box 29, Folder 1237, Yale University Archives.
96. Shaler to Eliot, 28 July 1892, Eliot Papers, UA I.5.150, Box 120, Folder 321.
97. Lurie, *Agassiz,* p. 271.
98. Jules Marcou, *A Little More Light on the United States Geological Survey* (Cambridge, Mass.: Privately printed, 1892), pp. 5–6.

99. Berg, "Shaler," p. 123.
100. Cloud, "The Improbable Democracy"; Darrah, *Powell*, p. 347.
101. See letters to Shaler from Walcott, Shaler Papers.
102. See, for example, reviews in *Critic* 26 (13 April 1895): 273; *Nation* 60 (17 January 1895): 60; *Popular Science Monthly* 46 (1895): 704.
103. See *Critic* 30 (3 April 1897): 233–34; *Dial* 22 (16 February 1897): 125–26.
104. *Science*, n.s., 8 (1898): 712–15; *Nature* 59 (1899): 604–5.
105. For this brief account of the Lawrence Scientific School, I have relied largely on Love, *Lawrence Scientific School;* M. W. Rossiter, "Lawrence Scientific School," in *Dictionary of American History*, vol. 4, pp. 119–20.
106. Shaler, *Autobiography* (1909), p. 328. When McKay died, his will stipulated inter alia that Shaler's daughters should receive $3,000 per year during their lifetime. As the various beneficiaries died, Harvard's share increased, and when the last payment was made in 1949, the total sum amounted to $15,766,755. See *Harvard Alumni Bulletin* 52 (1949): 155–56.
107. See *Harvard College Report for 1891–92* (Cambridge, Mass.: Harvard University, 1892), p. 90.
108. Davis, "Shaler," p. 2.
109. Shaler, *Autobiography* (1909), pp. 377, 379.
110. See Davis, "Shaler," p. 3.
111. Shaler to Eliot, 1 February 1898, Eliot Papers, Box 120, Folder 321.
112. On the basis of such a varied manifesto, W. A. Koelsch has recently maintained that Shaler "set a long-hidden agenda for twentieth century human geography." "Nathaniel Southgate Shaler," in *Geographers: Biobibliographical Studies*, ed. T. W. Freeman and Philippe Pinchemel (London: Mansell, 1979), vol. 3, p. 134.
113. *Catalogue of the Officers and Students of the Theological Seminary, Andover, Mass., 1891–92* (Andover: Draper, 1891), p. 25.
114. Shaler, "Use of Numbers" (1879); idem, "Natural History of Politics" (1879); idem, "Natural History of Sympathy" (1884).
115. *Nature* 54 (1896): 220–21.
116. Shaler, *Autobiography* (1909), p. 425.
117. "Professor Shaler Dead," *Harvard Bulletin* (11 April 1906).
118. Shaler, *Autobiography* (1909), pp. 396–97.
119. See Barbara M. Solomon, *Ancestors and Immigrants: A Changing New England Tradition* (Chicago: University of Chicago Press, 1956), p. 93.
120. Shaler, *Autobiography* (1909), p. 400. Shaler's close friendship with William James, Charles Peirce, and Chauncey Wright would have influenced his own disposition toward pragmatism. Various letters from these philosophers are extant in the Shaler Papers.
121. Davis, "Shaler," p. 3. Writing to Eliot in 1908 about the inscription to be appended to the bronze bust of Shaler placed in University Hall, Davis observed: "I wish something more characteristic might be said of him. *Adventurous* is characteristic, I know; but I should like some hint at his quick interest in any and every subject." Davis to Eliot, 8 January 1908, Eliot Papers, Box 246. Also in Shaler's memory, a fund was established in the 1920s to institute four scientific professorships at Berea College in Kentucky. See file on Shaler Memorial Fund, Shaler Papers, Harvard University Archives. Finally, in 1936, when a committee was appointed for the Harvard Tercentenary to select the most celebrated fifty men in the university's history, Shaler's was the twelfth name on the list.

Chapter 3

1. See J. A. Campbell and D. N. Livingstone, "Neo-Lamarckism and the Development of Geography in the United States and Great Britain," *Transactions of the Institute of British Geographers,* n.s., 8 (1983): 267–94. For the more general impact of evolutionary thought forms on geography, see David R. Stoddart, "Darwin's Impact on Geography," *Annals of the Association of American Geographers* 56 (1966): 683–98; idem, "'That Victorian Science': Huxley's *Physiography* and Its Impact on Geography," *Transactions of the Institute of British Geographers* 66 (1975): 17–40; idem, "Darwin's Influence on the Development of Geography in the United States, 1859–1914," in *Origins,* ed. Blouet, pp. 265–78; Jurgen Herbst, "Social Darwinism and the History of American Geography," *Proceedings of the American Philosophical Society* 105 (1961): 538–44.

2. See Peter J. Bowler, *The Eclipse of Darwinism: Anti-Darwinian Evolution Theories in the Decades around 1900* (Baltimore: Johns Hopkins University Press, 1983), p. 46; Ernst Mayr, "Agassiz, Darwin, and Evolution," *Harvard Library Bulletin* 13 (1959): 165–94; Edward Lurie, "Louis Agassiz and the Idea of Evolution," *Victorian Studies* 3 (1959): 87–108; Roy M. MacLeod, "Evolutionism and Richard Owen, 1830–1868: An Episode in Darwin's Century," *Isis* 56 (1965): 259–80; Coleman, *Biology in the Nineteenth Century,* p. 50.

3. Peter J. Bowler, *Evolution: The History of an Idea* (Berkeley: University of California Press, 1984), p. 120.

4. Ernst Mayr, "The Nature of the Darwinian Revolution," *Science* 176 (1972): 988.

5. Young, "Impact of Darwin," p. 28.

6. Charles Darwin, *On the Origin of Species by Means of Natural Selection: or, The Preservation of Favoured Races in the Struggle for Life* (London: Murray, 1859), p. 489.

7. See Derek Freeman, "The Evolutionary Theories of Charles Darwin and Herbert Spencer," *Current Anthropology* 15 (1974): 211–37; see also J. D. Y. Peel, *Herbert Spencer: The Evolution of a Sociologist* (London: Heinemann, 1971).

8. See Stephen J. Gould, *Ontogeny and Phylogeny* (Cambridge, Mass.: Harvard University Press, 1977); Bowler, *Evolution,* pp. 119–23.

9. Shaler, "Lateral Symmetry" (1861), p. 279. In 1865 Agassiz recorded his acceptance of Shaler's thesis that the valves of the brachiopoda were front and back, not dorsal and ventral. Louis Agassiz, "On Brachiopods," *Proceedings of the Boston Society of Natural History* 9 (1865): 68–69. This position was refuted by E. S. Morse. See Ralph W. Dexter, "Historical Aspects of Studies on the Brachiopoda by E. S. Morse," *Systematic Zoology* 15 (1966): 241–44.

10. Shaler, *Autobiography* (1909), pp. 181, 110.

11. Bowler, *Eclipse,* pp. 3–6; Vernon L. Kellogg, *Darwinism Today: A Discussion of Present-Day Scientific Criticism of the Darwinian Selection Theories, together with a Brief Account of the Principal Other Proposed Auxiliary and Alternative Theories of Species Forming* (New York: Holt, 1908); Vorzimmer, *Charles Darwin;* Hull, *Darwin and His Critics.*

12. William North Rice, "Darwinian Theory of the Origin of Species," *New Englander* 25 (1867): 607–33; Francis Bowen, "Remarks on the Latest Form of the Development Theory," *Memoirs of the American Academy of Arts and Science,* n.s., 8 (1860): 98–107.

13. Francis Darwin, *The Life and Letters of Charles Darwin,* 3 vols. (London: Murray, 1887), vol. 2, p. 232.
14. See Bowler, *Eclipse,* pp. 118–40; idem, "Edward Drinker Cope and the Changing Structure of Evolutionary Theory," *Isis* 68 (1977): 249–65; Edward J. Pfeifer, "The Genesis of American Neo-Lamarckism," *Isis* 56 (1965): 156–67; also George W. Stocking, Jr., "Lamarckianism in American Social Science: 1890–1915," *Journal of the History of Ideas* 23 (1962): 239–56.
15. See R. T. Jackson, "Hyatt, His Life and Work," Manuscript, Harvard University Archives, HUG 1461.541, p. 18. Also Alpheus Hyatt, "On the Parallelism between the Different Stages of Life in the Individual and Those in the Entire Group of the Molluscous Order Tetrabranchiata," *Memoirs of the Boston Society of Natural History* 1 (1866–67): 193–209.
16. Quoted in George Daniels, ed., *Darwinism Comes to America* (Waltham, Mass.: Blaisdell Publishing, 1968), p. 91. The original source is Edward Drinker Cope, "Evolution and Its Consequences," *Penn Monthly Magazine* 3 (1872): 222–36, 366–85, 429–38, 461–77.
17. See William H. Dall, "On a Provisional Hypothesis of Saltatory Evolution," *American Naturalist* 11 (1877): 135–37; Clarence King, "Catastrophism and Evolution," *American Naturalist* 11 (1877): 449–70; Joseph Le Conte, *Evolution: Its Nature, Its Evidences, and Its Relation to Religious Thought,* 2d ed. (New York: Appleton, 1897), pp. 73–74; Henry Fairfield Osborn, "Paleontological Evidence for the Transmission of Acquired Characteristics," *American Naturalist* 23 (1889): 561–66; John A. Ryder, "Proofs of the Effects of Habitual Use in the Modification of Animal Organisms," *Proceedings of the American Philosophical Society* 26 (1889):541–49; John Wesley Powell, "Competition as a Factor in Human Evolution," *American Anthropologist* 1 (1888): 297–323; Joel A. Allen, "The Influence of Physical Conditions in the Genesis of Species," *Radical Review* 1 (1877–78): 108–37.
18. See, for example, L. H. Bailey, "Neo-Darwinism and Neo-Lamarckism," *American Naturalist* 28 (1894): 661–78.
19. Campbell and Livingstone, "Neo-Lamarckism."
20. Eiseley, *Darwin's Century,* p. 217; Bowler, *Eclipse,* p. 66.
21. See Moore, *Post-Darwinian Controversies,* p. 231.
22. See Cynthia Eagle Russett, *Darwin in America: The Intellectual Response, 1865–1912* (San Francisco: Freeman, 1976), pp. 10–11. Also Greta Jones, *Social Darwinism and English Thought: The Interaction between Biological and Social Theory* (Sussex: Harvester Press, 1980), p. 82; Bowler, *Evolution,* pp. 278–82.
23. Thus, "Nathaniel Southgate Shaler," *Nation* 82 (1906): 318; Sidney Ratner, "Evolution and the Rise of the Scientific Spirit in America," *Philosophy of Science* 3 (1936): 104–22.
24. Shaler, "The Rattlesnake" (1872), pp. 36–37. Shaler's doubts persisted throughout his life. In 1894, for example, he outlined the difficulties in the Darwinian explanation of the loss of sight in certain fish. *Sea and Land* (1894), p. 104. And in 1902 he insisted that "the Darwinian hypothesis is still essentially unverified." "Faith in Nature" (1902), p. 288.
25. Charles Darwin, *The Expression of the Emotions in Man and Animals,* popular edition (London: Murray, 1904), p. 108, n. 31.
26. Shaler, "Origin of Domestic Cat" (1872), p. 162.

27. Shaler, *Interpretation* (1893), p. 146. This view is further emphasized when he observes: "The most important novelty which life introduces into the world consists in the principle of progressive accumulation of inheritances, so that the properties of the structure are determined, not as in the case of the physical unit, by laws dependent upon the primal conditions of matter, but by experiences won in the earlier life, either in that of the individual or in that of its ancestors since the beginning of the series to which it belongs." Ibid., pp. 150–51.

28. Shaler, "Faith in Nature" (1902), p. 283.

29. Shaler, *Neighbor* (1904), pp. 3–4.

30. Shaler, *Story of Continent* (1892), p. 73.

31. Shaler, "School Vacations" (1889), p. 828.

32. See D. W. Forrest, *Francis Galton: The Life and Work of a Victorian Genius* (New York: Elek, 1974); T. W. Freeman, *The Geographer's Craft* (Manchester: Manchester University Press, 1967), pp. 22–43.

33. Francis Galton, "Heredity, Talent, and Character," *Macmillan's Magazine* 12 (1865): 157–66, 318–27.

34. P. Froggatt and N. C. Nevin, "The 'Law of Ancestral Heredity' and the Mendelian-Ancestrian Controversy in England, 1889–1906," *Journal of Medical Genetics* 8 (1971): 5; see also Peter Vorzimmer, "Charles Darwin and Blending Inheritance," *Isis* 54 (1963): 371–90. Shaler's acceptance of blending inheritance is suggested by his comment that "by the intermingling of the people it [the state] has tended to average the results of local peculiarities, and so diminish their value in furnishing variations." "Natural History of Politics" (1879), p. 305.

35. F. Galton, "A Theory of Heredity," *Journal of the Anthropological Institute of Great Britain and Ireland* 4 (1875): 331.

36. Recognizing Galton's clear rejection of the inheritance of acquired characteristics, Darwin wrote to him to say that, since his work implied "that many parts are not modified by use and disuse during the life of the individual, I differ widely from you, as every year I come to attribute more and more to such an agency." Francis Darwin and A. C. Seward, eds., *More Letters of Charles Darwin: A Record of His Life and Work in a Series of Unpublished Letters,* 2 vols. (New York: Appleton, 1903), vol. 1, p. 360.

37. See F. B. Churchill, "The Weismann-Spencer Controversy over the Inheritance of Acquired Characteristics," *Proceedings of the International Congress of the History of Science* 15 (1978): 451–68.

38. Shaler, "European Peasants" (1893); idem, *Man and Earth* (1905), p. 155.

39. Shaler, *Nature and Man* (1891), pp. 263–83. See also discussion in chapter 4.

40. This idea, present in Galton's original formulations, was clarified and extended by Karl Pearson, leader of the biometricians in England, in his distinction between the law of ancestral heredity as applicable to blending inheritance and the law of reversion where the offspring completely resembles one parent or reverts to a more distant ancestor. See Froggatt and Nevin, " 'Law of Ancestral Heredity' "; Robert De Marrais, "The Double-Edged Effect of Sir Francis Galton: A Search for the Motives in the Biometrician-Mendelian Debate," *Journal of the History of Biology* 7 (1974): 141–74.

41. Shaler, *Interpretation* (1893), p. 171.

42. Shaler, "Critical Points" (1890), p. 12.

43. Ibid., pp. 13, 17.

44. Shaler, *Man and Earth* (1905), p. 156.
45. See Bowler, *Eclipse*, pp. 182–226; Peter J. Bowler, "Hugo De Vries and Thomas Hunt Morgan: The Mutation Theory and the Spirit of Darwinism," *Annals of Science* 35 (1978): 55–73; Garland E. Allen, "Hugo De Vries and the Reception of the Mutation Theory," *Journal of the History of Biology* 2 (1969): 55–87.
46. Shaler, *Interpretation* (1893), p. 294.
47. Shaler, *Individual* (1900), pp. 22–23.
48. See Robert M. Young, *Mind, Brain, and Adaptation in the Nineteenth Century: Cerebral Localization and Its Biological Context from Gall to Ferrier* (Oxford: Clarendon Press, 1970); Eiseley, *Darwin's Century*, pp. 314–24.
49. "The Origin of Intellect," *Edinburgh Review* 170 (1889): 359.
50. See Paul F. Boller, Jr., *American Thought in Transition: The Impact of Evolutionary Naturalism, 1865–1900* (Chicago: University of Chicago Press, 1969), pp. 42–43.
51. See Stocking, "Lamarckianism." The prevalence of Lamarckian thinking in psychology is recorded in Hamilton Cravens and John C. Burnham, "Psychology and Evolutionary Naturalism in American Thought, 1890–1940," *American Quarterly* 23 (1971): 635–57.
52. Herbert Spencer, *Principles of Psychology*, 2d ed., 2 vols. (London: Williams & Norgate, 1870), vol. 1, p. 422.
53. Robert J. Richards, "Lloyd Morgan's Theory of Instinct: From Darwinism to Neo-Darwinism," *Journal of the History of the Behavioral Sciences* 13 (1977): 12–32.
54. See Fred W. Voget, "Man and Culture: An Essay in Changing Anthropological Interpretation," *American Anthropologist* 62 (1960): 943–65. In chapter 4, I review a variety of anthropometric measures.
55. W J McGee, "The Trend of Human Progress," *American Anthropologist*, n.s., 1 (1899): 401–7.
56. George John Romanes, *Mental Evolution in Animals* (London: Kegan Paul, Trench, 1883).
57. Charles Darwin, *The Descent of Man and Selection in Relation to Sex*, 2d ed. (London: Murray, 1874), pp. 128–40.
58. See Eiseley, *Darwin's Century*, p. 297.
59. John W. Powell, "On the Evolution of Language, As Exhibited in the Specialization of the Grammatic Processes, the Differentiation of the Parts of Speech, and the Integration of the Sentence; from a Study of the Indian Languages," *Bureau of Ethnology, Annual Report* 1 (1881): 1–16; idem, "Competition."
60. Shaler, *Neighbor* (1904), p. 254.
61. Ibid., pp. 16, 17.
62. McGee, "Trend of Human Progress," p. 411.
63. Shaler, *Domesticated Animals* (1896), p. 133.
64. Ibid., p. 43. He made similar observations about the elephant. Ibid., p. 135.
65. Shaler, "Problem of Discipline" (1889), p. 25. Shaler applied the metaphor of elasticity or plasticity to the horse (*Domesticated Animals* [1896], p. 78), human beings ("Use of Numbers" [1879], p. 227), and plants ("Effects of Extraordinary Seasons" [1872], p. 2).
66. Shaler, *Domesticated Animals* (1896), pp. 45–48. He made similar judgments about domesticated birds. Ibid., pp. 154–56.
67. Philip P. Wiener, *Evolution and the Founders of Pragmatism* (Cambridge, Mass.: Harvard University Press, 1949), p. 109.

68. Thomas H. Huxley, *Man's Place in Nature and Other Anthropological Essays* (London: Macmillan, 1894), p. 134; Shaler, "Plant and Animal Intelligence" (1903).
69. Alfred Russel Wallace, *Contributions to the Theory of Natural Selection* (London: Macmillan, 1870), p. 316. The essay originally appeared as "The Origin of Human Races and the Antiquity of Man Deduced from the Theory of 'Natural Selection,' " *Journal of the Anthropological Society of London* 2 (1864): clvii–clxxxvii.
70. See Roger Smith, "Alfred Russel Wallace: Philosophy of Nature and Man," *British Journal for the History of Science* 6 (1972): 177–99.
71. See, for example, "Law of Fashion" (1888), p. 395.
72. "Natural History of Politics" (1879), p. 304.
73. Shaler, "Environment and Man" (1896), p. 738. Again, in 1893 he had noted that naturalists now recognized "the curious emancipation from the dominance of the body which man exhibits." *Interpretation* (1893), p. 188.
74. Shaler, *Individual* (1900), pp. 87–88.
75. Shaler, "Use of Numbers" (1879), p. 329. The parallel with Henry Drummond on this point is striking. "Nature is full of new departures," Drummond wrote; "but never since time began was there anything approaching in importance that period when the slumbering animal, Brain, broke into intelligence, and the Creature first felt that it had a Mind. From that dateless moment a higher and swifter progress of the world began. Henceforth, Intelligence triumphed over structural adaptation." *The Ascent of Man* (London: Hodder & Stoughton, 1894), p. 148.
76. Richard W. Burkhardt, Jr., *The Spirit of System: Lamarck and Evolutionary Biology* (Cambridge, Mass.: Harvard University Press, 1977), p. 177. See also Ernst Mayr, "Lamarck Revisited," *Journal of the History of Biology* 5 (1972): 55–94.
77. Bowler, "Cope and the Changing Structure," pp. 249–65. Other Lamarckians, like the psychologists James Mark Baldwin and G. Stanley Hall, shared this same emphasis. See Stocking, "Lamarckianism," p. 244.
78. Powell, "Competition," pp. 304–5.
79. Shaler, "Use and Limits" (1890), p. 163.
80. Shaler, "Natural History of Politics" (1879), p. 304.
81. Shaler, *Individual* (1900), pp. 234–35; Shaler, "Last Gift" (1889), p. 675.
82. Darwin, *Descent of Man*, p. 148.
83. Shaler, *Neighbor* (1904), pp. 21, 22.
84. Shaler, *Interpretation* (1893), p. 251. His full discussion of the subject is contained on pp. 232–77.
85. Thomas H. Huxley, *Evolution and Ethics and Other Essays* (London: Macmillan, 1898), p. 83. Yet Shaler did observe that, "in the case of the maxim 'the survival of the fittest', the half-conceived idea leads many persons to the pernicious corollary that whatever is, is right. . . . The naturalist alone . . . can never hope to show men the moral significance of their personality." Shaler, *Interpretation* (1893), pp. 230–31.
86. See Peter Kropotkin, *Mutual Aid* (London: Heinemann, 1902); Powell, "Competition." Henry Drummond, too, maintained, this time in a theistic context, that the animal struggle for existence was gradually replaced by a struggle for the life of others. Again, George Harris, professor of theology at Andover, where Shaler gave his 1891 lectures on the interpretation of nature, set out "to establish the harmony of personal and social morality with the facts of evolution." See Drum-

mond, *Ascent of Man,* pp. 275–341; George Harris, *Moral Evolution* (New York: Houghton Mifflin, 1896), p. iii.

87. Shaler, *Interpretation* (1893), pp. 248, 252.
88. See Milton Berman, *John Fiske: The Evolution of a Popularizer* (Cambridge, Mass.: Harvard University Press, 1961), p. 98; John Fiske, *Outlines of Cosmic Philosophy,* 2 vols. (London: Macmillan, 1874), vol. 2, pp. 344, 360, 369.
89. Shaler, *Individual* (1900), p. 130. Shaler had already called this caring a "spiritual quality." See *Domesticated Animals* (1896), p. 187.
90. Shaler, *Interpretation* (1893), p. 254.
91. Ibid., pp. 260–61. Herbert Spencer expressed a similar view when he described evolution as "an advance by degress from unconscious parental altruism to conscious parental altruism. . . [and] by degress from the altruism of the family to social altruism." *Principles of Ethics,* 2 vols. (London: Williams & Norgate, 1904), vol. 1, p. 204.
92. Shaler, *Domesticated Animals* (1896), p. 206.
93. Shaler, *Neighbor* (1904), p. 24. He had already pointed out that, as far as maternal love was concerned, survival advantage applied not to the creature *possessing* the quality but to the offspring *receiving* its benefit. Moreover, beyond even this expression of self-giving, Shaler found it "difficult to see how selective or hereditary action can have anything to do with advance in the altruistic motive." *Individual* (1900), pp. 264–65.
94. Shaler, *Interpretation* (1893), p. 266. Some tried to circumvent this problem by seeing the tribe rather than the individual as the basic unit of selection. So Spencer, Kropotkin, and Fiske all endorsed such solidarism; but as Michael Ruse has recently made clear, "Darwin opted firmly for hypotheses supposing selection always to work at the level of the individual rather than the group." "Charles Darwin and Group Selection," *Annals of Science* 37 (1980): 615.
95. Shaler, *Interpretation* (1893), p. 86.
96. See Burrow, *Evolution and Society;* Idus Murphree, "The Evolutionary Anthropologists. The Progress of Mankind, The Concepts of Progress and Culture in the Thought of John Lubbock, Edward B. Tylor, and Lewis H. Morgan," *Proceedings of the American Philosophical Society* 105 (1961): 265–300; Marvin Harris, *The Rise of Anthropological Theory: A History of Theories of Culture* (New York: Crowell, 1968); Fred W. Voget, *A History of Ethnology* (New York: Holt, Rinehart & Winston, 1975).
97. Richard Hofstadter, *Social Darwinism in American Thought, 1860–1915* (Boston: Beacon Press, 1955; first published 1944).
98. See Boller, *American Thought,* pp. 53–62. At the same time, Carnegie believed that the future of the human race depended on the practice of philanthropy.
99. See Robert C. Bannister, *Social Darwinism: Science and Myth in Anglo American Social Thought* (Philadelphia: Temple University Press, 1979); Jones, *Social Darwinism and English Thought;* Irvine Wyllie, "Social Darwinism and the Businessman," *Proceedings of the American Philosophical Society* 103 (1959): 629–35; Robert C. Bannister, " 'The Survival of the Fittest Is Our Doctrine': History or Histrionics?" *Journal of the History of Ideas* 31 (1970): 377–98.
100. Lester F. Ward, *Psychic Factors of Civilization* (Boston: Houghton Mifflin, 1893), p. 262. According to Henry Steele Commager, Ward viewed civilization as "the triumph of man over the blind forces of nature and the deliberate application of

human genius to the task of emancipating man from the tyranny of those forces." *The American Mind* (New Haven: Yale University Press, 1950), p. 207.

101. Shaler, "Environment and Man" (1896), p.726. Similarly, in "Natural History of Politics" (1879), p. 305, he noted: "It may be said that nothing can be more dangerous than the effort to explain a civilization by a reference to a single force."

102. Shaler, "Natural History of Politics" (1879), p. 304.

103. Shaler, *Citizen* (1904), p. 16.

104. Ibid.

105. Voget, *Ethnology,* p. 65.

106. Shaler, *Citizen* (1904), p. 17; idem, *Neighbor* (1904), p. 28.

107. Frank J. Sulloway, "Geographic Isolation in Darwin's Thinking: The Vicissitudes of a Crucial Idea," *Studies in the History of Biology* 3 (1979): 23–65. See also Ernst Mayr, "Isolation as an Evolutionary Factor," *Proceedings of the American Philosophical Society* 103 (1959): 221–30.

108. Shaler, "Nature and Man" (1890), p. 365.

109. Summarizing the views of Reclus, Ratzel, Semple, and Giddings in 1925, Franklin Thomas observed, "It is now generally admitted that the most important element in cultural progress is the contact of many cultures, while nothing breeds stagnation like isolation. Hence an environment which invites contact and provides easy access to and from other districts will promote psychic plasticity and cultural advancement, while one which produces isolation must of necessity lead to psychic stagnation and repetition." *The Environmental Basis of Society* (New York: Century, 1925), p. 7.

110. Shaler, "European Peasants" (1893), p. 649.

111. Shaler, *Citizen* (1904), p. 19.

112. See Walter Bagehot, *Physics and Politics; or, Thoughts on the Application of the Principles of "Natural Selection" and "Inheritance" to Political Society,* 11th ed. (London: Kegan Paul, Trench, Trubner, 1896), pp. 41–80.

113. Shaler, "Natural History of War" (1903), pp. 23–24.

114. Shaler, "Use of Numbers" (1879), p. 332.

115. See the discussion of this point in Hofstadter, *Social Darwinism,* p. 90.

116. See Voget, *Ethnology,* p. 82.

117. Wiener, *Evolution and Pragmatism,* p. 153.

118. See Henry Sumner Maine, *Ancient Law: Its Connection with the Early History of Society and Its Relation to Modern Ideas* (London: Murray, 1861).

119. Shaler, *Interpretation* (1893), p. 214.

120. Shaler, *Citizen* (1904), p. 123.

121. See the discussion in Boller, *American Thought,* pp. 148–84.

122. Shaler, *Citizen* (1904), pp. 125, 128.

123. Ibid., pp. 145–46, 122.

124. See Wilford A. Bladen and Pradyumna P. Karan, eds., *The Evolution of Geographic Thought in America: A Kentucky Root* (Dubuque, Iowa: Kendall/Hunt Publishing, 1983).

Chapter 4

1. See Manfred Büttner, "Kant and the Physico-Theological Consideration of the Geographical Facts," *Organon* 11 (1975): 231–49; idem, "Zum Gegenüber von

Naturwissenschaft (insbesondere Geographie) und Theologie im 18 Jahrhundert," *Philosophia Naturalis* 14 (1973): 95–123.

2. John William Draper, *History of the Conflict between Religion and Science* (London: King, 1875; first published 1874), p. 353.

3. See R. Hooykaas, *Philosophia Libera: Christian Faith and the Freedom of Science* (London: Tyndale Press, 1957). It is significant that Calvin did not treat the question of natural law in the section of his *Institutes of the Christian Religion* entitled "Predestination," but rather under that of "Providence."

4. See Merton, "Science, Technology, and Society." Also M. B. Foster, "The Christian Doctrine of Creation and the Rise of Modern Natural Science," *Mind* 43 (1934): 446–68; R. Hooykaas, "Science and Reformation," *Journal of World History* 3 (1956): 109–39.

5. For various perspectives see Ian G. Barbour, *Issues in Science and Religion* (New York: Harper & Row, 1971), pp. 44–55; John Dillenberger, *Protestant Thought and Natural Science: A Historical Interpretation* (London: Collins, 1961), pp. 50–74; Thomas F. Torrance, *Theological Science* (London: Oxford University Press, 1969), pp. 55–105; Stanley L. Jaki, "Science and Christian Theism: A Mutual Witness," *Scottish Journal of Theology* 32 (1979): 563–70; idem, *The Road of Science and the Ways to God* (Edinburgh: Scottish Academic Press, 1978); Alfred North Whitehead, *Science and the Modern World* (Cambridge: Cambridge University Press, 1926), pp. 13–14.

6. R. Hooykaas, *Religion and the Rise of Modern Science* (Edinburgh: Scottish Academic Press, 1972), pp. 98–160; Christopher Hill, *Intellectual Origins of the English Revolution* (Oxford: Oxford University Press, 1965); see also the literature review by Douglas S. Kemsley, "Religious Influences in the Rise of Modern Science: A Review and Criticism, Particularly of the 'Protestant-Puritan Ethic' Theory," *Annals of Science* 24 (1968): 199–226; Charles Webster, ed., *The Intellectual Revolution of the Seventeenth Century* (London: Routledge & Kegan Paul, 1974); Richard S. Westfall, *Science and Religion in Seventeenth-Century England* (New Haven: Yale University Press, 1958).

7. E. G. R. Taylor, *The Mathematical Practitioners of Tudor and Stuart England* (Cambridge: Cambridge University Press, 1954); Michael Hunter, *Science and Society in Restoration England* (Cambridge: Cambridge University Press, 1981). The latter's bibliographical essay represents a mine of source information.

8. Hunter, *Science and Society*, pp. 113–14.

9. Nathanael Carpenter, *Geography Delineated Forth in Two Bookes: Containing the Sphaericall and Topicall Parts Thereof,* Theatrum Orbis Terarum; the English Experience, no. 787 (Norwood, N.J.: Johnson, 1976; first published 1625). For further discussion see Hooykaas, *Philosophia Libera.* Carpenter's geography is described in J. N. L. Baker, "Nathanael Carpenter and English Geography in the Seventeenth Century," *Geographical Journal* 71 (1928): 261–71; E. G. R. Taylor, *Late Tudor and Early Stuart Geography, 1583–1650* (London: Methuen, 1934), pp. 136–37.

10. Manfred Büttner, "Bartholomäus Keckermann, 1572–1609," in *Geographers: Bio-bibliographical Studies,* ed. T. W. Freeman and Philippe Pinchemel (London: Mansell, 1977), vol. 2, pp. 73–79. Also idem, "Die Neuausrichtung der Geographie im 17 Jahrhundert durch Bartholomäus Keckermann, ein Betrag zur Geographie in ihren Beziehungen zu Theologie und Philosophie," *Geographische Zeitschrift* 63

(1975): 1–12. For Keckermann, "If God's intervention in nature is not or is no longer theologically interesting, geography can, at least as far as the physiogeographical branch is concerned, emancipate itself from theology. It no longer needs, as a servant of theology, to confirm the *Providentia,* but can construct its system on the basis of autonomous criteria, independent of theology and can set itself the goals of its own research." Quoted in Büttner, "Kant," p. 237.

11. Büttner, "Kant"; also idem, "Kant und die Uberwindung der Physiko-Theologischen Betrachtung der Geographisch-Kosmologischen Fakten," *Erdkunde* 29 (1975): 53–60.

12. See Glacken, *Traces,* pp. 415–26. Yi-Fu Tuan has suggested that the concept of the hydrological cycle in its standard form was the chief handmaiden of natural theology in this period. See *The Hydrologic Cycle and the Wisdom of God: A Theme in Geoteleology,* Department of Geography Research Paper no. 1 (Toronto: University of Toronto, 1968).

13. Ronald L. Numbers, *Creation by Natural Law: Laplace's Nebular Hypothesis in American Thought* (Seattle: University of Washington Press, 1977); John Hedley Brooke, "Nebular Contraction and the Expansion of Naturalism," *British Journal for the History of Science* 12 (1979): 200–211.

14. Dov Ospovat, "Perfect Adaptation and Teleological Explanation: Approaches to the Problem of the History of Life in the Mid-Nineteenth Century," *Studies in the History of Biology* 14 (1981): 193–230; idem, "God and Natural Selection: The Darwinian Idea of Design," *Journal of the History of Biology* 13 (1980): 169–94; idem, *The Development of Darwin's Theory: Natural History, Natural Theology, and Natural Selection, 1838–1859* (Cambridge: Cambridge University Press, 1981).

15. See John Hedley Brooke, "The Natural Theology of the Geologists: Some Theological Strata," in *Images of the Earth: Essays in the History of Environmental Sciences,* ed. L. J. Jordanova and Roy S. Porter (Chalfont St. Giles: British Society for the History of Science, 1979), pp. 39–64.

16. See, for example, W. F. Cannon, "The Problem of Miracles in the 1830's," *Victorian Studies* 4 (1960): 5–32; idem, "The Bases of Darwin's Achievement: A Revaluation," *Victorian Studies* 5 (1961): 109–34; Robert M. Young, "Darwin's Metaphor: Does Nature Select?" *Monist* 55 (1971): 442–503; Richard Yeo, "William Whewell, Natural Theology"; Young, "Natural Theology, Victorian Periodicals."

17. David N. Livingstone, "Natural Theology and Neo-Lamarckism: The Changing Context of Nineteenth-Century Geography in the United States and Great Britain," *Annals of the Association of American Geographers* 74 (1984): 9–28; idem, "History of Science and History of Geography."

18. *Encyclopaedia Britannica,* 14th ed., s.v. "Geography."

19. See Clarence J. Glacken, "Environment and Culture," in *Dictionary of the History of Ideas,* ed. Wiener, vol. 2, pp. 127–34.

20. As Lurie writes, "In an age alive and responsive to the idealism of Emerson, Agassiz gave a scientific demonstration of the spiritual quality underlying all material creation. Men might know all there was to know about the facts of nature, but if they did not appreciate the magnificence of the master plan fashioned by its Author . . . they could know the world only partially. . . . The religion of nature . . . demonstrated that the entire history of creation, beginning with the smallest radiated animals and ending with man, had been wisely ordained." *Agassiz,* p. 127.

21. Lucas, "Wilberforce and Huxley"; Gilley and Loades, "Huxley"; Moore, "1859 and All That."

22. Gertrude Himmelfarb, *Darwin and the Darwinian Revolution* (London: Chatto & Windus, 1959), pp. 314–39.

23. See James R. Moore, "Engines of Empire, Energies of Extinction: Reflections on the Crisis of Faith" (Paper presented at conference, the Victorian Crisis of Faith, Victoria College, University of Toronto, November 1984); see also idem, "On Revolutionising the Darwin Industry: A Centennial Retrospect," *Radical Philosophy* 37 (1984): 13–22.

24. Turner, "Victorian Conflict"; Young, "Natural Theology, Victorian Periodicals"; Frank Miller Turner, *Between Science and Religion: The Reaction to Scientific Naturalism in Late Victorian England* (New Haven: Yale University Press, 1974); T. W. Heyck, *The Transformation of Intellectual Life in Victorian England* (London: Croom Helm, 1982); Frank Miller Turner, "Rainfall, Plagues, and the Prince of Wales: A Chapter in the Conflict of Religion and Science," *Journal of British Studies* 13 (1974): 46–65.

25. Patrick Geddes, "Biology," in *Chambers's Encyclopaedia* (London: Chambers, 1925; first published 1882), vol. 2, pp. 157–64; Gilley and Loades, "Huxley," p. 290; D. W. Forrest, *Francis Galton: The Life and Work of a Victorian Genius* (New York: Elek, 1974); Francis Galton, *English Men of Science: Their Nature and Nurture* (London: Macmillan, 1874), p. 193.

26. See Alvar Ellegård, *Darwin and the General Reader: The Reception of Darwin's Theory of Evolution in the British Periodical Press, 1859–1872* (Göteborg: Göteborg Universitets Arsskrift, 1958), pp. 114–40.

27. See, for example, Charles D. Cashdollar, "The Social Implications of the Doctrine of Divine Providence: A Nineteenth-Century Debate in American Theology," *Harvard Theological Review* 71 (1978): 265–84, for various reconstructions of the providential tradition.

28. Charles Caverno, "Louis Agassiz and Charles Darwin: A Synthesis," *Bibliotheca Sacra* 73 (1916): 138.

29. I have in mind such figures as James McCosh and Alexander Winchell. See David N. Livingstone, "The Idea of Design: The Vicissitudes of a Key Concept in the Princeton Response to Darwin," *Scottish Journal of Theology* 37 (1984): 329–57; idem, *Darwin's Forgotten Defenders.*

30. Peter J. Bowler, "Darwinism and the Argument from Design: Suggestions for a Reevaluation," *Journal of the History of Biology* 10 (1977): 37; see also idem, *Fossils and Progress: Palaeontology and the Idea of Progressive Evolution in the Nineteenth Century* (New York: Science History Publications, 1976), p. 117; Thomas McPherson, *The Argument from Design* (London: Macmillan, 1972).

31. So, for example, Duke of Argyll [George John Douglas Campbell], *The Reign of Law* (London: Strahan, 1867).

32. John Passmore, "Darwin's Impact on British Metaphysics," *Victorian Studies* 3 (1959): 41–54; Numbers, *Creation*, pp. 105–18.

33. Shaler, "Chance or Design" (1889), p. 118.

34. Ibid., p. 119.

35. Ibid., p. 120. As he had already observed, "I have undertaken this study with every effort to clear my mind of prepossessions and to do the work with no prejudices whatever. I cannot be sure as to the measure of success at which I have

arrived in this endeavor, for the evident reason that no one can ascertain the extent to which he has cleared himself from the intellectual past on which all his thought immediately depends." Ibid., p. 119.

36. Peter A. Bertocci, "Creation in Religion," *Dictionary of the History of Ideas,* ed. Wiener, vol. 1, p. 576.

37. Henri Bergson, *Creative Evolution,* trans. Arthur Mitchell (London: Macmillan, 1911), pp. 140, 194.

38. Shaler, "Chance or Design" (1889), pp. 133, 122, 129.

39. Ibid., pp. 132–33.

40. Henry Ward Beecher, *Evolution and Religion* (New York: Fords, Howard & Hulbert, 1885), p. 115.

41. See Henry Drummond, *Natural Law in the Spiritual World* (London: Hodder & Stoughton, 1883); idem, *Ascent of Man.*

42. John Bascom, *Evolution and Religion; or, Faith as a Part of a Complete Cosmic System* (New York: Putnam's Sons, 1897); also idem, *Natural Theology* (New York: Putnam's Sons, 1880).

43. Joseph Le Conte, *Evolution: Its Nature, Its Evidences, and Its Relation to Religious Thought* (New York: Appleton, 1888). See also Lester D. Stephens, *Joseph Le Conte: Gentle Prophet of Evolution* (Baton Rouge: Louisiana State University Press, 1982), pp. 158–77.

44. See Irwin Reist, "Augustus Hopkins Strong and William Newton Clarke: A Study in Nineteenth Century Evolutionary and Eschatological Thought," *Foundations* 13 (1970): 26–43; Livingstone, *Darwin's Forgotten Defenders.*

45. See for example Minot Judson Savage, *The Religion of Evolution* (Boston: Lockwood, Brooks, 1876); idem, *Evolution and Religion from the Standpoint of One Who Believes Both: A Lecture Delivered in the Philadelphia Academy of Music, Seventh December, 1885* (Philadelphia: Buchanan, 1886).

46. See John Fiske, *The Idea of God As Affected by Modern Knowledge* (London: Macmillan, 1901).

47. See the useful summary in Moore, *Post-Darwinian Controversies,* pp. 220–24; Don Cupitt, "Darwinism and English Religious Thought," *Theology* 78 (1975): 125–31.

48. R. Hooykaas, "Catastrophism in Geology, Its Scientific Character in Relation to Actualism and Uniformitarianism," in *Philosophy of Geohistory,* ed. Claude C. Albritton (Stroudsburg, Pa.: Dowden, Hutchinson & Ross, 1975), p. 333. See also George Gaylord Simpson, "Uniformitarianism: An Inquiry into Principle, Theory, and Method in Geohistory and Biohistory," in ibid., pp. 256–309.

49. Shaler, "Critical Points" (1890), pp. 2, 4.

50. I discuss Shaler's gradualism in detail in chapter 8.

51. Shaler, "Critical Points" (1890), pp. 7, 11. When A. R. Wallace commented that volcanoes, "usually held to be blots on the fair face of nature, and even to be opposed to belief in a beneficent Creator, should now be proved to be really essential to the earth's habitability," similar conceptual moves were in operation. See his book *Man's Place in the Universe: A Study of the Results of Scientific Research in Relation to the Unity or Plurality of Worlds,* 4th ed. (London: Chapman & Hall, 1904), p. 185.

52. See Gillespie, *Charles Darwin,* p. 89; Stanley M. Guralnick, "Geology and Religion

before Darwin: The Case of Edward Hitchcock, Theologian and Geologist (1793–1864)," *Isis* 63 (1972): 529–43; Philip J. Lawrence, "Edward Hitchcock: The Christian Geologist," *Proceedings of the American Philosophical Society* 116 (1972): 21–34.

53. R. Hooykaas, *The Principle of Uniformity in Geology, Biology, and Theology* (Leiden: Brill, 1963), p. 92.

54. Shaler, *Nature and Man* (1891), pp. vi, vii. In a similar vein, George Perkins Marsh, with whose work Shaler was familiar, urged the votaries of science, "by demonstrating the God in and over nature, [to] give a loftier direction to the study of His works, and the investigation of the laws which He has imposed on the material creation." "The Study of Nature," *Christian Examiner* 68 (1860): 62. I discuss their respective attitudes to nature in my paper "Nature and Man in America: Nathaniel Southgate Shaler and the Conservation of Natural Resources," *Transactions of the Institute of British Geographers*, n.s., 5 (1980): 369–82.

55. *Catalogue of the Officers and Students of the Theological Seminary, Andover, Mass., 1891–92* (Andover: Andover Press, 1891), p.25. While indicating that Shaler was recommended for this lectureship, neither the trustee nor the faculty records mention the reasons for their choice. Letter to the author from Diana Yount, Special Collections Librarian, Andover Newton Theological Seminary, 5 September 1979. Previous Winkley Foundation lecturers included Professors T. W. Dwight and J. W. Burgess of Columbia College, President E. B. Andrews of Brown University, and Professor J. G. Schurman of Cornell University.

56. See Bruce L. Shelley, "Andover Controversy," in *The New International Dictionary of the Christian Church,* ed. J. D. Douglas (Exeter: Paternoster Press, 1974), p. 40. The final victory for liberal theology was complete when, in 1922, Andover Seminary—originally established as a direct assault on Unitarianism—merged with the Harvard Divinity School, the citadel of Unitarian orthodoxy.

57. Daniel Day Williams, *The Andover Liberals: A Study in American Theology* (New York: King's Crown Press, 1941), p. 86.

58. Shaler, *Interpretation* (1893), p. 16.

59. Ibid., pp. 46–47.

60. Ibid., p. 48.

61. Ibid., p.137. See also Shaler, *Outlines of Earth's History* (1898), pp. 1–8, 16–22. Similar viewpoints are discussed in Edward A. White, *Science and Religion in American Thought* (Stanford: Stanford University Press, 1952), pp. 33–47.

62. Shaler, *Interpretation* (1893), p.140; Rudolph Otto, *The Idea of the Holy: An Inquiry into the Idea of the Divine and Its Relation to the Rational* (Oxford: Oxford University Press, 1925). See also Linda H. Graber, *Wilderness as Sacred Space* (Washington, D.C.: Association of American Geographers, 1976).

63. Perry Miller, *Errand into the Wilderness* (London: Harper Books, 1956), p. 198. See also Schneider, "Religious Enlightenment."

64. See J. E. Carpenter, "Unitarianism," in *Encyclopaedia of Religion and Ethics,* ed. James Hastings (Edinburgh: Clark, 1921), vol. 12, pp. 525–57.

65. John B. Wilson, "Darwin and the Transcendentalists," *Journal of the History of Ideas* 26 (1965): 286–90; W. Owen Chadwick, "Religion and Science in the Nineteenth Century," in *Dictionary of the History of Ideas,* ed. Wiener, vol. 4, pp. 106–8. A useful synoptic review of philosophical trends is Kenneth R. Merrill,

"From Edwards to Quine: Two Hundred Years of American Philosophy," in *Issues and Ideas in America,* ed. Benjamin J. Taylor and Thurman J. White (Norman: University of Oklahoma Press, 1976), pp. 219–63.

66. Stow Persons, *Free Religion: An American Faith* (Boston: Beacon Press, 1963; first published 1946), p. 154.

67. Shaler, *Domesticated Animals* (1895), p. 263. This passage is strikingly parallel with the central thrust of Lynn White, Jr., "The Historical Roots of Our Ecologic Crisis," *Science* 155 (1967): 1203–7.

68. Shaler, *Interpretation* (1893), pp. 267–68.

69. Shaler, *Citizen* (1904), pp. 219–20.

70. See William James, *The Varieties of Religious Experience* (New York: Random House, 1929); also John Macquarrie, *Twentieth Century Religious Thought: The Frontiers of Philosophy and Theology, 1900–1970* (London: SCM Press, 1963), pp. 178–79.

71. Shaler, *Interpretation* (1893), p. 271.

72. Ibid., pp. 271–72. According to a ledger in the Shaler Papers compiled by J. B. Woodworth of papers written by Shaler during 1888–90, Shaler prepared a manuscript entitled "The Problem of Drunkenness." Woodworth, however, provided no publication details, and I have not been able to trace it. On the theme of alienation, Shaler commented in *Citizen* (1904), p. 294, that "this need of friendship, this struggle against the supreme evil of isolation, is most evident in man."

73. Shaler, *Interpretation* (1893), p. 273.

74. Shaler, "Last Gift," (1889), p. 676.

75. Shaler, *Citizen* (1904), p. 6.

76. Shaler, *Interpretation* (1893), pp. 274, 275.

77. Ibid., pp. 278–305.

78. Shaler to Horace Scudder, 6 August 1892, Horace Scudder Correspondence, bMS 801.4, Houghton Library, Harvard University.

79. Shaler to William James, 12 October 1900, James Papers, bMS Am 1092 (1009).

80. See Kuklick, *Rise of American Philosophy,* p. xxi.

81. Shaler, *Individual* (1900), p. xi.

82. Shaler's view was quoted by John Fiske, *Life Everlasting* (Boston: Houghton Mifflin, 1901), p. 30; and by Frank Ballard, *Haeckel's Monism False* (London: Kelly, 1905), pp. 366–67.

83. Shaler, *Individual* (1900), pp. 27, 242–43.

84. Ibid., pp. 207–8.

85. Shaler likewise disapproved of the similar investigations of A. R. Wallace. See ibid., p. 319.

86. See discussion in Turner, *Between Science and Religion,* pp. 104–33.

87. See R. E. D. Clark, "Myers, Frederic William Henry (1843–1901)," in *Dictionary of Christian Church,* ed. Douglas, p. 690.

88. Turner makes the point that current scientific theory seemed to encourage his speculations about invisible forces. The concept of cosmic ether widely accepted by physicists could be seen as compatible with Myers's "metetherial environment." Turner, *Between Science and Religion,* p. 120.

89. Shaler, "Human Personality" (1903), p. 1251.

90. Ibid., p. 1252. Shaler concluded his review with some personal reflections on his

own experience with a local medium, which underscored his deep-seated distaste for such research: "Moreover, it may well be doubted whether the interest in a future life will be enhanced by the kind of conviction which spiritism affords. In face of certain experiences with a well-known medium the present writer was forced to the conviction that he was in very unprofitable communication with a near friend among the dead. This experience brought no sense of pleasure or of elevation, but rather an immediate and abiding sense of degradation.... Thus against the contention that men need confidence in the hereafter, it may well be held that the majestic doubt that hangs over it is more inspiring than the cheap certainties that spiritism offers in its place." Ibid., pp. 1252–53.

91. Shaler, "Huxley's Life" (1901), p. 254. Shaler noted, however, that Huxley saw the value of retaining Bible reading in public schools, for he acknowledged it as a basis of morality and a source of literary culture. See Thomas Henry Huxley, *Science and Education* (London: Macmillan, 1893), p. 397.

92. Shaler, *Autobiography* (1909), p. 436.

93. Shaler to Charles Eliot, 16 January 1901, Eliot Papers, UA I.1.150, Box 120, Folder 321.

94. Shaler to William James, 12 October 1900, James Papers, bMS Am 1092 (1009). See also *Individual* (1900), pp. vii, viii.

95. Shaler to Eliot, 28 August, 1896, Eliot Papers, UA I.1.150, Box 120, Folder 321.

96. Kuklick, *Rise of American Philosophy,* p. 283.

97. Merrill, "From Edwards to Quine," p. 246. For Royce, the ultimate "community" was the Absolute Spirit.

98. James's letter is reproduced in Shaler, *Autobiography* (1909), pp. 436–38.

99. Moore, "Engines of Empire," pp. 25–26.

100. William Morris Davis, "The Faith We Need," Manuscript, Davis Papers, bMS Am 1798 (637), Houghton Library, Harvard University.

101. William Morris Davis, "The Greater Influence of the Brotherhood," Davis Papers, bMS Am 1798 (639).

102. William Morris Davis, "The Faith of Reverent Science," in Chorley, Beckinsale, and Dunn, *Davis,* pp. 785, 791.

103. D. H. Meyer, "American Intellectuals and the Victorian Crisis of Faith," *American Quarterly* 27 (1975): 585–603.

104. Shaler, *Interpretation* (1893), p. 136.

105. This very topic was made the subject of a symposium in the first issue of the *Nineteenth Century* in 1877.

Chapter 5

1. On the various intellectual and social contexts of racial ideology and immigration restriction in America, see Barbara Solomon, "The Intellectual Background of the Immigration Restriction Movement in New England," *New England Quarterly* 25 (1952): 47–59; idem, *Ancestors and Immigrants: A Changing New England Tradition* (Chicago: University of Chicago Press, 1965); Oscar Handlin, ed., *Immigration as a Factor in American History* (Englewood Cliffs, N.J.: Prentice-Hall, 1959); Thomas F. Gossett, *Race: The History of an Idea in America* (Dallas: Southern Methodist University Press, 1963); Milton M. Gordon, "Assimilation in America:

Theory and Reality," *Daedalus* 90 (1961): 263–85; idem, *Assimilation in American Life: The Role of Race, Religion, and National Origins* (New York: Oxford University Press, 1964); John S. Haller, Jr., *Outcasts from Evolution: Scientific Attitudes of Racial Inferiority, 1859–1900* (Chicago: University of Illinois Press, 1971); John Higham, *Strangers in the Land: Patterns of American Nativism, 1865–1925* (New York: Atheneum, 1973).

2. John S. Haller, Jr., "Nathaniel Southgate Shaler: A Portrait of Nineteenth-Century Academic Thinking on Race," *Essex Institute Historical Collections* 107 (1971): 173.

3. Harris, *Rise of Anthropological Theory*, pp. 83–87.

4. See A. I. Hallowell, "The Beginnings of Anthropology in America," in *Selected Papers from the American Anthropologist, 1888–1920*, ed. F. de Laguna (Evanston, Ill.: American Anthropologist, 1960), pp. 66–67. A few tried to use polygenism in the cause of slavery, but it must not be assumed that monogenists and polygenists were opposed, a priori, on the question of racial inferiority. Monogenists were often far from egalitarian and, relying on environmentalist explanations, saw no reason to view the black as other than inferior. Indeed, William Stanton has shown that, for all its theoretical attractions, polygenism never caught on as the ideology of slavocracy. William Stanton, *The Leopard's Spots: Scientific Attitudes toward Races in America, 1815–59* (Chicago: University of Chicago Press, 1960).

5. Lurie, *Agassiz*, pp. 256–65; Stephen Jay Gould, "Flaws in a Victorian Veil," *New Scientist*, 31 August 1978, pp. 632–33; Edward Lurie, "Louis Agassiz and the Races of Man," *Isis* 45 (1954): 227–42.

6. Haller, *Outcasts*, pp. 69–85.

7. Charles Darwin, *The Descent of Man and Selection in Relation to Sex* (London: Murray, 1901 ed.), p. 260. See also George W. Stocking, Jr., *Race, Culture, and Evolution: Essays in the History of Anthropology* (London: Free Press, 1968), p. 46.

8. Quoted in J. Hunt, "On the Application of the Principle of Natural Selection in Anthropology," *Anthropological Review* 4 (1866): 320.

9. Alfred R. Wallace, "The Origin of Human Races and the Antiquity of Man Deduced from the Theory of 'Natural Selection,'" *Journal of the Anthropological Society* 2 (1864): clviii–clxx. On the politicoracial ideology of James Hunt, founder and president of the Anthropological Society of London, see Ronald Rainger, "Race, Politics, and Science: The Anthropological Society of London in the 1860's," *Victorian Studies* 22 (1978): 51–70.

10. In *Neighbor* (1904), p. 252, Shaler defined a species as "an aggregate of kindred creatures in which the sympathies bind the individuals together so as to form a common mind."

11. Persons's suggestion that Shaler, in holding this position, departed from contemporary anthropological orthodoxy has been undermined by Stocking's demonstration of the persistence of polygenism in post-Darwinian anthropology. See Persons, *American Minds*, p. 290; Stocking, *Race, Culture, and Evolution*, pp. 42–68.

12. Shaler, "Nature of Negro" (1890), p. 24.

13. Paul Broca, *On the Phenomena of Hybridity in the Genus Homo*, ed. C. Carter Blake (London: Anthropological Society of London, 1864); see also Francis Schiller, *Paul Broca: Founder of French Anthropology, Explorer of the Brain* (Berkeley: University of California Press, 1979).

14. Shaler, *Neighbor* (1904), p. 162.

15. Ibid., p. 160.

16. Shaler, "Science and African Problem" (1890), p. 37.

17. Stocking, "Lamarckianism." Lamarckian assumptions underlay the sociological writings of such figures as Herbert Spencer, Lewis Henry Morgan, John Wesley Powell, Lester Frank Ward, and W J McGee.

18. See Robert L. Carneiro, "Herbert Spencer's *The Study of Sociology* and the Rise of Social Science in America," *Proceedings of the American Philosophical Society* 118 (1974): 540–54; Murphree, "Evolutionary Anthropologists"; Irving Goldman, "Evolution and Anthropology," *Victorian Studies* 3 (1959): 55–75; Derek Freeman, "The Evolutionary Theories of Charles Darwin and Herbert Spencer," *Current Anthropology* 15 (1974): 211–21; Leslie A. White, "The Concept of Evolution in Cultural Anthropology," in *Evolution and Anthropology: A Centennial Appraisal,* ed. Betty J. Meggers (Washington, D.C.: Anthropological Society of Washington, 1959), pp. 106–25.

19. See Churchill, "Weismann-Spencer Controversy."

20. See Haller, *Outcasts,* pp. vii–xi; Stocking, *Race, Culture, and Evolution,* pp. 42–68.

21. Shaler, *Neighbor* (1904), p. 28. The savage, he added, "must slay when he can and propitiate when he cannot slay, and all with hatred in his heart."

22. A short account of Morgan's work is to be found in Leslie A. White, "Lewis Henry Morgan: Pioneer in the Theory of Social Evolution," in *An Introduction to the History of Sociology,* ed. Harry Elmer Barnes (Chicago: University of Chicago Press, 1948); Merwyn S. Garbarino, *Sociocultural Theory in Anthropology: A Short History* (New York: Holt, Rinehart & Winston, 1977), pp. 27–29.

23. Shaler, *Neighbor* (1904), p. 28.

24. Shaler, "Science and African Problem" (1890), p. 40; see also Persons, *American Minds,* pp. 289–95.

25. See Rainger, "Race, Politics, and Science"; John W. Burrow, "Evolution and Anthropology in the 1860's: The Anthropological Society of London, 1863–1871," *Victorian Studies* 7 (1963): 137–54; George W. Stocking, "What's in a Name? The Origins of the Royal Anthropological Institute, 1837–1871," *Man: The Journal of the Royal Anthropological Institute* 6 (1971): 369–90.

26. Voget, *Ethnology,* pp. 41–112, 165–319; Harris, *Rise of Anthropological Theory,* pp. 108–216.

27. T. D. Stewart, "The Effect of Darwin's Theory of Evolution on Physical Anthropology," in *Evolution and Anthropology,* ed. Meggers, pp. 21–22.

28. Hallowell, "Beginnings of Anthropology," p. 58; see also L. E. Hoyme, "Physical Anthropology and Its Instruments," *Southwestern Journal of Anthropology* 9 (1953): 408–30.

29. See John S. Haller, Jr., "Concepts of Race Inferiority in Nineteenth-Century Anthropology," *Journal of the History of Medicine and Allied Sciences* 25 (1970): 40–51; idem, "The Physician versus the Negro: Medical and Anthropological Concepts of Race in the Late Nineteenth Century," *Bulletin of the History of Medicine* 44 (1970): 154–67.

30. Various anthropometric measurements are summarized in Gossett, *Race,* pp. 69–83. See also Stephen Jay Gould, *The Mismeasure of Man* (New York: Norton, 1981). Among other instruments was Peter A. Browne's "trichometer," designed for the analysis of human hair.

31. Alfred C. Haddon, *History of Anthropology* (London: Watts, 1910), pp. 88–98.

32. Stanford B. Hunt, "The Negro as a Soldier," *Anthropological Review* 7 (1869): 40–54.

33. Shaler, "Science and African Problem" (1890), p. 41.

34. Shaler, *General Account of Kentucky* (1876), pp. 21–24; Shaler, "Scotch Element" (1896).

35. Shaler, *Nature and Man* (1891), pp. 271–83.

36. See discussions and reviews in Hofstadter, *Social Darwinism,* pp. 170–200; idem, *The Progressive Historians* (New York: Vintage Books,1970), pp. 30–43; Boller, *American Thought,* pp. 199–223; Jurgen Herbst, *The German Historical School in American Scholarship: A Study in the Transfer of Culture* (Ithaca: Cornell University Press, 1965).

37. See Ray Allen Billington, *Genesis of the Frontier Thesis: A Study in Historical Creativity* (San Marino, Calif.: Huntington Library, 1971), pp. 3–8, 28.

38. Hofstadter, *Social Darwinism,* p. 172.

39. Persons, *American Minds,* p. 283.

40. See John Higham, *Send These to Me: Jews and Other Immigrants in Urban America* (New York: Atheneum, 1975), p. 46.

41. See discussions in Louis L. Snyder, *The Idea of Racialism: Its Meaning and History* (New York: Van Nostrand Reinhold, 1962), pp. 86–87; Milton Berman, *John Fiske: The Evolution of a Popularizer* (Cambridge, Mass.: Harvard University Press, 1961), pp. 199–219.

42. Higham, *Send These to Me,* p. 46.

43. Shaler, *Autobiography* (1909), p. 378.

44. Shaler, *Nature and Man* (1891), p. 151.

45. Shaler, *United States* (1894), p. 1270.

46. Ibid., p. 1272.

47. Shaler, *Nature and Man* (1891), p. 270. His discussion of the American family and its derivations appears in *United States* (1894), pp. 1274–76.

48. Shaler, "European Peasants" (1893), p. 655.

49. Quoted in Harris, *Rise of Anthropological Theory,* p. 41.

50. See Glacken, *Traces.*

51. Douglas A. Feldman, "The History of the Relationship between Environment and Culture in Ethnological Thought: An Overview," *Journal of the History of the Behavioral Sciences* 11 (1975): 67–81; see also June Helm, "Ecological Approach in Anthropology," *American Journal of Sociology* 67 (1962): 630–39.

52. See George Tatham, "Geography in the Nineteenth Century," in *Geography in the Twentieth Century,* ed. Griffith Taylor (New York: Philosophical Library, 1951), pp. 28–69; Marvin Mikesell, "Friedrich Ratzel," in *International Encyclopedia of the Social Sciences,* ed. D. L. Sills (New York: Macmillan and Free Press, 1968), vol. 13, pp. 327–29; O. H. K. Spate, "Environmentalism," ibid., vol. 5, pp. 93–97.

53. Hanno Beck, "Moritz Wagner als Geograph," *Erdkunde* 7 (1953): 125–28; also Carl O. Sauer, "The Formative Years of Ratzel in the United States," *Annals of the Association of American Geographers* 61 (1971): 245–54; William Coleman, "Science and Symbol in the Turner Frontier Hypothesis," *American Historical Review* 72 (1966): 22–49.

54. Harriet Wanklyn, *Friedrich Ratzel: A Biographical Memoir and Bibliography* (Cambridge: Cambridge University Press, 1961); Robert E. Dickinson, *The Makers of*

Modern Geography (London: Routledge & Kegan Paul, 1969), p. 69; David R. Stoddart, "Organism and Ecosystem as Geographical Models," in *Models in Geography,* ed. Richard J. Chorley and Peter Haggett (London: Methuen, 1969), pp. 517–18; Preston James, *All Possible Worlds: A History of Geographical Ideas* (Indianapolis: Bobbs-Merrill, 1972), p. 240; Harold H. Sprout, "Political Geography," in *Encyclopedia of Social Sciences,* ed. Sills, vol. 6, pp. 116–23.

55. Shaler, *Nature and Man* (1891), p. vii. Although Coleman observes that "Shaler's book [*Nature and Man in America*] popularized the central argument of Ratzel's theory," I have found no direct quotation of Ratzel in Shaler's writings. Coleman, "Science and Symbol," p. 40. Nevertheless his insistence that "the life of the people is peculiar in proportion to the measure of their isolation and the length of time for which it had endured" is clearly consonant with Ratzel's views. *Nature and Man* (1891), p. 161. On geographical determinism more generally, see Thomas, *Environmental Basis;* Robert S. Platt, "Determinism in Geography," *Annals of the Association of American Geographers* 38 (1948): 126–28; Paul Claval, *Essai sur l'évolution de la géographie humaine* (Paris: Belles Lettres, 1976), pp. 41–52; Gordon R. Lewthwaite, "Environmentalism and Determinism: A Search for Clarification," *Annals of the Association of American Geographers* 56 (1966): 1–23.

56. Shaler, *United States* (1894), p. 10. More generally, Shaler noted in 1890 that the "variations of man and beast in the various stations of the world are essentially due to the influence of the climatal conditions which surround them." Shaler, "African Element" (1890), p. 660.

57. Shaler, *Nature and Man* (1891), pp. 168–69.

58. J. Russell Smith, *Industrial and Commercial Geography* (New York: Holt, 1913), p. 8.

59. Thomas, *Environmental Basis,* p. 148; O. H. K. Spate, "Toynbee and Huntington: A Study in Determinism," *Geographical Journal* 108 (1952): 406–28; Ellen Churchill Semple, *Influences of Geographic Environment: On the Basis of Ratzel's Anthropogeography* (New York: Holt, 1911), p. 65.

60. Shaler, *Nature and Man* (1891), p. 164.

61. Ibid., pp. 156, 165. Although less disposed toward the importance of isolation—perhaps because of less Darwinian influence—Reclus in a similar vein noted that "lakes, swamps, and fens, while in early periods of civilization they may have served an important function as a protective agency, ultimately become a chief cause of social stagnation." Quoted in Thomas, *Environmental Basis,* p. 170.

62. Shaler, *Nature and Man* (1891), pp. 267–68. Haller has noted that Shaler regarded the "American-type of man" as somewhat thinner and more angular than his European cousin. Haller, "Shaler." Shaler, however, only introduced these "supposed" differences to dismiss them. His own investigations revealed rather different distinctions between European and American. *Nature and Man* (1891), p. 264.

63. Ibid., p. 283.

64. For the sake of consistency in my reference to blacks, I have used Shaler's own terminology. Nothing pejorative is intended by the language I have used.

65. Haller, "Shaler," p. 179; Shaler, "Economic Future of New South" (1890), p. 260.

66. See R. L. Morrill and O. F. Donaldson, "Geographical Perspectives on the History of Black America," *Economic Geography* 48 (1972): 1–23. For other accounts of blacks in America see Rayford Whittingham Logan, *The Negro in American Life*

and Thought: The Nadir, 1877–1901 (New York: Dial Press, 1954); Robert Cruden, *The Negro in Reconstruction* (Englewood Cliffs, N.J.: Prentice-Hall, 1969); James M. McPherson et al., *Blacks in America: Bibliographical Essays* (New York: Doubleday, 1971); George M. Frederickson, *The Black Image in the White Mind: Debates on Afro-American Character and Destiny, 1817–1914* (New York: Irvington, 1971).

67. Solomon, *Ancestors and Immigrants,* pp. 13–14.

68. Shaler, "Our Negro Types" (1900), p. 44; Shaler, *Neighbor* (1904), p. 135.

69. Gould, "Flaws." So fierce was Agassiz's revulsion that he asserted that the "production of half-breeds is as much a sin against nature, as incest in a civilized community is a sin against purity of character." Ibid., p. 633.

70. On the idea that heredity is transmitted through blood, T. Dobzhansky, writing as late as 1946, could say: "Even now this notion continues to bedevil the thinking of many biologists and anthropologists." "The Genetic Nature of Differences among Men," in *Evolutionary Thought in America,* ed. Stow Persons (New York: Braziller, 1956), p. 89.

71. Shaler, "Science and African Problem" (1890), p. 37.

72. Shaler, "An Ex-Southerner" (1870), p. 57.

73. Gordon, *Assimilation,* pp. 84–114.

74. Shaler, "Science and African Problem" (1890), p. 37.

75. Shaler, *Neighbor* (1904), pp. 183–84.

76. Shaler, "Transplantation" (1900), p. 514.

77. Shaler, "African Element" (1890), p. 667.

78. Shaler, "Nature of Negro" (1890), p. 25; also *Neighbor* (1904), p. 190; "An Ex-Southerner" (1870), p. 57.

79. Shaler, "Negro since Civil War" (1900), p. 29.

80. Shaler, *Neighbor* (1904), p. 135.

81. Ibid., p. 138. See also "Negro Problem" (1884).

82. Shaler, "Nature of Negro" (1890), p. 35. Shaler felt that no inalienable right would be infringed "by a requirement that the voter should prove either that he earns a fair wage or holds a certain amount of property." *Neighbor* (1904), p. 332. More generally, he believed that the role given to blacks by the Fourteenth and Fifteenth Amendments to the United States Constitution had been premature: "We mocked the African with the gift of the franchise," he reported. "Future of Negro" (1900), p. 148. For British perspectives on black suffrage during the American Civil War era, see J. E. Cairnes, "The Negro Suffrage," *Macmillan's Magazine* 12 (1865): 334–43; C. MacKay, "The Negro and the Negrophilists," *Blackwood's Edinburgh Magazine* 99 (1866): 581–97.

83. Shaler, "Nature of Negro" (1890), pp. 33–34; idem, "Negro Problem" (1884), pp. 701–2. Shaler also noted that, while the religious impulse was highly developed in the race, there was no corresponding level of morality. Indeed this divorce between the spiritual and the ethical was a final indication that the black race was as yet "unfit for an independent place in a civilized state." Ibid., p. 702.

84. Shaler, *Neighbor* (1904), p. 133.

85. Shaler, "African Element" (1890), p. 671.

86. Shaler, "Transplantation" (1900), pp. 522–23.

87. Shaler, *Neighbor* (1904), p. 187.

88. Shaler, *Citizen* (1904), p. 226.

89. Thus Shaler affirmed in 1904: "There are those who hold a man to be a mere

receptacle into which we may by the process of education pour so much as we will of that distillate of experience we term knowledge. The teacher, if he has learned the most obvious truth of his trade, knows that to have any value instruction must be educative, it must awaken inherited latencies, capacities that are in the stock and which all his resources can in no wise create." *Neighbor* (1904), p. 135. On Spencer's theories of education, see R. F. Butts and L. A. Cremin, *A History of Education in American Culture* (New York: Holt, 1959).

90. Shaler, "Science and African Problem" (1890), p. 42; idem, "Negro Problem" (1884), p. 704.

91. For a description of black lynching, see Gossett, *Race,* p. 269.

92. See Shaler, "American Quality" (1901); idem, *Neighbor* (1904), pp. 150–52.

93. For a review of scientific conceptions of the American Indian during this period, see Reginald Horsman, "Scientific Racism and the American Indian in the Mid-Nineteenth Century," *American Quarterly* 27 (1975): 152–68.

94. See discussion in Murphree, "Evolutionary Anthropologists"; Hallowell, "Beginnings of Anthropology."

95. Shaler, *Story of Continent* (1892), p. 153.

96. John Wesley Powell, "The North American Indians," in *United States,* ed. Shaler (1894), pp. 190–272.

97. Shaler, *Story of Continent* (1892), pp. 156, 161, 158.

98. Powell, "North American Indians," p. 196. In his *Systems of Consanguinity and Affinity of the Human Family* (1871), Morgan asserted: "It is impossible to overestimate the influence of property upon the civilization of mankind. It was the germ, and is still the evidence, of his progress from barbarism, and the ground of his claim to civilization." Quoted in Murphree, "Evolutionary Anthropologists," p. 294.

99. In his analysis he drew substantially on the works of Theodore Reinach, *Textes d'auteurs grecs et romains relatifs au Judaism* (Paris: Société des Etudes Juives, 1895); idem, *Histoires des Israelites depuis l'époque de leur dispersion jusqu'à nos jours* (Paris: n.p., 1885).

100. Shaler discussed the Jewish question in chap. 6 of *The Neighbor* (1904).

101. The difficulty in establishing somatic criteria was highlighted by Carl C. Seltzer, "The Jew—His Racial Status," *Harvard Medical Alumni Bulletin,* April 1939, pp. 10–11; *The Race Question in Modern Science* (New York: UNESCO, 1956), pp. 31–36.

102. See Ben Halpern, "America Is Different," in *The Jews: Social Patterns of an American Group,* ed. M. Sklare (Glencoe, Ill.: Free Press, 1958), pp. 23–39.

103. See Higham, *Send These to Me,* pp. 138–44. Here he estimates that, from numbering less than fifteen thousand in the late 1830s, the Jewish population had swollen to a quarter of a million by 1887, with a major concentration of a fifth of American Jewry in New York City. According to Solomon, "The presence of Jews emerging from ghettos into polite society became a recurring topic of conversation among Lowell, Norton, and others of the Cambridge circle of the 1870s." *Ancestors and Immigrants,* p. 18. See also Bernard D. Weinryb, "Jewish Immigration and Accommodation to America," in *Jews,* ed. Sklare, pp. 4–22.

104. Shaler, *Neighbor* (1904), pp. 107–8, 116.

105. Higham, *Send These to Me,* p. 122.

106. Karl B. Raitz, "Themes in the Cultural Geography of European Ethnic Groups in the United States," *Geographical Review* 69 (1979): 79–94.
107. Higham, *Strangers,* p. 88.
108. Gordon, "Assimilation," p. 267; Higham, *Strangers,* p. 55.
109. Shaler, *Nature of Intellectual Property* (1878), pp. 34, 38.
110. Shaler, "Immigration Problem" (1888); idem, "European Peasants" (1893).
111. Shaler, "Scotch Element" (1896), pp. 515, 516.
112. Quoted in E. Estyn Evans, "The Scotch-Irish in the New World: An Atlantic Heritage," *Journal of the Royal Society of Antiquaries of Ireland* 95 (1965): 41.
113. *Chicago Record-Herald,* 25 September 1901, quoted in E. N. Saveth, *American Historians and European Immigrants, 1875–1925* (New York: Columbia University Press, 1948), pp. 128–29.
114. Francis A. Walker, *Restriction of Immigration,* Immigration Restriction League, Publication no. 33 (New York, 1899).
115. Charles E. Rosenberg, "The Bitter Fruit: Heredity, Disease, and Social Thought in Nineteenth-Century America," *Perspectives in American History* 8 (1974): 189. Francis Galton, "Eugenics: Its Definition, Scope, and Aims," *American Journal of Sociology* 10 (1904): 1–6. In answering H. G. Wells's comment that "eugenics" was merely a new word for the more popular term "stirpiculture," Galton replied that he himself had coined both terms. Shaler employed the latter expression. See his *Man and Earth* (1905), p. 155. A comprehensive introduction to the literature of eugenics is available in Lyndsay A. Farrall, "The History of Eugenics: A Bibliographical Review," *Annals of Science* 36 (1979): 111–23; see also Kenneth M. Ludmerer, *Genetics and American Society: A Historical Appraisal* (Baltimore: Johns Hopkins University Press, 1972).
116. Nathan Allen, *The Intermarriage of Relations* (New York: Appleton, 1869), p. 25. Similarly, according to H. S. Pomeroy, "Heredity is one of the exact sciences." *The Ethics of Marriage* (New York: Funk & Wagnalls, 1888), p. 107.
117. Francis Galton, *Hereditary Genius: An Inquiry into Its Laws and Consequences* (London: Macmillan, 1869). On Galton more generally see D. W. Forrest, *Francis Galton: The Life and Work of a Victorian Genius* (New York: Elek, 1974). Historical studies of the eugenics movement include Mark Haller, *Eugenics: Hereditarian Attitudes in American Thought* (New Brunswick, N.J.: Rutgers University Press, 1963); Michael Freeden, "Eugenics and Progressive Thought: A Study in Ideological Affinity," *Historical Journal* 22 (1979): 645–71; Donald MacKenzie, "Eugenics in Britain," *Social Studies of Science* 6 (1976): 449–532. Eugenics was thoroughly compatible with socialist opinion, as is clear from the study by Diane Paul, "Eugenics and the Left," *Journal of the History of Ideas* 45 (1984): 567–90.
118. Rosenberg, "Bitter Fruit," pp. 221–22; Gossett, *Race,* p. 157; Richard Dugdale, *"The Jukes": A Study in Crime, Pauperism, Disease, and Heredity* (New York: Putnam's Sons, 1877). See also R. C. Olby, "Human Genetics, Eugenics, and Race: Essay Review," *British Journal for the History of Science* 10 (1977): 156–59.
119. Gossett, *Race,* p. 157; Bannister, *Social Darwinism,* pp. 137–63; Rosenberg, "Bitter Fruit," p. 224.
120. Haller, *Eugenics,* p. 40.
121. Shaler, "Immigration Problem II" (1888), p. 2. In "European Peasants" (1893), p. 651, Shaler observed: "The researches of Mr. Francis Galton, and of the other investigators who have followed his admirable methods of inquiry, have clearly

shown that the inheritance of qualities in man is as certain as among the lower animals. The cases are indeed rare where persons of conspicuous ability have been born of parents of inferior capacity."

122. Quoted in G. R. Searle, *Eugenics and Politics in Britain, 1900–1914* (Leiden: Noord-hoff International Publishing, 1976), p. 34; see also idem, *The Quest for National Efficiency, 1899–1914* (Oxford: Oxford University Press, 1971).

123. Francis Galton, *Essays in Eugenics* (London: Eugenics Education Society, 1909), p. 24.

124. Shaler, *Man and Earth* (1905), p. 159.

125. Arthur R. Schultz, "Immigration," in *Dictionary of American History*, vol. 3, pp. 332–41; Victor Greene, "Immigration Restriction," in ibid., pp. 342–43.

126. Gordon, "Assimilation," p. 272.

127. See Lee A. Benson, "The Historical Background of Turner's Frontier Essay," *Agricultural History* 25 (1951): 59–82; James C. Malin, "Space and History: Reflections on the Closed-Space Doctrines of Turner and Mackinder and the Challenge of Those Ideas by the Air Age," *Agricultural History* 18 (1944): 65–74, 107–26.

128. Solomon, *Ancestors and Immigrants,* passim.

129. Immigration Restriction League of Boston, Papers, US 10583.9.8, Houghton Library, Harvard University.

130. Norton Papers, bMA Am 1088 (6539), Houghton Library, Harvard University.

131. Shaler, "Summer's Journey" (1873), pp. 710, 712–13.

132. Shaler, *United States* (1894), p. 1283.

133. Solomon, *Ancestors and Immigrants,* p. 115. This bill, however, was vetoed in March 1897 by President Grover Cleveland.

134. Persons, *American Minds,* p. 294.

135. See Vogel, *Ethnology,* p. 75.

136. Shaler, *Citizen* (1904), p. 299.

137. Shaler, *Neighbor* (1904), p. 274.

138. Shaler, *Citizen* (1904), p. 299.

139. Rosenberg, "Bitter Fruit," p. 235.

Chapter 6.

1. Robert E. Dickinson, *Regional Concept: The Anglo-American Leaders* (London: Routledge & Kegan Paul, 1976), p. 195; Coleman, "Science and Symbol," p. 40; Billington, *Genesis of Frontier Thesis,* p. 97.

2. Thus too Walter Berg's doctoral dissertation on Shaler, for although he concedes that Shaler "was not a radical geographical determinist belonging in the camp of Buckle and Semple" (p. 282) and even suggests that the label "possibilist" might be an appropriate designation for some of his work, he concludes that Shaler "was a carrier of many ideas on the geographical interpretation of American history . . . a trustee of the past." "Shaler," p. 299.

3. A useful review of this perspective is provided in Morris Cohen, *American Thought: A Critical Sketch* (Glencoe: Free Press, 1954), pp. 53–55.

4. See Eric H. Warmington, ed., *Greek Geography* (London: Dent, 1934); Arnold J. Toynbee, *A Study of History* (London: Oxford University Press, 1934); Ellsworth Huntington, *Civilization and Climate* (New Haven: Yale University Press, 1915);

idem, *Mainsprings of Civilization* (New York: Wiley, 1945). Even more recently John E. Chappell, Jr., has defended a revisionist geographical determinism in "Environmental Causation," in *Themes in Geographic Thought,* ed. Harvey and Holly, pp. 163–86.

5. See Glacken, *Traces;* Spate, "Environmentalism"; idem, "Toynbee and Huntington: A Study in Determinism," *Geographical Journal* 118 (1952): 406–28.

6. George Bancroft, *History of the United States,* 14th ed., 10 vols. (Boston: Little, Brown, 1854), vol. 1, p. 126; see also Hofstadter, *Progressive Historians,* p. 15.

7. Quoted in Dean Moor, "The Paxton Boys: Parkman's Use of the Frontier Hypothesis," *Mid-America* 36 (1954): 216. Moor notes (p. 217) that Parkman "did not place his chief reliance on racial heredity and self-determined growth. He looked for the influence of environment."

8. Carl Ortwin Sauer, "On the Background of Geography in the United States," in *Heidelberger Studien zur Kulturgeographie: Festgabe für Gottfried Pfeifer,* Heidelberger Geographische Arbeiten, no. 15 (Wiesbaden, 1966), p. 67.

9. Arnold Guyot, *The Earth and Man,* rev. ed. (New York: Scribner's Sons, 1897; first published 1849), p. 16.

10. Clarence J. Glacken, "Changing Ideas of the Habitable World," in *Man's Role in Changing the Face of the Earth,* ed. William L. Thomas (Chicago: University of Chicago Press, 1956), p. 80.

11. This long phrase was the title for the second chapter of the first volume. For Buckle, climate, soil, and physiography not only affected social history but also produced distinctive mental characteristics. See Thomas Henry Buckle, *History of Civilization in England,* 3 vols. (London: Longmans, Green, 1894; first published 1867–70), vol. 1, pp. 39–41.

12. Quoted in Berg, "Shaler," p. 255. Writing in the *History of the Conflict between Religion and Science* (London: King, 1875), p. 4, Draper observed: "The topographical configuration of Greece gave an impress to her political condition. It divided her people into distinct communities having conflicting interests, and made them incapable of centralization."

13. See Gordon R. Lewthwaite, "Environmentalism and Determinism: A Search For Clarification," *Annals of the Association of American Geographers* 56 (1966): 1–23.

14. Thus Sauer, "Formative Years of Ratzel."

15. Spate, "Environmentalism," p. 94.

16. Derek Gregory, "Environmental Determinism," in *The Dictionary of Human Geography,* ed. R. J. Johnston (Oxford: Blackwell, 1981), p. 103.

17. Arthur M. Schlesinger, *New Viewpoints in American History* (New York: Macmillan, 1922), p. 45; Oscar Handlin et al., *Harvard Guide to American History* (Cambridge, Mass.: Harvard University Press, 1954), p. 18.

18. Berg, "Shaler," p. 236.

19. Shaler, "Earthquakes" (1869), p. 682.

20. On such a map, he speculated, the most densely shaded areas would include southern Italy, Sicily, Syria, Persia, the Malayan Archipelago from Singapore to Manila, most of Japan, and the Chinese Empire; in the New World, Mexico, Central America, and the northern and western shoreline of South America.

21. Shaler, "Earthquakes" (1869), p. 682. Perhaps the most marked instance of such environmental retardation had occurred among the Icelandic peoples, "who, although at first . . . seemed to develop an intellectual activity proportionate to

the intensity of the movements of the physical world about them, are now reduced far below the position of the people of their race on the main-land." Ibid., p. 683.

22. Ibid., p. 684. In a subsequent article Shaler pointed out that the "style of the structures erected by the ancient Mexicans and Peruvians is as well suited for the resistance of earthquake shocks as that of modern fortifications for the protection of their occupants against projectiles." Shaler, "Earthquakes of American Continents" (1869), p. 461.

23. Shaler, "Great Earthquakes of Old World" (1869), p. 150.

24. Shaler, "Earthquakes of American Continents" (1869), p. 467.

25. Shaler, "An Ex-Southerner" (1870), pp. 54, 55.

26. Shaler, "Summer's Journey" (1873), p. 711. Another regional cameo, "A Winter Journey in Colorado" (1881), similarly displays how impossible it is to typify Shaler as an environmental determinist. Here, commenting on the fertility of the Mississippi Valley, he noted (pp. 46–47) the future impact that human society would exert on that environment: "Undoubtedly, the energy that men bring with them to this land of monotonous fertility, together with the protective influences of institutions, literature, and travel, will secure them from the effects which the stranger feels there, and in time art will come to diversify that which nature has so dismally uniformed."

27. Shaler, "Reelfoot Lake" (1878), p. 218.

28. Shaler, "Outline of Geology of Boston" (1880).

29. According to Koelsch, Shaler's contribution was unique in escaping the annotative spirit of Winsor's editorial pen. See William A. Koelsch, " 'A Profound Though Special Erudition': Justin Winsor as Historian of Discovery," *Proceedings of the American Antiquarian Society* 93 (1983): 63.

30. Compare Guyot, *Earth and Man*, pp. 221–22.

31. Shaler, "Physiography" (1884), p. xi.

32. Shaler referred to Green's book *The Making of England*.

33. Shaler, "Physiography" (1884), p. xiv.

34. Shaler, "Origin and Nature of Soils" (1892), p. 338.

35. Ibid.

36. Similarly geographical in origin was tetanus: "Thus in certain districts in and adjacent to Long Island, New York, the disease known as tetanus, or lockjaw, is of unusually common occurrence among men and animals. It is the opinion of experts in medical science that this malady is caused by some species of soil-inhabiting bacterian which invests this part of the country." Ibid., p. 342.

37. Ibid., p. 343.

38. Advertisement on pp. 365–66 of William Barrows, *Oregon: The Struggle for Possession* (Boston: Houghton Mifflin, 1884). Other contributors to the series included philosopher Josiah Royce, novelist John E. Locke, and jurist Thomas M. Cooley.

39. Shaler, *Kentucky* (1884), pp. v, viii.

40. Thomas D. Clark maintained that many Kentucky historians followed Shaler's views about slavery in the state. See Clark's book *A History of Kentucky* (New York: Prentice-Hall, 1937), p. 278.

41. Shaler, *Kentucky* (1884), pp. 394, 408.

42. Ibid., pp. 24, 233, 232–33.

43. Shaler, "Peculiarities of the South" (1890), p. 486.
44. See John B. McMaster, *A History of the People of the United States from the Revolution to the Civil War*, 8 vols. (New York: Appleton, 1883–1913); John W. Caughey, *Hubert Howe Bancroft, Historian of the West* (Berkeley: University of California Press, 1946).
45. Shaler, *Nature and Man* (1891), p. 13.
46. Ibid., pp. 65, 146.
47. Ibid., p. 149.
48. Ibid., p. 211.
49. Ibid., pp. 180, 210.
50. William Z. Ripley, "Geography as a Sociological Study," *Political Science Quarterly* 10 (1896): 649. Also idem, *The Races of Europe: A Sociological Study* (London: Kegan Paul, Trench, Trubner, 1899), p. 11. This work, it is interesting to note, was also the outcome of a series of lectures at the Lowell Institute.
51. [Hugh Robert Mill?] "Review of *Nature and Man in America*," *Proceedings of the Royal Geographical Society*, n.s., 14 (1892): 428.
52. Shaler, *Story of Continent* (1892), p. 166. As he further observed: "Thus when a people have certain definite characteristics imprinted upon them in the cradle land, the region where their infancy was spent, they are likely to preserve these features long after they have been separated from their birthplace. Great as are the influences of these early conditions on the character of a people, they are in a measure affected by the country in which their lives are lived; gradually the nature about them determines their habits into new channels, and modifies their ways of thinking and acting." Ibid., p. 167.
53. Shaler, "Geological History of Harbors" (1893), pp. 101–2, 103.
54. Shaler, *United States* (1894), pp. 1–2.
55. Shaler, "Environment and Man" (1896), p. 736.
56. Berg, "Shaler," p. 283.
57. Shaler, *United States* (1894), pp. 35, 1272.
58. Frederick J. Turner, "Geographical Interpretations of American History," *Journal of Geography* 3 (1905): 34–37.
59. Albert Perry Brigham, *Geographic Influences in American History* (Boston: Ginn, 1903), p. ix. In a 1904 address, Brigham acknowledged that "Francis Parkman must be given the pioneer in this splendid field" and that "Fiske and McMaster are not far behind." "Geography and History in the United States," in *Report of the Eighth International Geographic Congress Held in the United States, 1904* (Washington, D.C.: Government Printing Office, 1905), p. 964. Again Shaler received no mention. But see idem, *Geographic Influences*, pp. 44, 48, 103, 206, 323.
60. Richard Elwood Dodge, "Albert Perry Brigham," *Annals of the Association of American Geographers* 20 (1930): 56.
61. Ellen Churchill Semple, *American History and Its Geographic Conditions* (Boston: Houghton Mifflin, 1903), pp. 36, 46, 76, 115, 123, 226, 285. Besides, like Shaler, Semple was a native of Kentucky, and her paper "The Anglo Saxons of the Kentucky Mountains: A Study in Anthropogeography," *Geographical Journal* 17 (1901): 588–623, parallels Shaler's subscription to the Teutonic theory. A short biographical sketch is provided in Charles C. Colby, "Ellen Churchill Semple," *Annals of the Association of American Geographers* 23 (1933): 229–40. By limiting investigation only to the sources of Semple's later *Influences of Geographic Environ-*

ment, Numa Broc does not do justice to the influence of Shaler on Semple's thinking. See Numa Broc, "Les classiques de Miss Semple: Essai sur les sources des *Influences of Geographic Environment,* 1911," *Annales de géographie* 497 (1981): 87–102. In this volume Semple referred to Shaler's work on pp. 69 and 396.

62. L. Rodwell Jones and P. W. Bryan, *North America: An Historical, Economic, and Regional Geography* (London: Methuen, 1924), p. vii.

63. W. M. Davis, "The Progress of Geography in the United States," *Annals of the Association of American Geographers* 14 (1924): 169.

64. J. Russell Smith, *North America: Its People and the Resources, Development, and Prospects of the Continent as an Agricultural, Industrial, and Commercial Area* (London: Bell & Sons, 1924), p. vi. See also Virginia M. Rowley, *J. Russell Smith: Geographer, Educator, and Conservationist* (Philadelphia: University of Pennsylvania Press, 1964), p. 63.

65. Thomas, *Environmental Basis,* pp. 176–79.

66. See John Fiske, *The Discovery of America,* 2 vols. (Boston: Houghton Mifflin, 1896), vol. 1, p. 243; William A. Koelsch, ed., *Lectures on the Historical Geography of the United States As Given in 1933 by Harlan H. Barrows,* Department of Geography, Research Paper no. 77 (Chicago: University of Chicago, 1962). On Barrows's environmentalism see John E. Chappell, Jr., "Harlan Barrows and Environmentalism," *Annals of the Association of American Geographers* 61 (1971): 198–201.

67. Sauer, "Formative Years of Ratzel," p. 245.

68. For the origins of the term "frontier," see Fulmer Mood, "Notes on the History of the Word *Frontier,*" *Agricultural History* 22 (1948): 78–83; John T. Juricek, "American Usage of the Word 'Frontier' from Colonial Times to Frederick Jackson Turner," *Proceedings of the American Philosophical Society* 110 (1966): 10–34.

69. Roderick Nash, *Wilderness and the American Mind* (New Haven: Yale University Press, 1967); Hans Huth, *Nature and the American: Three Centuries of Changing Attitudes* (Berkeley: University of California Press, 1957); J. Wreford Watson, "The Image of Nature in America," in *The American Environment,* ed. W. R. Mead (London: Athlone Press, 1974), pp. 1–20.

70. Poem by Michael Wigglesworth (1662), "God's Controversy with New England," quoted in Henry Nash Smith, *Virgin Land: The American West as Symbol and Myth* (New York: Vintage Books, 1950), p. 4.

71. Smith, *Virgin Land;* John F. Davis, "Constructing the British View of the Great Plains," in *Images of the Plains: The Role of Human Nature in Settlement,* ed. Brian W. Blouet and Merlin P. Lawson (Lincoln: University of Nebraska Press, 1975), pp. 181–85.

72. Fred L. Pattee, ed., *The Poems of Philip Freneau, Poet of the American Revolution,* 3 vols. (Princeton: Princeton University Press, 1902), vol. 1, p. 73.

73. See Arthur A. Ekirch, Jr., *Man and Nature in America* (New York: Columbia University Press, 1963), pp. 16–17.

74. See G. Malcolm Lewis, "Three Centuries of Desert Concepts in the Cis-Rocky Mountain West," *Journal of the West* 4 (1965): 457–68; idem, "First Impressions of the Great Plains and Prairies" in *The American Environment: Perceptions and Policies,* ed. J. Wreford Watson and Timothy O'Riordan (London: Wiley, 1976), pp. 37–45; Ralph C. Morris, "The Notion of a Great American Desert East of the Rockies," *Mississippi Valley Historical Review* 13 (1926–27): 190–200; Herman R.

Friis, "The Role of the United States Topographical Engineers in Compiling a Cartographic Image of the Plains Region," in *Images of the Plains,* ed. Blouet and Lawson, pp. 59–74.

75. G. Malcolm Lewis, "Early American Exploration and the Cis-Rocky Mountain Desert, 1803–1823," *Great Plains Journal* 5 (1965): 1–11; also idem, "Regional Ideas and Reality in the Cis-Rocky Mountain West," *Transactions of the Institute of British Geographers* 38 (1966): 135–50.

76. G. Malcolm Lewis, "William Gilpin and the Concept of the Great Plains Region," *Annals of the Association of American Geographers* 56 (1966): 33–51. In this attitude Gilpin reflected the influence of de Tocqueville, whom he had met in Washington in the early 1830s.

77. See Walter M. Kollmorgen, "The Woodsman's Assaults on the Domain of the Cattleman," *Annals of the Association of American Geographers* 59 (1969): 215–39.

78. See G. Malcolm Lewis, "Changing Emphases in the Description of the Natural Environment of the American Great Plains Area," *Transactions of the Institute of British Geographers* 30 (1962): 75–90; Dan E. Clark, "Great Plains," in *Dictionary of American History,* vol. 3, pp. 220–21.

79. Stemming from the work of Walker in 1874, the historical map became increasingly important in the teaching of American history. See Lester J. Cappon, "The Historical Map in American Atlases," *Annals of the Association of American Geographers* 69 (1979): 622–34.

80. See Fulmer Mood, "The Development of Frederick Jackson Turner as a Historical Thinker," *Publications of the Colonial Society of Massachusetts* 34 (1943): 283–352; idem, "The Concept of the Frontier, 1871–1898," *Agricultural History* 19 (1945): 24–30. Carl Sauer emphasized Walker's importance in American geography, extolling in particular his organization of the massive census of 1880, which he believed to be textually "the greatest single contribution to the historical geography of the United States." Sauer, "Background of Geography," pp. 59–71.

81. Francis A. Walker, "The Indian Question," *North American Review* 116 (1873): 329–88; Henry Gannett, "The Settled Area and the Density of Our Population," *International Review* 12 (1882): 70–77.

82. Frederick Jackson Turner, "The Significance of the Frontier in American History," *Proceedings of the State Historical Society of Wisconsin* 41 (1894): 79–112.

83. Gilman M. Ostrander, however, has argued that, in effect, "Turner's frontier thesis was based upon an implicit assumption of the validity of this racial determinism as it was then most influentially formulated in the Germanic germ theory of history." "Turner and the Germ Theory," *Agricultural History* 23 (1958): 258.

84. Coleman, "Science and Symbol"; Robert F. Berkhofer, Jr., "Space, Time, Culture, and the New Frontier," *Agricultural History* 38 (1964): 21–30; Rudolf Freund, "Turner's Theory of Social Evolution," *Agricultural History* 19 (1945): 78–87; Ray A. Billington, *Genesis of Frontier Thesis.*

85. Shaler, "Improvement of Native Pasture-Lands" (1883), p. 186.

86. John Wesley Powell, *Report on the Lands of the Arid Region of the United States* (Washington, D.C.: Government Printing Office, 1878).

87. James C. Malin, *The Grassland of North America: Prolegomena to Its History, with Addenda and Postscript* (Gloucester, Mass.: Smith, 1967; first published 1947), p. 199.

88. Shaler, "Improvement of Native Pasture-Lands" (1883), p. 186.

89. The six regions he identified were (1) the eastern lowlands between the shore and the Appalachians, (2) the lowlands of the Gulf states, (3) the Mississippi Valley, (4) the Appalachian Mountains, (5) the Cordilleras, and (6) the Pacific shoreland fringe.
90. Shaler, "Physiography" (1884), p. vii.
91. Shaler, "Forests of North America" (1887), p. 579.
92. Shaler, *Nature and Man* (1891), pp. 250–63. Some of this material had already appeared in the 1890 issue of *Scribner's Magazine*.
93. Shaler, *United States* (1894), p. 168.
94. James C. Malin, *Essays on Historiography* (Lawrence, Kans.: James C. Malin, 1948), p. 61. In his monograph "The Origin and Nature of Soils" (1892), Shaler viewed forest soils as the norm against which other soil types were to be judged. Here prairie soils were subsumed under the rubric of "Certain Peculiar Soil Conditions."
95. Shaler, *United States* (1894), pp. 145–46.
96. Carl Sauer supported this explanation in "The Agency of Man on the Earth," in *Man's Role,* ed. Thomas, pp. 350–66. More recently, however, Rostlund has suggested that Shaler's view of Indian firing of woodlands is not applicable to the southeast region. See Erhard Rostlund, "The Geographical Range of the Historic Bison in the Southeast," *Annals of the Association of American Geographers* 50 (1960): 395–407. See also John A. Jakle, "Salt on the Ohio Valley Frontier, 1770–1820," *Annals of the Association of American Geographers* 59 (1969): 687–709.
97. Shaler, "Physiography" (1884), pp. xiv, xv.
98. Among these problems were early perceptions of the prairies as a virtually woodless desert with severe winter climates, the danger of fires, an apparent lack of water, and primitive transportation methods. See Carl O. Sauer, "Conditions of Pioneer Life in the Upper Illinois Valley," in *Land and Life: A Selection from the Writings of Carl Ortwin Sauer,* ed. John Leighly (Berkeley: University of California Press, 1963), pp. 11–22.
99. Shaler, "Origin and Nature of Soils" (1892), p. 339.
100. Shaler, *United States* (1894), pp. 46–47.
101. Ibid., p. 1270. The importance of transportation is stressed in Lee A. Benson, "The Historical Background of Turner's Frontier Essay," *Agricultural History* 25 (1951): 59–82.
102. Shaler, *United States* (1894), pp. 134–35.
103. Malin, *Grassland of North America,* p. 201.
104. Shaler, *Story of Our Continent* (1892), pp. 236–40.
105. Billington, *Genesis of Frontier Thesis,* p. 272.
106. Coleman, "Science and Symbol," p. 41.
107. A letter from Chamberlin to Shaler in 1887 indicates the closeness of their friendship: "Nothing would give me more pleasure than to take a vacation with you next summer in your charming retreat on Martha's Vineyard." T. C. Chamberlin, Office of the President, University of Wisconsin, to Shaler, 8 December 1887, Shaler Papers, Box File HUG 1784.10.
108. See Robert H. Block, "Frederick Jackson Turner and American Geography," *Annals of the Association of American Geographers* 70 (1980): 31–42.
109. These factors were recognized in the later writings of Walter Prescott Webb, who sharply differentiated between the plains and woodland frontiers, although

he used the distinction to explain the failure of early settlement beyond the ninety-eighth meridian. Nevertheless, as Tobin shows, Webb's concern "was simply that the experience of the Anglo-American beyond the ninety-eighth meridian was not merely a continuation of the processes Turner had discussed but something qualitatively different." Gregory M. Tobin, *The Making of a History: Walter Prescott Webb and the Great Plains* (Austin: University of Texas Press, 1976), p. 111.

110. See the discussion in H. Roy Merrens, "Historical Geography and Early American History," *William and Mary Quarterly*, 3d ser., 22 (1965): 529–48.

111. Malin, *Historiography*, p. 92.

112. See Fulmer Mood, "The Historiographic Setting of Turner's Frontier Essay," *Agricultural History* 17 (1943): 153–55; Gene M. Gressley, "The Turner Thesis — a Problem in Historiography," *Agricultural History* 32 (1958): 227–49.

113. Shaler, *Kentucky* (1884), p. 398.

114. Shaler, *United States* (1894), p. 43. Ostensibly Turner pursued this line of reasoning, but without Shaler's overt racism, by contending that America meant generation not degeneration. As we have seen, however, Turner was distinctly opposed to the coming of southern and eastern Europeans, believing that their presence was detrimental to the structure of American society. See Gossett, *Race*, pp. 292–93.

115. Shaler, *United States* (1894), pp. 44–45.

116. Shaler, "Forests of North America" (1887), p. 561. The myth of a monolithic eastern forest in Shaler's thinking was challenged in Erhard Rostlund, "The Myth of a Natural Prairies Belt in Alabama: An Interpretation of Historical Records," *Annals of the Association of American Geographers* 47 (1957): 392–411.

117. Later Sauer was also to stress the importance of this agricultural acquisition. See Carl O. Sauer, "The Settlement of the Humid East," in *Yearbook of Agriculture* (Washington, D.C.: Government Printing Office, 1941), pp. 157–66.

118. Frederick Jackson Turner, *Frontier and Section: Selected Essays of Frederick Jackson Turner*, ed., with intro. and notes, Ray A. Billington (Englewood Cliffs, N.J.: Prentice-Hall, 1961), p. 39.

119. Shaler, *United States* (1894), pp. 408–9.

120. Shaler, *Citizen* (1904), p. 321.

121. Such quotations show the inadequacy of Gulley's summary of Shaler's views when he concluded that for Shaler, family, church, and political organization were inherited from the Old World. Clearly that was only half the story. See J. L. M. Gulley, "The Turnerian Frontier: A Study in the Migration of Ideas," *Tijdschrift voor economische en sociale geografie* 50 (1959): 68. The extracts from Shaler's work appear in *Citizen* (1904), p. 30.

122. Shaler, *United States* (1894), p. 1308.

123. Ibid., p. 1271.

Chapter 7

1. Donald Worster, "Introduction," in *American Environmentalism: The Formative Period, 1860–1915*, ed. Donald Worster (New York: Wiley, 1973), p. 2.

2. See David Cushman Coyle, *Conservation: An American Story of Conflict and Accomplishment* (New Brunswick, N.J.: Rutgers University Press, 1957); John R. Ross,

"Man over Nature: Origins of the Conservation Movement," *American Studies* 16 (1975): 49–62.

3. See, for example, Samuel P. Hays, *Conservation and the Gospel of Efficiency: The Progressive Conservation Movement, 1890–1920* (Cambridge, Mass.: Harvard University Press, 1959); David Lowenthal, *George Perkins Marsh, Versatile Vermonter* (New York: Columbia University Press, 1958); Whitney R. Cross, "W J McGee and the Idea of Conservation," *Historian* 15 (1953): 148–62; Martin Nelson McGeary, *Gifford Pinchot: Forester/Politician* (Princeton: Princeton University Press, 1960); Donald C. Swain, *Wilderness Defender: Horace M. Albright and Conservation* (Chicago: University of Chicago Press, 1970); S. B. Sutton, *Charles Sprague Sargent and the Arnold Arboretum* (Cambridge, Mass.: Harvard University Press, 1970).

4. Thus, for example, J. R. Whitaker, "World View of Destruction and Conservation of Natural Resources," *Annals of the Association of American Geographers* 30 (1940): 143–62. Introductions to the literature of conservation are available in Gordon B. Dodds, "The Historiography of American Conservation: Past and Prospects," *Pacific Northwest Quarterly* 56 (1965): 75–81, and the compendious survey by Ronald J. Fahl, *North American Forest and Conservation History: A Bibliography* (Santa Barbara: A.B.C.-Clio Press, 1977).

5. Smith, *North America*, p. vi.

6. Lewis Mumford, *The Brown Decades: A Study of the Arts in America, 1865–1895* (New York: Dover Publications, 1971; first published 1931), p. 29.

7. Malin, *Historiography*, pp. 45–92; idem, *Grassland of North America*, pp. 197–202.

8. See especially the essays by William L. Thomas, "Introductory: About the Symposium, About the People, About the Theme," pp. xxi–xxxviii; Clarence J. Glacken, "Changing Ideas of the Habitable World," pp. 70–102; James C. Malin, "The Grassland of North America: Its Occupance and the Challenge of Continuous Reappraisals," pp. 350–66.

9. Worster, in *American Environmentalism*, ed. Worster, p. 209; Koelsch, "Shaler," pp. 133–39. The neglect of Shaler's warning about the ill effects of destroying wildlife is noted in H. Wayne Morgan, "American History as Experiment," in *Issues and Ideas*, ed. Taylor and White, pp. 5–18.

10. Peter N. Carroll, *Puritanism and the Wilderness: The Intellectual Significance of the New England Frontier, 1629–1700* (New York: Columbia University Press, 1969).

11. Nash, *Wilderness*, p. 24.

12. Watson, "Image of Nature in America," pp 1–16; Smith, *Virgin Land*, pp. 1–13.

13. See Peter J. Schmitt, *Back to Nature: The Arcadian Myth in Urban America* (New York: Oxford University Press, 1969). Paul Shepard, in *Man in the Landscape: A Historic View of the Esthetics of Nature* (New York: Knopf, 1967), p. 64, argues that the kind of emotional attachment to wilderness typical of the back-to-nature movement was "a sort of intellectual creation . . . which is impossible except in a world of ideas whose survival depends on the city." See also Huth, *Nature and the American*.

14. See Miller, *Errand into Wilderness*; Ahlstrom, *Religious History*, pp. 388–402; Schneider, "Religious Enlightenment."

15. See Lowenthal, *Marsh*; McGeary, *Pinchot*.

16. The whole Hetch Hetchy debate is recounted in Nash, *Wilderness*, chap. 10; quotation from p. 171.

17. See, for example, David Lowenthal, "Is Wilderness 'Paradise Enow'? Images of Nature in America," *Columbia University Forum* 7 (1964): 34–40.
18. See R. P. McIntosh, "Ecology since 1900," in *Issues and Ideas,* ed. Taylor and White, pp. 353–72.
19. An excellent introduction to the major currents of Western environmental thought is provided in John Passmore, *Man's Responsibility for Nature* (London: Duckworth, 1974).
20. See, for instance, Frank F. Darling, "Man's Responsibility for the Environment," in *Biology and Ethics,* ed. F. J. Ebling, Symposium of the Institute of Biology, no. 18 (London: Academic Press, 1969), pp. 117–22; also Theodore Roszak, "Ecology and Mysticism," *Humanist* 86 (1971): 134–36.
21. See Gina Bari Kolata, "Theoretical Ecology: Beginnings of a Predictive Science," *Science* 183 (1974): 400–401, 450.
22. Herbert London, "American Romantics: Old and New," *Colorado Quarterly* 18 (1969): 11.
23. Although the modern rediscovery of Marsh is attributed to Lewis Mumford, whose essay *The Brown Decades* reputedly prompted Carl Sauer to resurrect Marsh's contributions, Shaler and other early conservationists were familiar with it during the latter part of the nineteenth century. In a letter to the author, 18 December 1980, John Leighly informs me that he had introduced Marsh's writings to Sauer in the late 1930s.
24. Shaler, "History of Operations" (1877), p. 113; idem, *Annual Report* (1877), p. 38.
25. Lowenthal, *Marsh,* p. 377.
26. See discussion in Morton White, *Science and Sentiment in America: Philosophical Thought from Jonathan Edwards to John Dewey* (New York: Oxford University Press, 1973), pp. 144–67.
27. In August 1869 Charles Peirce, assisted by Shaler, set up equipment for observing an eclipse of the sun at Bardstown, Kentucky. Extant correspondence reveals a close friendship between them. See Charles S. Peirce Papers, CSP-L-401, Houghton Library, Harvard University.
28. Berg, "Shaler," pp. 60–61.
29. This attitude is clear, for example, in his utilitarian approach to the study of Kentucky's marl and lime deposits as sources of future agricultural value.
30. See Frank N. Egerton, "Ecological Studies and Observations before 1900," in *Issues and Ideas,* ed. Taylor and White, pp. 311–51.
31. Shaler, *Autobiography* (1909), p. 350.
32. Shaler, "Faith in Nature" (1902), p. 281. This concern is also noted by Schmitt, *Back to Nature,* p. 141.
33. Shaler, "Faith in Nature" (1902), p. 304.
34. See for example Theodore Roszak, *The Making of a Counter Culture: Reflections on the Technocratic Society and Its Youthful Opposition* (London: Faber, 1970).
35. See Donald Fleming, "Roots of the New Conservation Movement," *Perspectives in American History* 6 (1972): 7–91.
36. See Suk-Han Shin, "American Conservation Viewpoints," *Environmental Conservation* 4 (1977): 273–77.
37. Shaler, "Forests of North America" (1887), p. 561.
38. Ibid., p. 580.

39. Charles R. Van Hise, *The Conservation of Natural Resources in the United States* (New York: Macmillan, 1910), p. 222.
40. Lewis W. Moncrief, "The Cultural Basis for Our Environmental Crisis," *Science* 170 (1970): 508–12. Ekirch maintains that, in the short span of American history, the cycle of primitivism to civilization had been compressed and laid bare for study. See Arthur A. Ekirch, Jr., *Man and Nature in America* (New York: Columbia University Press, 1963), p. 6.
41. See the discussion of the controversy in chapter 4.
42. Shaler, *Interpretation* (1893), pp. 132–33.
43. Shaler, "Humanism in Study of Nature" (1885). Such affirmations continue to be echoed in the writings of those ecological activists who appeal to the nature-affirming reverence for all life that is characteristic of many Eastern religions. See the discussion in Gabriel Fackre, "Ecology and Theology," *Religion in Life* 40 (1971): 210–24. The destruction of the environment in the East, its religion notwithstanding, is charted in Yi-Fu Tuan, "Discrepancies between Environmental Attitude and Behaviour: Examples from Europe and China," *Canadian Geographer* 12 (1968): 176–91. A recent review of trends is provided in Robin W. Doughty, "Environmental Theology: Trends and Prospects in Christian Thought," *Progress in Human Geography* 5 (1981): 234–48.
44. Shaler, "Annual Report" (1887), p. 35.
45. See Howard W. Lull, "Forest Influences: Growth of a Concept," *Journal of Forestry* 47 (1949): 700–5.
46. Raphael Zon, "Climate and the Nation's Forests," in *The Yearbook of Agriculture* (Washington, D.C.: United States Department of Agriculture, 1941), pp. 477–98; H. G. Wilm, "The Status of Watershed Management Concepts," *Journal of Forestry* 44 (1946): 968–71; Joseph Kittredge, *Forest Influences* (New York: McGraw-Hill, 1948).
47. See, for example, Van Hise, *Conservation*, pp. 4–5. McGee was secretary to the Inland Waterways Commission of 1907, the report of which emphasized that the control of water necessitated the preservation of forests. Gifford Pinchot wrote "Relation of Forests to Stream Control," *Annals of the American Academy of Political and Social Science* 31 (1908): 219–28.
48. Glacken, "Changing Ideas," p. 84.
49. Malin, *Grassland of North America*, p. 199.
50. Shaler, "Origin and Nature of Soils" (1892), p. 220.
51. Ibid., p. 219. Almost twenty years later, in 1910, Van Hise was to remark: "Frequently the idea is held that the thing is most useful which is dearest; but on the contrary, those useful things which are cheap and abundant are the most valuable to civilization. For instance, the soil, which is far more abundant and cheaper than fuel or any metal, is the most valuable of our resources. *Conservation*, p. 62.
52. See A. A. Rode, *Soil Science*, trans. A. Gourevitch (Jerusalem: Israel Program for Scientific Translations, 1962).
53. See James G. Cruickshank, *Soil Geography* (Newton Abbott: David & Charles, 1972), pp. 24–25.
54. C. C. Nikiforoff, "Reappraisal of the Soil," *Science* 129 (1959): 196.
55. Shaler, "Origin and Nature of Soils" (1892), p. 305.
56. See *Life and Work of C. F. Marbut* (Columbia, Mo.: Soil Science Society of America,

1942). Marbut worked with Shaler on the glacial brick clays of Rhode Island. See *Annual Report of the United States Geological Survey, 1895–96.*

57. George Perkins Marsh, *Address Delivered before the Agricultural Society of Rutland County, 30th September, 1847* (Rutland, Vt.: Agricultural Society of Rutland County, 1848). This address is reprinted in Barbara Guttmann Rosenkrantz and William A. Koelsch, eds., *American Habitat: A Historical Perspective* (New York: Free Press, 1973), p. 343.

58. Shaler, *Man and Earth* (1905), p. 126.

59. Shaler, "Economic Aspects of Soil Erosion" (1896), p. 374. As he had already observed, "Man himself is, through his arts, particularly those of agriculture, one of the great agents of change, and . . . through these interferences with the course of nature the operation of many forces had been greatly increased in energy." Ibid., p. 328.

60. Clarence J. Glacken, "The Origins of the Conservation Philosophy," *Journal of Soil and Water Conservation* 11 (1956): 65.

61. Shaler, "Origin and Nature of Soils" (1892), pp. 332–33. The conservationist thrust of these proposals is further attested in his later observation that the "true aim, therefore, of conservative agriculture, such as is to maintain the soil in shape to be useful to man for an indefinitely long future, is to bring about and keep the balance between the processes of rock decay and erosion in fitting adjustment." *Man and Earth* (1905), p. 123.

62. See C. E. Kellogg, "We Seek; We Learn" in *The Yearbook of Agriculture* (Washington, D.C.: The United States Department of Agriculture, 1957), pp. 2–11.

63. On Shaler's death, Roosevelt wrote in the *Harvard Crimson* (no. 49, 12 April 1906): "I am greatly shocked at the death of Dean Shaler and mourn his loss. I not only feel for him the affectionate remembrance of scholar toward instructor, but the remembrance of the friendship and regard I grew to feel in constantly growing measure for him after I left College."

64. In 1910, Pinchot wrote: "The waste of soil is among the most dangerous of all wastes now in progress in the United States. In 1896, Professor Shaler, than whom no one has spoken with greater authority on this subject, estimated that in the upland regions of the states south of Pennsylvania three thousand square miles of soil had been destroyed as the result of forest denudation, and that destruction was then proceeding at the rate of one hundred square miles of fertile soil per year." *The Fight for Conservation* (Seattle: University of Washington Press, 1967; first published 1910), p. 9. This comment is especially significant, for as Gerald D. Nash's introduction to this 1967 reprint points out, Pinchot lacked historical perspective and "gave little credit to others who were important in furthering conservation in his own day, especially those who disagreed with his views." Ibid., p. xxiii. Besides this reference to Shaler, Pinchot referred in this essay to only two other individuals, namely, Theodore Roosevelt and President Hadley of Yale.

65. T. C. Chamberlin, "Soil Wastage," in *Proceedings of a Conference of Governors in the White House* (Washington, D.C.: Government Printing Office, 1909), pp. 75–83.

66. Carl O. Sauer, "Theme of Plant and Animal Destruction in Economic History," in *Land and Life*, ed. Leighly, p. 151.

67. See William W. Speth, "Carl Ortwin Sauer on Destructive Exploitation," *Biological Conservation* 11 (1977): 273–77.

68. Shaler, *Man and Earth* (1905), p. 82. Shaler, of course, did not restrict his discussion of irrigation to North America but spoke also of conditions in India and China. In view of his racist sentiments, too, it is worth noting that he believed that the Aryan race possessed "no native sense of the use of water on the land" and had learned the techniques of irrigation from Arabs and other desert dwellers. Ibid., p. 77.

69. See John H. Paterson, *North America: A Geography of Canada and the United States,* 4th ed. (London: Oxford University Press, 1970), p. 243. A short, useful introduction to the irrigation question in the United States may be found in Robert G. Dunbar, "Irrigation," in *Dictionary of American History,* vol. 3, pp. 477–78.

70. See Nash, *Wilderness,* chap. 10.

71. Shaler's close friendship with Powell has already been noted in chapter 1, and his agreement with Powell on the irrigation question was evident when he wrote: "It is reckoned by Major Powell, director of the United States Geological Survey, that by storing the water from the winter rains in the valleys above the headwaters of the rivers, it will be possible to irrigate at least 180,000 square miles of this desert land, a region more than four times the size of the state of Illinois." Shaler, *Story of Continent* (1892), p. 172.

72. See Stegner, *Beyond Hundredth Meridian,* pp. 309–11; Darrah, *Powell,* pp. 299–314.

73. Shaler, *Man and Earth* (1905), pp. 85–86.

74. Shaler, "Preliminary Report on Sea-Coast Swamps" (1885), p. 380.

75. Shaler, "Fresh-Water Morasses" (1890), p. 303. Shaler also believed that the American Swamp Cypress provided a useful timber source that had not been exploited. Shaler, "American Swamp Cypress" (1883).

76. Shaler, *Man and Earth* (1905), pp. 93, 146.

77. Shaler, "Future of Precious Metal Mining" (1880), p. 774. A shortened version of this paper appeared in the *Kansas City Review of Science and Industry* for 1880.

78. His *Report on the Resources of the Region Adjacent to the Proposed Cincinnati and Southeastern Railway* (1884), for example, contained detailed estimates of the quantity and quality of various coal types to be found between Cincinnati and Pound Gap in the Ohio Valley, along with observations on the region's timber and agricultural resources.

79. Shaler, *Man and Earth* (1905), p. 52.

80. Shaler, "Rock Gases" (1890), p. 639.

81. Shaler, *Man and Earth* (1905), p. 41.

82. Van Hise, *Conservation,* p. 18.

83. See chapter 3.

84. Lawrence Scientific School Report, Eliot Papers, UA I.5.150, Box 246, Folder: Shaler, Nathaniel S.

85. The first forestry school was started at Cornell University in 1898, and Yale followed in 1900. In 1903, forestry curricula were established in the University of Michigan, Michigan State College, the University of Maine, and the University of Minnesota; in 1904, three additional institutions inaugurated forestry schools: Iowa State College, Harvard University, and the University of Nebraska. See Ralph Sheldon Hosmer, "The Progress of Education in Forestry in the United States," *Empire Forestry Journal* 2 (1923): 1–24; Henry Clepper, "Forestry Education in America," *Journal of Forestry* 54 (1956): 455–57; George A. Garratt, "Gifford

Pinchot and Forestry Education," *Journal of Forestry* 63 (1965): 597–600. For a general introduction to forestry in America, see Frank J. Harmon, "Forestry," in *Dictionary of American History,* vol. 3, pp. 72–76.

86. Shaler to the President and Fellows of Harvard College, 28 March 1903, Eliot Papers, UA I.5.150, Box 120, Folder 321.

87. Shaler, "Landscape as Means of Culture" (1898), pp. 777, 779.

88. Shaler, however, was not prepared to go so far as Muir in his celebration of romantic primitivism. Shaler's matter-of-fact language in discussing environment sharply contrasts with that used by Muir in a letter to Asa Gray: "Before filling your sack I witnessed one of the most glorious of our mountain sunsets—not one of the assembled mountains seemed remote—all had ceased their labor of beauty & gathered around their Parent sun to receive the evening blessing & waiting angels could not be more solemnly hushed. The sun himself seemed to have reached a higher life as if he had died & only his soul were glowing with rayless bodiless *Light,* & as Christ to his disciples so this departing sun-soul said to every precious heart—to every pine & weed, to every stream & mountain, My Peace I give unto you." John Muir, Yosemite Valley, to Asa Gray, 18 December 1872, Archives, Gray Herbarium Library, Harvard University.

89. Shaler, "Landscape as Means of Culture" (1898), p. 778.

90. Shaler, *Story of Continent* (1892), p. 134.

91. So Passmore, for example, who distinguishes problems in ecology (scientific queries) from ecological problems (social evaluations). See Passmore, *Responsibility for Nature,* pp. 43–45.

92. In this concession Shaler anticipated the later recognition that the "environmental crisis brings to the surface a public demand for collective responsibility—for protecting collective environmental property. . . . Current demands for action seem to reflect a conviction that the aggregate of private decisions does not automatically ensure general welfare." See Samuel Z. Klausner, "Thinking Social-Scientifically about Environmental Quality," *Annals of the American Academy of Political and Social Science* 389 (1970): 9–10. In a similar spirit Shaler himself had written: "It may be charged that the legislation which established these reservations is, in its tendencies, socialistic, but the most inveterate enemy of that political theory, if he be open to reason, will not be disposed to contend against such action. He will have to acknowledge that these gifts to the community are very helpful to its best interests, and that they could not have been secured by private or corporate endeavor or even by the action of individual States." Shaler, "Proposed Appalachian Park" (1901), p. 774.

93. Shaler, *Man and Earth* (1905), p. 230.

94. See, for example, Shaler, *United States* (1894), pp. 511, 517.

95. Shaler, *Man and Earth* (1905), p. 1.

96. Shaler, *Citizen* (1904), p. 277.

97. Shaler, *Man and Earth* (1905), p. 184. He had already proposed that the mountain region of western North Carolina should be preserved as a national park because of its unsurpassed broad-leaved forest. See Shaler, "Proposed Appalachian Park" (1901).

98. Perceptually, this position would seem to represent an urban response to countryside, for, as Shepard observes, "The beauty of the farmed land is seldom felt

by the farmer, or, if felt, seldom articulated. Farmland is admired not by those who work it, but by those who live in the city and travel through the countryside." Shepard, *Man in Landscape,* p. 131.

99. Shaler, *Man and Earth* (1905), p. 181.

100. Shaler drew on a paper by H. Langford Warren entitled "A Plea for Esthetic Considerations in Building Roads." See Shaler, *Highways* (1896), p. 111.

101. Ibid.

102. Ibid., pp. 111–12, 113–14, 120.

103. For other efforts of a similar nature, see Kenneth Robert Olwig, "Historical Geography and the Society/Nature 'Problematic': The Perspective of J. F. Schouw, G. P. Marsh, and E. Reclus," *Journal of Historical Geography* 6 (1980): 29–45.

104. Shaler, *Man and Earth* (1905), p. 172.

105. See Morton and Lucia White, "The American Intellectual versus the American City," *Daedalus* 90 (1961): 166–79; also idem, *The Intellectual versus the City* (Cambridge, Mass.: Harvard University Press, 1962).

106. Shaler, *Man and Earth* (1905), p. 189.

107. Harold M. Rose, "Conservation in the United States," in *Conservation of Natural Resources,* 3d ed., ed. Guy-Harold Smith (New York: Wiley, 1965), p. 13.

108. The reality of this distinction between romantic and scientific approaches to the study of the earth is demonstrated in W. M. Davis's comments on John Muir in a letter to President Eliot: "In reply to your note of the 19th, about Mr. Muir, I should say that he does not project among geographers enough to warrant you in distinguishing him as proposed. He is a great lover of nature, a good woodsman, and an active explorer. He may be a good botanist, but he is not a scientific geographer or geologist." W. M. Davis to President Eliot, Harvard University, 22 December 1894, Eliot Papers, Box 106, Folder 68.

109. Fleming, "Roots of New Conservation Movement," p. 75.

110. Lewis Mumford, *The Golden Day: A Study in American Literature and Culture* (New York: Dover Publications, 1968; first published 1926), p. 103.

111. Frank Erisman, "The Environmental Crisis and Present-Day Romanticism: The Persistence of an Idea," *Rocky Mountain Social Science Journal* 10 (1973): 7.

Chapter 8

1. Shaler, *Autobiography* (1909), p. 245.

2. See Young, "Historiographic and Ideological Contexts"; Brooke, "Natural Theology of Geologists."

3. See Gordon L. Davies, *The Earth in Decay: A History of British Geomorphology, 1578–1878* (New York: Macdonald, 1969); Edward Battersby Bailey, *James Hutton—the Founder of Modern Geology* (Amsterdam: Elsevier, 1967). Hutton, of course, did not mean that God had not created the universe, only that that doctrine was irrelevant to his study. See Frank F. Cunningham, *The Revolution in Landscape Science* (Vancouver: Tantalus Research, 1977), pp. 53–65.

4. See Young, "Natural Theology, Victorian Periodicals"; idem, "Impact of Darwin."

5. Shaler, *Outlines of Earth's History* (1898), pp. 8, 6, 22.

6. In 1898, for example, he observed that "from the beginning of organic life in the

remote past to the present day one kind of animal or plant has been in a natural and essentially gradual way converted into the species which was to be its successor." Ibid, p. 15.

7. See the general discussion in R. Hooykaas, "The Parallel between the History of the Earth and the History of the Animal World," *Archives internationales d'histoire des sciences* 10 (1957): 1–18.

8. Shaler, *Outlines of Earth's History* (1898), p. 394.

9. Quoted in R. Hooykaas, "Geological Uniformitarianism and Evolution," *Archives internationales d'histoire des sciences* 19 (1966): 3.

10. Shaler, "Critical Points" (1890), pp. 3, 4.

11. Ibid., pp. 17, 18.

12. In Hutton's case, for example, an acceptance of the principle that all past changes on the globe had been brought about by the slow agency of existing causes did not rule out certain aspects of catastrophism. Hooykaas, "Parallel," p. 3.

13. R. Hooykaas, "Catastrophism," p. 314. See also idem, "The Principle of Uniformity in Geology, Biology, and Theology," *Journal of the Transactions of the Victoria Institute* 88 (1956): 101–16; idem, *The Principle of Uniformity in Geology, Biology, and Theology* (Leiden: Brill, 1963).

14. George Gaylord Simpson, "Uniformitarianism: An Inquiry into Principle, Theory, and Method in Geohistory and Biohistory," in *Philosophy of Geohistory*, ed. Albritton, p. 264.

15. Walter F. Cannon, "The Uniformitarian-Catastrophist Debate," *Isis* 51 (1960): 38–55.

16. See Stephen Jay Gould, "Is Uniformitarianism Necessary?" *American Journal of Science* 263 (1965): 223–28.

17. Simpson, "Uniformitarianism," p. 271.

18. Hooykaas, "Catastrophism," p. 314.

19. Ibid., p. 326.

20. Charles G. Higgins, "Theories of Landscape Development: A Perspective," in *Theories of Landform Development,* ed. W. N. Melhorn and R. C. Flemal (New York: State University of New York, 1975), pp. 14, 15. Higgins in fact goes on to argue that the main reason for the absence of a satisfactory post-Davisian theory of landscape evolution stems from a general failure to "recognize that in many parts of the world the gross forms of the landscape are relics formed by processes no longer operating there." His own preference for some form of actualistic catastrophism is plain when he writes: "It is most likely that future geomorphic theories will reflect a broadly uniformitarian view of Earth history, at least to the extent of denying any major roles in it to cosmic collisions and other adventitious calamities. However, the relative importance of ordinary catastrophes in land sculpture is still disputed. . . . [Nevertheless] distinctive long-lasting landforms have clearly resulted from rare catastrophic floods and rockfalls. The part played by such catastrophic events in long-term development of landscape needs further study." Ibid., pp. 19, 24.

21. See J. William Dawson, "Some Recent Discussions in Geology," *Bulletin of the American Geological Society* 5 (1894): 101–16.

22. W. M. Davis, "Bearing of Physiography on Uniformitarianism" [Abstract], *Bulletin of the Geological Society of America* 7 (1896): 8. In this connection it is interesting to note that Chorley, Beckinsale, and Dunn write: "Davis subconsciously looked

at landscapes as if they were tectonically stationary whereas Penck viewed them eternally unstable. Paradoxically Davis was at heart a traditional catastrophist: if landforms moved he preferred them to move briefly with celerity. Penck, on the other hand, was a uniformitarian—the events may be catastrophic but in the past they went as indefinitely as they do today." Davis, p. 538.

23. See Ronald C. Flemal, "The Attack on the Davisian System of Geomorphology: A Synopsis," *Journal of Geological Education* 19 (1971): 3–13.

24. David R. Stoddart, "Darwin's Influence on the Development of Geography in the United States, 1859–1914," in *Origins,* ed. Blouet, p. 272.

25. Shaler, *Autobiography* (1909), pp. 190–91.

26. Stephen G. Brush, "Nineteenth-Century Debates about the Inside of the Earth: Solid, Liquid, or Gas?" *Annals of Science* 36 (1979): 228. See also Richard J. Chorley, A. J. Dunn, and Robert P. Beckinsale, *The History of the Study of Landforms,* vol. 1, *Geomorphology before Davis* (London: Methuen, 1964), p. 154.

27. Chorley, Dunn, and Beckinsale, *Geomorphology before Davis,* p. 344.

28. William Hopkins, "On the Thickness and Constitution of the Earth's Crust," *Philosophical Transactions of the Royal Society of London* 132 (1842): 43–55.

29. William Thomson, "On the Rigidity of the Earth," *Proceedings of the Glasgow Philosophical Society* 5 (1862): 169–70.

30. Earlier American theories are discussed in Joseph Barrell, "The Growth of Knowledge of Earth Structure," *American Journal of Science,* 4th ser., 46 (1918): 133–70.

31. Brush, "Nineteenth-Century Debates about Inside of Earth," p. 242.

32. Shaler, "Preliminary Notice concerning Elevation of Continental Masses" (1865), p. 239.

33. Clarence Edward Dutton, "On Some of the Greater Problems of Physical Geology," *Bulletin of the Philosophical Society of Washington* 11 (1889): 51–64.

34. Brush, "Nineteenth-Century Debates about Inside of Earth," p. 242.

35. Shaler, "On the Formation of Mountain Chains" (1866), p. 10.

36. Ibid., pp. 11, 12.

37. George P. Merrill, *The First One Hundred Years of American Geology* (New Haven: Yale University Press, 1924), p. 464.

38. Osmond Fisher, "On the Effect upon the Ocean-Tides of a Liquid Substratum beneath the Earth's Crust," *Philosophical Magazine* 14 (1882): 213–15.

39. See James D. Dana, "On Some Results of the Earth's Contraction from Cooling, Including a Discussion of the Origin of Mountains, and the Nature of the Earth's Interior. Part I," *American Journal of Science,* 3d ser., 5 (1873): 423–43; idem, "Part II, The Condition of the Earth's Interior, and the Connection of the Facts with Mountain-Making. Part III, Metamorphism," ibid., 6 (1873): 6–14; idem, "Part IV, Igneous Ejections, Volcanoes," ibid., pp. 104–15.

40. Dana, "Part I," pp. 423, 424; idem, "Part III," p. 11. In the third edition of his *Manual of Geology* (1880), he claimed that the arguments against fluidity had not been substantiated.

41. Shaler, "Earthquakes" (1869), p. 678; idem, "Production of Cape Hatteras" (1871), pp. 110–21.

42. Shaler, "Notes on Age of Cumberland Gap" (1877), pp. 385, 391. Here he claimed to have isolated at least four distinct periods of elevation, the central anticline being the oldest and the other monoclinal mountains having been separately formed and only recently uplifted.

43. Joseph Le Conte, *Elements of Geology* (New York: Appleton, 1888), p. 93.
44. Shaler, "Sedimentary to Volcanic Rocks" (1879); he alluded to it again in 1887, in rather more detail the following year in an article subsequently reprinted as a chapter of *Aspects of the Earth* (1889), and again in 1902. See Shaler, "Stability of Earth" (1887); idem, "Volcanoes" (1888); idem, "Nature of Volcanoes" (1902).
45. Shaler, "Volcanoes" (1888), p. 223.
46. Israel C. Russell, *Volcanoes of North America* (New York: Macmillan, 1904), p. 326.
47. Shaler, "Crenitic Hypothesis" (1888), p. 281.
48. Shaler, "Pleistocene Distortions" (1894), p. 202.
49. Shaler, "Relation of Mountain-Growth to Formation of Continents" (1894), p. 204.
50. James Geikie, *Mountains: Their Origin, Growth, and Decay* (Edinburgh: Oliver & Boyd, 1913), p. 201.
51. Merrill, *First One Hundred Years of American Geology*, p. 386.
52. Shaler, *Geology of Narragansett Basin* (1899), p. 16. Here he insisted there was no "real similarity between these hypotheses." For him, the "steps of action which are postulated are as follows: First, the excavation in ancient and compact rocks, in their nature good transmitters of thrusts, of a trough or basin such as is likely to be formed in the estuarine section of a considerable river; second, the filling in of this basin by sediments accumulated during a downward oscillation of the area in which the basin lies; third, the development of compression strains, such as are involved in rock folding, the relief being afforded by the folding of these stratified deposits." Twenty years earlier, it should perhaps be noted, Shaler had spoken of a similar process in a discussion of the passage from sedimentary to volcanic rocks in the Brighton district of Boston. See Shaler, "Sedimentary to Volcanic Rocks" (1879), pp. 130–31.
53. Chorley, Dunn, and Beckinsale, *Geomorphology before Davis*, pp. 196, 273. For an account of the early reception of the glacial theory in Britain, see B. Hansen, "The Early History of Glacial Theory in British Geology," *Journal of Glaciology* 9 (1970): 135–41.
54. Shaler, "Position and Character of Glacial Beds" (1866), p. 29.
55. Shaler, "Movements in Changes of Level" (1868), pp. 135–36.
56. Shaler, "Absence of Distinct Evidences of Glacial Action" (1868). Shaler accounted for such lack of evidence by suggesting that, with a barrier to ice movement, the ice would build up to the height of the barrier and would move over it but that there would be no movement of the basal layer. See also Shaler, "Parallel Ridges of Glacial Drift" (1870); idem, "Glacial Moraines of Charles River Valley" (1870).
57. Shaler, "Glacial Moraines of Charles River Valley" (1870), p. 278.
58. Shaler, "Geology of Island of Aquidneck" (1872), p. 611.
59. See Chorley, Dunn, and Beckinsale, *Geomorphology before Davis*, pp. 197–205.
60. Shaler, "Geology of Island of Aquidneck" (1872), p. 611.
61. See John T. Andrews, ed., *Glacial Isostasy* (Pa.: Dowden, Hutchinson & Ross, 1974), p. 20. See also Richard Foster Flint, "Introduction: Historical Perspectives," in *The Quaternary of the United States*, ed. H. E. Wright, Jr., and David G. Frey (Princeton: Princeton University Press, 1965), pp. 3–11; T. F. Jamieson, "On the Causes of the Depression and Re-Elevation of the Land during the Glacial Period," *Geological Magazine* 9 (1882): 400–7.

62. Shaler, "Recent Changes of Level" (1874), p. 339.
63. Shaler, "Some Phenomena of Elevation" (1874), p. 288.
64. James D. Dana to Asa Gray, 14 August 1878, Gray Herbarium Library.
65. As Andrews summarizes Shaler's contribution: "Shaler's conclusion is that it is the mass of the ice load that causes the depression of the continents. He uses an analogy of a weight placed on a sheet of lake ice—the ice is depressed around the weight and an elevated area occurs around the 'sunken point.' This may be the first suggestion of the existence of a forebulge! Shaler thus envisages the crust as being rigid but having an underlying region where loads are accommodated by the outward flow of material. This is basically the concept that is still in use today." Andrews, *Glacial Isostasy*, p. 20.
66. W. B. Wright, *The Quaternary Ice Age* (London: Macmillan, 1937), p. 404.
67. Thomas F. Jamieson, "On the History of the Last Geological Changes in Scotland," *Quarterly Journal of the Geological Society of London* 21 (1865): 178.
68. Baron Gerard de Geer, "On Pleistocene Changes of Level in Eastern North America," *Proceedings of the Boston Society of Natural History* 25 (1892): 461.
69. Review in *Geological Magazine* 9 (1882): 34–38.
70. It may be noted in passing that Shaler believed that, even "without any profound changes of climate, very extensive retreats and advances of the ice might be brought about." Shaler, "Recent Advances and Retrogressions" (1881), p. 166. Again in 1890 he insisted that "the facts . . . militate against any hypothesis which seeks to account for the glacial period on the supposition that the climate in the glacial regions was cooler than at present." Shaler, "Glacial Climate" (1890), p. 465.
71. Indeed Croll's development of Adhémar's astronomical theory was to be revived in the early decades of the twentieth century by the Yugoslavian astronomer Milankovitch, whose own account, in turn, was resurrected during the 1970s. See John Imbrie and Katherine Palmer Imbrie, *Ice Ages: Solving the Mystery* (London: Macmillan, 1979); H. H. Lamb, *Climate: Present, Past, and Future*, vol. 1, *Fundamentals and Climate Now* (London: Methuen, 1972), pp. 30–37, and vol. 2, *Climatic History and the Future* (London: Methuen, 1977), pp. 312–13.
72. While Shaler did not provide bibliographical details, it seems likely that he had in mind J. Thomson, "Theoretical Considerations on the Effect of Pressure in Lowering the Freezing Point of Water," *Transactions of the Royal Society of Edinburgh* 16, pt. 5 (1849): 575–80.
73. Shaler, "Motion of Continental Glaciers" (1875), pp. 132, 130.
74. These theories were summarized in Shaler and Davis, *Illustrations of Earth's Surface* (1881), pp. 151–52.
75. J. K. Charlesworth, *The Quaternary Era with Special Reference to Its Glaciation*, 2 vols. (London: Arnold, 1957), vol. 1, pp. 108–25. Indeed pressure melting and slippage over a water layer continue to be regarded as accurate descriptions of processes at the ice-bedrock interface that contribute to the basal shifting of glaciers. See David E. Sugden and Brian S. John, *Glaciers and Landscape: A Geomorphological Approach* (London: Arnold, 1976), pp. 28–30.
76. Shaler and Davis, *Illustrations of Earth's Surface* (1881), pp. 118, 119.
77. This skull, reported by J. D. Whitney to have been found in 1886 in Calaveras County, California, has been shown to correspond in type with those of modern Indian inhabitants of the district. Shaler's ideas on the antiquity of the last glacial

period, incidentally, were adopted by S. Laing, *Human Origins* (London: Chapman & Hall, 1902), p. 277.

78. W. Upham, "Drumlins and Marginal Moraines of Ice-Sheets," *Bulletin of the Geological Society of America* 7 (1896): 27. At this early stage, Shaler's terminology was that of James Hall. See August Böhm, *Geschichte der Moränenkunde* (Vienna: Lechner, 1901), pp. 198, 200.

79. Shaler, "Parallel Ridges of Glacial Drift" (1870), pp. 199–200.

80. Other early theories of drumlin formation included erosion by rain and streams, extraglacial retreat waters, and reticulated subglacial drainage.

81. Shaler, "Geology of Martha's Vineyard" (1888), p. 321.

82. Shaler, "Geology of Cape Ann" (1889), pp. 550–51. By thus invoking a double glacial advance, Shaler found it possible to provide an explanation for what he termed "frontal" or "shoved" moraines. By these features he understood only such end moraines as were "pushed forward by an ice sheet" or "that which has been urged forward in the advance of the ice sheet as the soil is carried onward in front of a scraper." He did not take into account, however, the possibility that ice shearing and overriding could take place, even though in 1870 he had thought it impossible for a glacier to move a thickness of one hundred feet or more of crushed boulder, sand, and silt at its base and had specified a "lifting motion" inherent in glacial movement evident from drift material high up in the glacier. Shaler, "Geology of Martha's Vineyard" (1888), p. 308; idem, "Geology of Cape Ann" (1889), p. 546; idem, "Parallel Ridges of Glacial Drift" (1870), p. 199. Also Shaler found drumlins to be grouped in belts "suggestive of morainic lines," which led him to speculate a possible derivation from glacial remodeling of earlier moraines, despite the obvious difficulty of reconciling their different orientations.

83. R. S. Tarr, "The Origin of Drumlins," *American Geologist* 8 (1894): 393–407; G. H. Barton, "Remarks on Drumlins," *Proceedings of the Boston Society of Natural History* 26 (1893): 23–25; idem, "Original Origin of Channels on Drumlins," *Bulletin of the Geological Society of America* 6 (1895): 8–13.

84. Charlesworth, *Quaternary Era,* p. 397. See also Böhm, *Geschichte,* p. 206.

85. See Clifford Embleton and Cuchlaine A. M. King, *Glacial Geomorphology,* 2d ed. (London: Arnold, 1975), p. 427. Also Charlesworth, *Quaternary Era,* p. 399.

86. Shaler, "Origin of Kames" (1884), pp. 38–39.

87. Charlesworth, *Quaternary Era,* pp. 417, 418.

88. Shaler, "Action of Glaciers" (1889), p. 408.

89. Shaler, "Conditions of Erosion beneath Deep Glaciers" (1893), p. 212.

90. Shaler, Woodworth, and Foerste, *Geology of Narragansett Basin* (1899), p. 72.

91. W. H. Hobbs, *Earth Features and Their Meaning: An Introduction to Geology* (New York: Macmillan, 1926), p. 306; Richard Foster Flint, *Glacial Geology and the Pleistocene Epoch* (New York: Wiley, 1947), p. 120; Charlesworth, *Quaternary Era,* p. 366.

92. According to Clapp, the "first suggestion of the possible complexity of the Glacial Period came in 1889, when Shaler published the probability that southern New England at least has been subjected to two ice-advances, separated by a retreat of considerable duration. His conclusions were based on the relations of glacial and interglacial beds at Martha's Vineyard and Nantucket." Frederick G. Clapp, "Complexity of the Glacial Period in Northeastern New England," *Bulletin of the Geological Society of America* 18 (1907–8): 507.

93. Shaler, Woodworth, and Marbut, "Glacial Brick Clays" (1896), p. 1003.
94. Shaler, "Value of Saliferous Deposits" (1890), p. 582.
95. Shaler, "Antiquity of Last Glacial Period" (1891), p. 261.
96. Shaler, "Movements in Changes of Level" (1868); idem, "Production of Cape Hatteras" (1871); idem, "Phosphate Beds of South Carolina" (1870).
97. See Richard J. Chorley, "Diastrophic Background to Twentieth-Century Geomorphological Thought," *Bulletin of the Geological Society of America* 74 (1963): 953–70.
98. Eduard Suess, *The Face of the Earth*, trans. Hertha B. C. Sollas under the direction of W. J. Sollas, 5 vols. (Oxford: Clarendon Press, 1904–24), vol. 2, p. 22.
99. Shaler, "Some Phenomena of Elevation" (1874), pp. 288–89.
100. Shaler, "Fresh-Water Morasses" (1890), p. 331.
101. Shaler, "Changes of Sealevel" (1895), p. 152.
102. Shaler, "Geology of Richmond Basin" (1899), p. 416.
103. Chorley, "Diastrophic Background," p. 956. See T. C. Chamberlin, "Diastrophism as the Ultimate Basis of Correlation," *Journal of Geology* 17 (1909): 685–93.
104. Shaler, Woodworth, and Foerste, *Geology of Narragansett Basin* (1899), p. 43.
105. Douglas W. Johnson, *The New England-Acadian Shoreline* (New York: Wiley, 1925), p. xv.
106. Shaler, "Phosphate Beds of South Carolina" (1870), p. 226.
107. Shaler, "Beaches and Tidal Marshes" (1895), pp. 151–53. Davis followed Shaler in relating the offshore bar to emergent shorelines. See D. W. Johnson, ed., *The Geographical Essays of William Morris Davis* (Boston: Ginn, 1909), p. 710.
108. Douglas W. Johnson, *Shore Processes and Shoreline Development* (New York: Wiley, 1919), p. 380.
109. Shaler, "Geology of Coast of Maine" (1875), p. 883. This view was also later adopted by Hubbard, Daly, and Andrews. See Johnson, *Shore Processes*, p. 181.
110. Thus Richard J. Pike and Wesley J. Rozema, "Spectral Analysis of Landforms," *Annals of the Association of American Geographers* 65 (1975): 514
111. See, for example, Shaler, "Caverns and Cavern Life" (1887). This piece reappeared as a chapter in *Aspects of the Earth* (1889). Also Shaler, *Outlines of Earth's History* (1898), pp. 250–63.
112. See James D. Dana, *Manual of Geology*, 4th ed. (New York: Ivison, 1895), p. 130; Ralph Stockman Tarr, *College Physiography* (New York: Macmillan, 1931), p. 93.
113. Shaler, "Caverns and Cavern Life" (1887), p. 450.
114. Shaler, "Antiquity of Caverns" (1876), p. 5.
115. Israel C. Russell, *River Development* (New York: Putnam's Sons, 1907), p. 90.
116. Tarr, *College Physiography*, p. 91.
117. Shaler, "Rivers and Valleys" (1888), pp. 147, 150.
118. W J McGee, "The Classification of Geographic Forms by Genesis," *National Geographic Magazine* 1 (1888): 27–36; W. M. Davis, "The Rivers and Valleys of Pennsylvania," *National Geographic Magazine* 1 (1889): 183–253. Davis himself was later to note the differing traditions of interpretation between the European and American schools: "In regard to the origin of plains of denudation, there prevail in Europe and this country two schools of belief: the European school, attributing denuded plains to marine action, follows Ramsay and Richthofen;

the American school, looking to subaërial denudation for the same result, follows Powell, Dutton, and others." "Plains of Marine and Subaërial Denudation," *American Geologist* 10 (1896): 96.

119. Chorley, Beckinsale, and Dunn, *Davis,* p. 160. See also Sheldon Judson, "Davis, William Morris," in *Dictionary of Scientific Biography* (New York: Scribner's, 1971), vol. 3, pp. 592–96.

120. Shaler, "Spacing of Rivers" (1899), p. 270.

121. Ibid., pp. 272, 273.

122. W. M. Davis, "Peneplains of Central France and Brittany," *Bulletin of the Geological Society of America* 12 (1901): 481.

123. Rhodes W. Fairbridge, "Gipfelflur," in *Encyclopedia of Geomorphology* (New York: Reinhold, 1968), pp. 426–27. See also Albrecht Penck, "Die Gipfelflur der Alpen," *Sitzberichte der Preussischen Akademie der Wissenschaften* (Berlin) 17 (1919): 256–68; R. A. Daly, "The Accordance of Summit Levels among the Alpine Mountains: The Fact and Its Significance," *Journal of Geology* 13 (1905): 105–25; Robert P. Beckinsale, "Penck, Albrecht," in *Dictionary of Scientific Biography,* vol. 10, pp. 501–6.

124. Flemal, "Attack on Davisian System," p. 8.

125. Shaler, "Spacing of Rivers" (1899), p. 275.

Chapter 9

1. Shaler's popularity as a teacher is further illustrated by David Starr Jordan's story that any great noise at Harvard was usually attributed to "student applause at 'one of Shaler's jokes,' even a clap of thunder thus being accounted for occasionally." *The Days of a Man; Being Memories of a Naturalist, Teacher, and Minor Prophet of Democracy,* 2 vols. (New York: World Book, 1922), vol. 1, p. 189.

2. H. Philip Bacon, "Fireworks in the Classroom: Nathaniel Southgate Shaler as a Teacher," *Journal of Geography* 54 (1955): 350.

3. Eliot Papers, UA I.5.150, Box 246, Folder: Shaler Memorial Fund; idem, Folder: Shaler, Nathaniel S.

4. Shaler Papers, HUG 1784, circular dated 20 April 1925. In an earlier circular, dated 15 June 1922, Davis appealed for the establishment of a memorial professorship to Shaler in the following way: "And Berea in particular deserves to be the beneficiary of the memorial, for it was a former student of Berea who last spring cast the deciding vote in the Kentucky Legislature, whereby the State was saved from stultifying itself by refusing to permit the teaching of evolution in its schools and colleges. No one who ever heard Shaler lecture can forget that he was a thorough-going evolutionist. How deeply would he have been mortified had his beloved Kentucky voted to stop the teaching of evolution!" Davis's proposal found favor with Bishop William Lawrence and with Theodore Roosevelt. Their letters are preserved in the Shaler Papers, HUG 1784, Folder: General.

5. Helen R. Fairbanks and Charles P. Berkey, *Life and Letters of R. A. F. Penrose, Jr.* (New York: Geological Society of America, 1952), pp. 106, 104. When completed, the bust was placed in the Faculty Room of University Hall; requests for replicas were received from the governor of Kentucky and from the Harvard Club of New York.

6. Raphael Pumpelly, *My Reminiscences* (New York: Holt, 1918), p. 123.

7. H. M. Knox, *Introduction to Educational Method* (London: Oldbourne, 1961), p. 36.
8. Lurie, *Agassiz*, p. 9.
9. A useful synopsis of Rousseau's educational philosophy is provided in Robert R. Rusk, *Doctrines of the Great Educators*, 4th ed. (London: Macmillan, 1969), pp. 157–207; see also Jean Jacques Rousseau, *Emile; or, On Education* (London: Everyman's Library, 1966; first published 1762), pp. 131–34.
10. Rusk, *Great Educators*, p. 215; Karl A. Sinnhuber, "Carl Ritter, 1779–1859," *Scottish Geographical Magazine* 75 (1959): 153–63; see also E. Plewe, "Carl Ritter," in *Encyclopaedia of the Social Sciences*, ed. Sills, vol. 13, pp. 517–20.
11. Lurie, *Agassiz*, p. 9.
12. J. B. Woodworth, manuscript on life of Shaler, Shaler Papers, HUG 1784.96.
13. Shaler, *Outlines of Earth's History* (1898), pp. 24–25.
14. Shaler, "Relations of Geologic Science to Education" (1896), p. 322.
15. Shaler, *Outlines of Earth's History* (1898), p. 25.
16. Shaler, "Rivers and Valleys" (1888), p. 131; idem, *Sea and Land* (1894), p. 2.
17. My discussion of Huxley's *Physiography* and the educational philosophy it represents draws on D. R. Stoddart, " 'That Victorian Science': Huxley's *Physiography* and Its Impact on Geography," *Transactions of the Institute of British Geographers* 66 (1975): 17–40.
18. Huxley, *Science and Education*, pp. 123, 108–9.
19. Thomas H. Huxley, *Physiography: An Introduction to the Study of Nature* (London: Macmillan, 1887), p. vii. Similarly, Peter Kropotkin noted that the geographies of the time were, with few exceptions, "a collection of information, too abstract, too incoherent, too wide, and too superficial at the same time, to be of any use in education." "The Teaching of Physiography," *Geographical Journal* 2 (1893): 350.
20. Shaler, *First Book in Geology* (1884), pp. iii–iv.
21. Quoted in Stoddart, " 'That Victorian Science,' " p. 23. See also J. W. Redway, "What Is Physiography?" *Educational Review* 10 (1895): 352–63; J. S. Keltie, "Report to the Council of the Royal Geographical Society," in *Report of the Proceedings of the Society in Reference to the Improvement of Geographical Education* (London: Royal Geographical Society, 1886).
22. Patrick Geddes, "Nature Study and Geographical Education," *Scottish Geographical Magazine* 18 (1902): 525–36. Geddes notes that the method was incorporated in the Scottish curriculum only in 1899 and in England only in 1900. He presented a paper "Facilities for Nature Study" at the Nature Study Exhibition of 1902. Details are recorded in *Nature* 66, no. 1709 (July 31 1902): 326.
23. J. H. Cowham, *The School Journey: A Means of Teaching Geography, Physiography, and Elementary Science* (London: Westminster School Book Depot, 1900). See also Stanley H. Beaver, "The Le Play Society and Field Work," *Geography* 47 (1962): 225–40.
24. Samuel Eliot Morison, *Three Centuries of Harvard* (Cambridge, Mass.: Harvard University Press, 1937), p. 355.
25. *The Organization and Progress of the Anderson School of Natural History, Report of the Trustees for 1873* (Cambridge, Mass.: Harvard University, 1874), p. 11. W. M. Davis observed regarding Agassiz's instruction at the Lawrence Scientific School: "Yet how extraordinary was the scheme of education here outlined. It was as if boys of 18 were let loose in the open field of investigation; it was truly 'actual

business from the start,' with very little systematic training; it was an excessive reaction from textbook study." "Professor Shaler and the Lawrence Scientific School," *Harvard Engineering Journal* 5 (1906): 130.

26. So, for instance, Harry G. Good, *A History of Western Education* (New York: Macmillan, 1950), p. 448; Edgar W. Knight, *Education in the United States* (Boston: Ginn, 1941), p. 337; Charles F. Thwing, *A History of Education in the United States since the Civil War* (Boston: Houghton Mifflin, 1910), pp. 246–49.

27. Manuscript "Notice concerning the Summer School," Eliot Papers, UA I.5.150, Box 120, Folder 321.

28. See Samuel Eliot Morison, ed., *The Development of Harvard University since the Inauguration of President Eliot, 1869–1929* (Cambridge, Mass.: Harvard University Press, 1930), p. 308.

29. W. W. Willoughby, "The History of Summer Schools in the United States," *Annual Report of the Commissioner of Education for the Year 1891–1892* (Washington, D.C.: Government Printing Office, 1894), p. 898.

30. Shaler, *Autobiography* (1909), p. 272. Berg notes that Shaler's role was recognized in the press at the time, citing the *New York Tribune* (9 July 1873); "The Anderson School of Natural History," *Nation* (11 September 1873); and "Penikese Island," *Harper's Weekly* (9 August 1873). See Berg, "Shaler," pp. 26–58.

31. Shaler to Louis Agassiz, 6 July 1868, reprinted in *Autobiography* (1909), p. 248. Again, Agassiz got the credit for the first endeavors to bring school and museum together. See Harry G. Good and James D. Teller, *A History of American Education,* 3d ed. (New York: Macmillan, 1973), p. 210.

32. James Lee Love, "Summer Courses of Instruction," in *Annual Report of the President and the Treasurer of Harvard College, 1906–1907* (Cambridge, Mass.: Harvard University, 1908), p. 318.

33. *Annual Report of the Trustees of the Museum of Comparative Zoology at Harvard College* (Cambridge, Mass.: Harvard University, 1870), p. 7.

34. These accounts appeared under the title "The Summer's Journey of a Naturalist" (1873).

35. Shaler, "Notice concerning the Summer School." See also *Annual Report of the Trustees of the Museum of Comparative Zoology at Harvard College* (Cambridge, Mass.: Harvard University, 1872), p. 30.

36. Shaler to Alexander Agassiz, 12 April 1873, Agassiz Papers, bAG 766.10.1.

37. *Anderson School of Natural History,* p. 13; Berg, "Shaler," p. 42.

38. Shaler, "Summer Schools" (1893), p. 456. See also Alexander Agassiz, "Abandonment of Penikese," *Popular Science Monthly* 42 (1892): 123; E. Ray Lankester, "An American Sea-Side Laboratory," *Nature* 21 (1880): 498.

39. Willoughby, "History of Summer Schools," pp. 901–3.

40. See Sally Gregory Kohlstedt, "From Learned Society to Public Museum: The Boston Society of Natural History," in *The Organization of Knowledge in Modern America, 1860–1920,* ed. Alexandra Oleson and John Voss (Baltimore: Johns Hopkins University Press, 1979), pp. 386–406; Ralph W. Dexter, "From Penikese to the Marine Biological Laboratory at Woods Hole—the Role of Agassiz's Students," *Essex Institute Historical Collections* 110 (1974): 151–61. The standard history is Frank R. Lillie, *The Woods Hole Marine Biological Laboratory* (Chicago: University of Chicago Press, 1944).

41. Shaler, chairman of Committee on Summer School of Arts and Sciences, to President and Fellows of Harvard, 25 October 1902, Eliot Papers, UA I.5.150, Box 120, Folder 321.
42. See Chorley, Beckinsale, and Dunn, *Davis,* p. 61. At the close of this summer session with Shaler, Davis was offered the position of assistant in geology at Harvard. The following year, 1876, Davis went as Shaler's assistant in the summer school excursion across eastern Tennessee into the mountains of North Carolina. See Albert Perry Brigham, "William Morris Davis," *Geographen Kalender* 7 (1909): 1–73. Also Preston E. James and Geoffrey J. Martin, eds., *The Association of American Geographers: The First Seventy-Five Years, 1904–1979* (Washington, D.C.: Association of American Geographers, 1979), pp. 11–16.
43. See "Harvard Summer School of Geology" (1876). In a letter to the editor of *Nature* entitled "Geology in America" (1875), Shaler remarked, "In our schools it is still worse: geology is taught in the air, not on the earth. The student never gets into the field for practical work, and the science remains for him a thing of names and shadows."
44. See Preston E. James and Cotton Mather, "The Role of Periodic Field Conferences in the Development of Geographical Ideas in the United States," *Geographical Review* 67 (1977): 446–61.
45. *Annual Report of the President and Treasurer of Harvard College, 1906–1907* (Cambridge, Mass.: Harvard University Press, 1907), p. 320. Shaler's devotion to the task is witnessed by his performing the duties of chairman and teacher without salary. Ibid., p. 317.
46. Berg, "Shaler," p. 54.
47. Shaler to Charles W. Eliot, 23 February 1900, Eliot Papers, UA I.5.150, Box 120, Folder 321.
48. Precisely, the public gave $71,145.33. See *Annual Report of the President and Treasurer of Harvard College, 1899–1900* (Cambridge, Mass.: Harvard University Press, 1900), p. 38. See also letter, Shaler to President and Fellows of Harvard, 25 September 1900, Eliot Papers, UA I.5.150, Box 120, Folder 321.
49. Shaler to Charles W. Eliot, 7 September 1901, Eliot Papers.
50. "School Vacations" (1889), p. 831. See also *Autobiography* (1909), p. 370.
51. Geddes, "Nature Study," p. 536.
52. Shaler, "School Vacations" (1889), pp. 831–32.
53. James S. Ross, *Groundwork of Educational Theory* (London: Harrap, 1942), p. 138.
54. Burton J. Bledstein, *The Culture of Professionalism: The Middle Class and the Development of Higher Education in America* (New York: Norton, 1976), p. 87.
55. Shaler, "Individualism in Education" (1891), p. 84.
56. Shaler, *Outlines of Earth's History* (1898), p. 403.
57. Shaler, "Individualism in Education" (1891), p. 89.
58. Shaler, *Autobiography* (1909), pp. 375–76.
59. R. H. Quick, *Some Thoughts concerning Education by John Locke* (Cambridge: Cambridge University Press, 1895), p. 43; Herbert Spencer, *Education: Intellectual, Moral, and Physical* (New York: Allison [c. 1880]), p. 288.
60. Shaler, "Individualism in Education" (1891), p. 88.
61. The other participants in the discussion were Presidents Bartlett of Dartmouth,

Angell of Michigan, Adams of Cornell, Hyde of Bowdoin, and Davis of California and Principal Dawson of McGill. See "Discipline in American Colleges," *North American Review* 149 (1889): 1–29.

62. Shaler, "Discipline in American Colleges" (1889), p. 13. See also idem, "Problem of Discipline" (1889).

63. Shaler, "Election of Studies" (1898), p. 418.

64. Shaler, "College Examinations" (1891), pp. 98, 100.

65. Rusk, *Great Educators*, p. 165.

66. Shaler, "Athletic Problem in Education" (1889), pp. 82, 83. Also *Annual Report of the President and Treasurer of Harvard College, 1892–1893* (Cambridge, Mass.: Harvard University Press, 1893), p. 13.

67. Shaler, "Humanism in Study of Nature" (1885), p. 66.

68. Shaler, "Value of Geological Science to Man" (1894), p. 173.

69. Shaler, *Autobiography* (1909), p. 367.

70. John D. Pulliman, "Shifting Patterns of Educational Thought," in *Issues and Ideas,* ed. Taylor and White, pp. 161–87. For Edwards, "epistemology begins with man's love of self, moves to the love of nature revealed through scientific laws. . . and attains its highest level in the contemplation of the nature of God." Ibid., p. 164.

71. Shaler, "Faith in Nature" (1902), p. 303; also "Chance or Design" (1889), p. 133.

72. Shaler, *Autobiography* (1909), p. 374.

73. See Carlton H. Bowyer, *Philosophical Perspectives for Education* (Glenview, Ill.: Scott, Foresman, 1970), p. 278. See also Merrill, "From Edwards to Quine"; W. B. Gallie, *Peirce and Pragmatism* (Harmondsworth: Penguin Books, 1952).

74. John S. Brubacher, *Modern Philosophies of Education,* 3d ed. (New York: McGraw-Hill, 1962), p. 315.

75. John Dewey, *The Quest for Certainty* (New York: Minton, Balch, 1929), p. 315. Underlying his support for the experimental method of teaching was his conviction that "we have no right to call anything knowledge except where our activity has actually produced certain physical changes in things, which agree with and confirm the conception entertained." *Democracy and Education* (New York: Macmillan, 1916), p. 393.

76. Thus, for example, John L. Childs, "Experimentalism and American Education," *Record* 44 (1943): 539–43.

77. Shaler, "Relations of Science to Industry" (1895), pp. 25, 27.

78. Shaler, "March of Invention" (1895).

79. Shaler, "Conquest of the Under Earth" (1895).

80. Shaler, "Use and Limits" (1890), pp. 161, 162.

81. Ibid., p. 167.

82. Bledstein, *Culture of Professionalism.* See also Laurence R. Veysey, *The Emergence of the American University* (Chicago: University of Chicago Press, 1965); Robert A. McCaughey, "The Transformation of American Academic Life: Harvard University, 1821–1892," *Perspectives in American History* 8 (1974): 239–332. For the impact on geography see Gary S. Dunbar, "Credentialism and Careerism in American Geography, 1890–1915," in *Origins,* ed. Blouet, pp. 71–88.

83. See Joseph F. Kett, "Adolescence and Youth in Nineteenth-Century America," *Journal of Interdisciplinary History* 2 (1971): 285–86.

84. See Hugh Hawkins, *Between Harvard and America: The Educational Leadership of Charles W. Eliot* (New York: Oxford University Press, 1972).

85. Bledstein, *Culture of Professionalism*, pp. 318–19.
86. Alexandra Oleson and John Voss, "Introduction," in *Organization of Knowledge*, ed. Oleson and Voss, p. x.
87. Prior to the publication of his article, Shaler wrote to Horace Scudder, editor of *Atlantic Monthly*, observing that "President Walker is in a somewhat excited state of mind concerning the advance in the Lawrence Scientific School." Shaler to Scudder, 17 May 1893, Horace Scudder Correspondence, bMS 801.4. In this letter Shaler also suggested that Walker's rejoinder to his article should appear in a subsequent issue of the journal rather than immediately following his own. He offered to let Walker see the manuscript so that he could prepare his response.
88. "Relations of Academic and Technical Instruction" (1893), pp. 260, 261.
89. In another letter to Scudder he proposed writing a piece, as he later did, called "The Transmission of Learning through the University." He hoped eventually to produce a booklet "Universities and Their Place in Modern Society." Such a booklet, however, never appeared. Shaler to Scudder, 21 September 1893, Scudder Correspondence.
90. Shaler, "Relations of Academic and Technical Instruction" (1893), pp. 262, 265.
91. Shaler to Scudder, undated, quoted in Berg, "Shaler," p. 185.
92. Walker to Scudder, 17 May 1893, Scudder Correspondence.
93. Francis A. Walker, "The Technical School and the University," *Atlantic Monthly* 72 (1893): 393. This article appeared in the September issue, Shaler's having appeared in August.
94. Davis, "Shaler and Lawrence Scientific School," provides the following figures for students admitted to the school: 27 in 1888, 44 in 1889, 53 in 1890, 65 in 1891 (Shaler's first year as dean), and 122 in 1892. By 1900, the number of new students had risen to 187, the total enrollment being 507.
95. Alexander Agassiz to Charles W. Eliot, 3 March 1902, Eliot Papers, Box 100, Folder 3, Alexander Agassiz.
96. Alexander Agassiz to Charles W. Eliot, 27 October 1901, Eliot Papers. On 11 October he had written complaining that he had gone into Davis's "room and found him instructing a class of women—I take it Radcliffe. You may remember I protested against Davis teaching elementary work and was told it was only temporary." The previous year, on 17 August 1900, he had written to Eliot: "It is becoming very evident that a mistake has been made in appointing Davis Sturgis Hooper Professor. . . . [He] is crazy for power and promotion, and I told Woodworth and Henshaw to drop all discussions with him." For some reason, the major study of Davis by Chorley, Beckinsale, and Dunn makes no reference to this continued feud between Agassiz and Davis and even comments that, after 1899, there is "an absence of reports of friction with Agassiz." *Davis*, p. 253.
97. Davis to Charles W. Eliot, 6 October 1903, Eliot Papers, Box 106, Folder 68.
98. Shaler to Horace Scudder, 6 August 1892, Scudder Correspondence.
99. See Love, *Lawrence Scientific School;* Rossiter, "Lawrence Scientific School."
100. Quoted in Russett, *Darwin in America*, p. 11.
101. See Walter Humes, "Evolution and Educational Theory in the Nineteenth Century," in *The Wider Domain of Evolutionary Thought*, ed. David Oldroyd and Ian Langham (Dordrecht: Reidel Publishing, 1983), pp. 27–56.
102. Quoted in Russell, *Darwin in America*, p. 11. See also Lester D. Stephens, "Joseph

Le Conte on Evolution, Education, and the Structure of Knowledge," *Journal of the History of the Behavioral Sciences* 12 (1976): 103–19; idem, "Joseph Le Conte's Evolutional Idealism: A Lamarckian View of Cultural History," *Journal of the History of Ideas* 39 (1978): 465–80; idem, *Le Conte: Gentle Prophet*.

103. Paul Buck, "Introduction," in *Social Sciences at Harvard, 1860–1920: From Inculcation to the Open Mind*, ed. Paul Buck (Cambridge, Mass.: Harvard University Press, 1965), p. 11.

104. See Arthur G. Powell, "The Education of Educators at Harvard, 1891–1912," in *Social Sciences at Harvard*, ed. Buck, pp. 223–74.

105. Shaler, "Transmission of Learning" (1894), p. 115.

106. Ibid., pp. 118–19.

107. On the widespread use of the organic metaphor, see D. C. Phillips, "Organicism in the Late Nineteenth and Early Twentieth Centuries," *Journal of the History of Ideas* 31 (1970): 413–32; Bannister, *Social Darwinism;* Roger Cooter, "The Power of the Body: The Early Nineteenth Century," in *Natural Order: Historical Studies of Scientific Culture*, ed. Barry Barnes and Steven Shapin (Beverly Hills: Sage Publications, 1979), pp. 73–92. For the use of the analogy in geography, see Stoddart, "Organicism and Ecosystem as Geographical Models"; David N. Livingstone and Richard T. Harrison, "Meaning through Metaphor: Analogy as Epistemology," *Annals of the Association of American Geographers* 71 (1981): 95–107.

108. Lester Frank Ward, it may be noted in passing, was likewise committed to education as a long-term instrument for the improvement of humanity and was equally reluctant to surrender Lamarckism. Nevertheless he did concede, in an article "The Transmission of Culture," first published in 1891, that acquired knowledge per se cannot be transmitted. The capacity to acquire such knowledge, however, was another matter, and he urged that talents obviously running in families were to be explained by inheritance and not by natural selection. The article is reprinted in Lester Frank Ward, *Glimpses of the Cosmos*, 6 vols. (New York: Putnam's Sons, 1913–18), vol. 4, pp. 246–52.

109. Shaler, "Direction of Education" (1895), p. 391.

Epilogue

1. C. S. Lewis, *The Abolition of Man; or, Reflections on Education with Special Reference to the Teaching of English in the Upper Forms of Schools* (London: Blis, 1943), p. 40.

Bibliography

The Works of Nathaniel Southgate Shaler

1861

"Lateral Symmetry in Brachiopoda." *Proceedings of the Boston Society of Natural History* 8: 274–79.
"On the Geology of Anticosti Island, Gulf of Saint Lawrence." *Proceedings of the Boston Society of Natural History* 8: 285–87.

1865

"List of the Brachiopoda from the Island of Anticosti, Sent by the Museum of Comparative Zoology to Different Institutions in Exchange for Other Specimens, with Annotations." *Bulletin of the Museum of Comparative Zoology* 1. 61–70.
"Preliminary Notice of Some Opinions concerning the Mode of Elevation of Continental Masses." *Proceedings of the Boston Society of Natural History* 10: 237–39.

1866

"Notes on the Position and Character of Some Glacial Beds Containing Fossils at Gloucester, Massachusetts." *Proceedings of the Boston Society of Natural History* 11: 27–30.
"On the Formation of Mountain Chains." *Proceedings of the Boston Society of Natural History* 11: 8–15.
"On the Formation of the Excavated Lake Basins of New England." *Proceedings of the Boston Society of Natural History* 10: 358–66.
"On the Modifications of Ocean Currents in Successive Geological Periods." *Proceedings of the Boston Society of Natural History* 10: 269–302.

341

1868

"Considerations concerning the Absence of Distinct Evidences of Glacial Action in the Valley of the Yukon River, Alaska." *Proceedings of the Boston Society of Natural History* 12: 145–59.

"On the Disappearance of the Cane from the Central Part of the Ohio Valley." *Proceedings of the Boston Society of Natural History* 12: 136–37.

"On the Formation of Mountain Chains." *Geological Magazine* 5: 511–17.

"On the Nature of the Movements Involved in the Changes of Level of Shorelines." *Proceedings of the Boston Society of Natural History* 12: 128–36.

"Report on the Collection of Fossil Remains in General." *Annual Report of the Museum of Comparative Zoology*, pp. 41–44.

1869

"Earthquakes." *Atlantic Monthly* 23: 676–85.

"Earthquakes of the American Continents." *Atlantic Monthly* 24: 461–69.

"Earthquakes of the Western United States." *Atlantic Monthly* 24: 549–59.

"Great Earthquakes of the Old World." *Atlantic Monthly* 24: 140–50.

"Note on the Concentric Structure of Granitic Rocks." *Proceedings of the Boston Society of Natural History* 12: 289–93.

"Note on the Occurrence of the Remains of *Tarandus Rangifer* Gray at Big Bone Lick, in Kentucky." *Proceedings of the Boston Society of Natural History* 13: 167.

"On Changes in Geographical Distribution of the American Buffalo." *Proceedings of the Boston Society of Natural History* 13: 136.

"On the Relations of the Rocks in the Vicinity of Boston." *Proceedings of the Boston Society of Natural History* 13: 172–77. (Abstract, *American Naturalist* 5 [1871]: 278.)

1870

"California Earthquakes." *Atlantic Monthly* 25: 351–60.

"An Ex-Southerner in South Carolina." *Atlantic Monthly* 26: 53–61.

"Father Blumhardt's Prayerful Hotel." *Atlantic Monthly* 26: 712–17.

"Note on the Glacial Moraines of the Charles River Valley, near Watertown." *Proceedings of the Boston Society of Natural History* 13: 277–79.

"On the Parallel Ridges of Glacial Drift in Eastern Massachusetts, with Some Remarks on the Glacial Period." *Proceedings of the Boston Society of Natural History* 13: 196–204.

"On the Phosphate Beds of South Carolina." In *United States Coast Survey Report for 1870*, pp. 182–89. Washington, D.C.: Government Printing Office.

"On the Phosphate Beds of South Carolina." *Proceedings of the Boston Society of Natural History* 13: 222–36.

"The Time of the Mammoths." *American Naturalist* 4: 148–66.

1871

"On the Causes Which Have Led to the Production of Cape Hatteras." *Proceedings of the Boston Society of Natural History* 14: 110–21. (Abstract, *American Naturalist* 5: 178–81.)

"Sources of Boulders in Ohio, Kentucky." *Proceedings of the Boston Society of Natural History* 16: 386.

1872

"Effects of Extraordinary Seasons on the Distribution of Animals and Plants." *American Naturalist* 6: 1–3.

"Geology of Martha's Vineyard and Nantucket." *Proceedings of the Boston Society of Natural History* 15: 219.

"Note on the Origin of Our Domestic Cat." *Proceedings of the Boston Society of Natural History* 15: 159–62.

"On Elongation of Pebbles." *Proceedings of the Boston Society of Natural History* 15: 2.

"On the Effects of the Vertical Position in Man." *Proceedings of the Boston Society of Natural History* 15: 188–91.

"On the Geology of the Island of Aquidneck and the Neighboring Parts of the Shores of Narragansett Bay." *American Naturalist* 6: 518–28, 611–21, 751–60.

"The Rattlesnake and Natural Selection." *American Naturalist* 6: 32–37.

1873

"Mixed Populations of North Carolina." *North American Review* 116: 150–66.

"Notes on the Right and Sperm Whales." *American Naturalist* 7: 1–4.

"The Summer's Journey of a Naturalist." *Atlantic Monthly* 31: 707–18; 32: 181–89, 349–57.

1874

"Martha's Vineyard." *Atlantic Monthly* 34: 732–40.

"The Moon." *Atlantic Monthly* 34: 270–78.

"Notes on Some of the Phenomena of Elevation and Subsidence of the Continents." *Proceedings of the Boston Society of Natural History* 17: 288–92.

"On the Geology of the Region about Richmond, Virginia." *Proceedings of the American Academy of Arts and Sciences* 4: 307–8.

"Recent Changes of Level on the Coast of Maine, with Reference to Their Origin and Relation to Other Similar Changes." *Memoirs of the Boston Society of Natural History* 2: 321–40.

Report of Special Committee. Boston: Wright & Potter. (Report of experts to a committee of the state board of education on a proposed scientific survey of Massachusetts.)

1875

"Geology in America." *Nature* 12: 5–6.

"Note on Some Points Connected with Tidal Erosion." *Proceedings of the Boston Society of Natural History* 17: 465–66.

"Note on the Geological Relations of Boston and Narragansett Bays." *Proceedings of the Boston Society of Natural History* 17: 488–90.

"On the Antiquity of the Caverns and Cavern Life of the Ohio Valley." *Memoirs of the Boston Society of Natural History* 2: 355–63.

"On the Cause and Geological Value of Variation in Rainfall." *Proceedings of the Boston Society of Natural History* 18: 176–82.

"Propositions concerning the Motion of Continental Glaciers." *Proceedings of the Boston Society of Natural History* 18: 126–33.

Question Guide to the Environs of Boston: Designed for the Use of Beginners in Geology in the Classes of Harvard University. Pt. 1, *Somerville and Cambridge.* Cambridge, Mass.: Sever.

"Remarks on the Geology of the Coast of Maine, New Hampshire, and That Part of Massachusetts North of Boston." In *United States Coast Survey, Coast Pilot for the Atlantic Seaboard, Gulf of Maine, and Its Coast from Eastport to Boston,* pp. 883–88. Washington, D.C.: Government Printing Office.

"Some Considerations on the Possible Means Whereby a Warm Climate May Be Produced within the Arctic Circle." *Proceedings of the Boston Society of Natural History* 17: 332–37.

"A State Survey for Massachusetts." *Atlantic Monthly* 35: 357–63.

1876

A General Account of the Commonwealth of Kentucky Prepared by the Geological Survey of the Commonwealth for the Centennial Exhibition at Philadelphia, 1876. Cambridge, Mass.: Wilson & Son.

"The Harvard Summer School of Geology." *American Naturalist* 10: 29–31.

"Introduction to Report on the Botany of Barren and Edmonson Counties by John Hussey." *Geological Survey of Kentucky,* 2d ser., 1: 27–58.

"On the Age of the Bison in the Ohio Valley." Appendix to *The American Bison, Living and Extinct,* by J. A. Allen, pp. 232–36. Memoirs of the Kentucky Geological Survey 1, pt. 2. Cambridge, Mass.: Harvard University Press.

On the Antiquity of the Caverns and Cavern Life of the Ohio Valley. Memoirs of the Kentucky Geological Survey 1, pt. 1. Cambridge, Mass.: Harvard University Press. (Abstract, *American Journal of Science,* 3d ser., 13 [1877]: 226–27.)

On the Fossil Brachiopods of the Ohio Valley. Memoirs of the Kentucky Geological Survey 1, pt. 3. Cambridge, Mass.: Harvard University Press.

With Lucien Carr. *On the Prehistoric Remains of Kentucky.* Memoirs of the Kentucky Geological Survey 1, pt. 4. Cambridge, Mass.: Harvard University Press.

With A. R. Crandall. "Report on the Forest Timber of Greenup, Carter, Boyd, and Lawrence Counties." *Geological Survey of Kentucky,* 2d ser., 1: 1–26.

1877

"Annual Report of N. S. Shaler for the Year 1876." *Geological Survey of Kentucky,* 2d ser., 3: 283–315.

"Annual Report of N. S. Shaler, State Geologist, for the Year 1877." *Geological Survey of Kentucky,* 2d ser., 3: 365–414.

"Description of the Preliminary Topographical and Geological Maps of Kentucky, Edition of 1877." *Geological Survey of Kentucky,* 2d ser., 3: 347–64.

"A General Account of the Commonwealth of Kentucky, Prepared by the Geological Survey of the Commonwealth." *Geological Survey of Kentucky,* 2d ser., 2: 361–468. (Previously published in 1876.)

"General Report of the Geological Survey of Kentucky." *Geological Survey of Kentucky*, 2d ser., 3: 1–30.

"History of the Operations of the Survey in 1874 and 1875." *Geological Survey of Kentucky*, 2d ser., 3: 31–127.

"How to Change the North American Climate." *Atlantic Monthly* 40: 724–31.

"Notes on the Age and Structure of the Several Mountain Axes in the Neighborhood of Cumberland Gap." *American Naturalist* 11: 385–92.

"Notes on the Investigations of the Kentucky Geological Survey during the Years 1873, 1874, and 1875." *Geological Survey of Kentucky*, 2d ser., 3: 129–240.

"On the Existence of the Allegheny Division of the Appalachian Range within the Hudson Valley." *American Naturalist* 11: 627–28.

"On the Origin of the Galena Deposits of the Upper Cambrian Rocks of Kentucky." *Geological Survey of Kentucky*, 2d ser., 2: 277–92.

"Report on the Unfinished Work of the Survey of the Commonwealth under the Direction of Dr. David Dale Owen." *Geological Survey of Kentucky*, 2d ser., 3: 415–20.

"The Transportation Routes of Kentucky and Their Relation to the Economic Resource of the Commonwealth." *Geological Survey of Kentucky*, 2d ser., 3: 316–46.

1878

"Mammoth Cave." In *Johnson's New Universal Cyclopaedia: A Scientific and Popular Treasury of Useful Knowledge*, editors-in-chief Frederick A. P. Barnard and Arnold Guyot. 4 vols. New York: Johnson & Son.

"Reelfoot Lake." *Atlantic Monthly* 47: 216–22.

"The Silver Question Geologically Considered." *Atlantic Monthly* 46: 620–29.

Thoughts on the Nature of Intellectual Property, and Its Importance to the State. Boston: Osgood.

1879

"The Natural History of Politics." *Atlantic Monthly* 43: 302–10.

"Notes on Certain Evidences of a Gradual Passage from Sedimentary to Volcanic Rocks Shown in the Brighton District of Boston." *Proceedings of the Boston Society of Natural History* 20: 129–33.

"Notes on the Submarine Coast-Shelf, or Hundred-Fathom Detrital Fringe." *Proceedings of the Boston Society of Natural History* 20: 278–82.

"On the Improvement of the Rivers of Kentucky." *Bulletin of the Kentucky Geological Survey* 2: 13–21.

"Petroleum." *Bulletin of the Kentucky Geological Survey* 1: 5–20.

"Sleep and Dreams." *International Review* 6: 234–37.

"The Use of Numbers in Society." *Atlantic Monthly* 44: 326–33.

With J. R. Proctor. "On the Importance of Improvement in the Navigation of the Kentucky River to the Mining and Manufacturing Interests of Kentucky." *Bulletin of the Kentucky Geological Survey* 3: 22–45.

1880

"Future of Precious Metal Mining in the United States." *Atlantic Monthly* 45: 765–74.
"The Future of Weather Foretelling." *Atlantic Monthly* 46: 645–51.
"General Report on the Building Stones of Rhode Island, Massachusetts, and Maine."
    Tenth Census Report, pt. 10: 107–15.
"Mica Mines of New England." *Tenth Census Report*, pt. 15: 833–36.
"Outline of the Geology of Boston and Its Environs." In *The Memorial History of
    Boston*, edited by Justin Winsor, vol. 1, pp. 1–8. 4 vols. Boston: Osgood.
"Precious Metal Mining in the United States." *Kansas City Review of Science and
    Industry* 4: 95–96.
*Preliminary Report concerning the Resources of the Country Adjacent to the Line of the
    Proposed Richmond and Southwestern Railway.* Cambridge, Mass.: Wheeler.
"Proposition concerning the Classification of Lavas, Considered with Reference to
    the Circumstances of Their Extrusion." *Anniversary Memoirs of the Boston Society
    of Natural History*, pp. 1–15. (Published in celebration of the fiftieth anniversary
    of the Society's foundation.)
Summary of the Work of the Geological Survey for the Years 1878–1879. Frankfort: Kentucky
    Geological Survey.

1881

"Great Kanawha, West Virginia, Iron Ores and Coals: The Black-Band Iron and Coal
    Company." *Virginias* 2: 154–55.
The Island of Campobello: Preliminary Report. Cambridge, Mass.: Wheeler.
"On the Recent Advances and Retrocessions of Glaciers." *Proceedings of the Boston
    Society of Natural History* 21: 162–67.
*Report on the Resources of the Country Traversed by the Virginia, Kentucky, and Ohio
    Railroad, and the Paris, Georgetown, and Frankfort Railroad, in the States of Virginia
    and Kentucky.* Cambridge, Mass.: Wheeler.
"The Value of University Records." *Harvard Register* 3: 200–201.
"A Winter Journey in Colorado." *Atlantic Monthly* 47: 46–55.
With W. M. Davis. *Illustrations of the Earth's Surface.* Vol. 1, *Glaciers.* Boston: Osgood.

1882

"Hurricanes." *Atlantic Monthly* 49: 330–36.

1883

"American Swamp Cypress." *Science* 2: 38–40.
"The Floods of the Mississippi Valley." *Atlantic Monthly* 51: 653–60.
"Improvement of the Native Pasture-Lands of the Far West." *Science* 1: 186–87.
The Kentucky Union Railway Company. Lexington: Kentucky Union Railway
    Company.
Report on the Croton Magnetic Iron Mines. Cambridge, Mass.: Wheeler.

1884

A First Book in Geology: Designed for the Use of Beginners. Boston: Ginn, Heath. (Translated into Polish by Henryk Wernic as *Dzieje Ziemi, Czyli Poczatki Geologii* [Warsaw: Slosarski i Siemeradski, 1888]; translated into German by C. von Karczewska as *Elementarbuch der Geologie für Anfänger* [Dresden: Schultze, 1903].)

Kentucky: A Pioneer Commonwealth. Boston: Houghton, Mifflin.

"The Natural History of Sympathy." *Christian Register,* 1 May, pp. 280–81. (Lecture delivered before the students of the Harvard Divinity School.)

"The Negro Problem." *Atlantic Monthly* 54: 696–709.

"On the Origin of Kames." *Proceedings of the Boston Society of Natural History* 23: 36–44.

"The Red Sunsets." *Atlantic Monthly* 53: 475–82.

Report on the Resources of the Region Adjacent to the Proposed Cincinnati and Southwestern Railway. Cambridge, Mass.: Wheeler.

1885

"Annual Report, Atlantic Coast Division." In *Sixth Annual Report of the United States Geological Survey, 1884 85,* pp. 18 22. Washington, D.C.: Government Printing Office.

"Humanism in the Study of Nature." *Science* 6: 64–66.

"Preliminary Report on Sea-Coast Swamps of the Eastern United States." In *Sixth Annual Report of the United States Geological Survey, 1884–85,* pp. 353–98. Washington, D.C.: Government Printing Office.

1886

"Preliminary Report on the Geology of the Cobscock Bay District, Maine." *American Journal of Science,* 3d ser., 32: 35–60. (Abstract, *American Naturalist* 20: 969.)

"Race Prejudices." *Atlantic Monthly* 58: 510 18.

Report of the Tremont Street Conglomerate Quarry of Boston, Mass. Cambridge, Mass.: n.p.

"The Swamps of the United States." *Science* 7: 232–33.

Use of a Topographical and Geological Survey to the People. Cambridge, Mass.: n.p. (Letter and report to Hon. Samuel J. Randall.)

With W. M. Davis and T. W. Harris. *A Series of Twenty-Five Colored Geological Models and Twenty-Five Photographs of Important Geological Objects, Each Accompanied by Letter-Press Description.* Boston: Heath.

1887

"Caverns and Cavern Life." *Scribner's Magazine* 2: 449–72.

"The Earthquake in the Riviera." *Epoch,* March, pp. 84–85.

"Field Geology." *Popular Science Monthly* 21: 80–82, 94–96.

"Fluviatile Swamps of New England." *American Journal of Science,* 3d ser., 33: 210–21.

"Forests of North America." *Scribner's Magazine* 1: 561–80.

"The Instability of the Atmosphere." *Scribner's Magazine* 2: 197–221.
"The Natural Gas Supply." *Forum*, May, pp. 305–12.
"Notes on the *Taxodium Distichium*, or Bald Cypress." *Memoirs of the Museum of Comparative Zoology* 16: 1.
"One of the Needs of American Universities." *Harvard Monthly* 3: 171–75.
"On the Original Connection of the Eastern and Western Coal-Fields of the Ohio Valley." *Memoirs of the Museum of Comparative Zoology* 16: 2.
"The Stability of the Earth." *Scribner's Magazine* 1: 259–79.

1888

"Animal Agency in Soil-Making." *Popular Science Monthly* 32: 484–87.
"Annual Report, Atlantic Coast Division." In *Seventh Annual Report of the United States Geological Survey, 1885–'86*, pp. 61–65. Washington, D.C.: Government Printing Office.
"The Crenitic Hypothesis and Mountain Building." *Science* 11: 280–81.
Directions for the Teaching of Geology: A Manual for the Teacher, to Accompany "The First Book in Geology." Boston: Heath.
"The Immigration Problem Historically Considered." *America* 1, no. 30: 1–2.
"The Immigration Problem Historically Considered. Second Paper." *America* 1, no. 31: 1–2.
"Introduction" to *Nature and Origin of Deposits of Phosphate of Lime* by R. A. F. Penrose, Jr. United States Geological Survey Bulletin no. 46. Washington, D.C.: Government Printing Office.
"The Law of Fashion." *Atlantic Monthly* 61: 386–98.
"On the Geology of the Cambrian District of Bristol County, Massachusetts." *Bulletin of the Museum of Comparative Zoology* 16: 13–26.
"Origin of the Divisions between the Layers of Stratified Rocks." *Proceedings of the Boston Society of Natural History* 23: 308–19.
"Report on the Geology of Martha's Vineyard." In *Seventh Annual Report of the United States Geological Survey, 1885–'86*, pp. 297–363. Washington, D.C.: Government Printing Office.
"Rivers and Valleys." *Scribner's Magazine* 4: 131–55.
"Volcanoes." *Scribner's Magazine* 3: 201–26.

1889

"The Action of Glaciers." *Chautauquan* 10: 405–9.
"Annual Report: Atlantic Coast Division." In *Ninth Annual Report of the United States Geological Survey, 1887–88*, pp. 71–74. Washington, D.C.: Government Printing Office.
Aspects of the Earth: A Popular Account of Some Familiar Geological Phenomena. New York: Scribner's Sons.
"The Athletic Problem in Education." *Atlantic Monthly* 63: 79–88.
"The Cause of Geographic Conditions." *Chautauquan* 10: 148–52.
"Chance or Design." *Andover Review* 12: 117–33.
"The Common Roads." *Scribner's Magazine* 6: 473–83.
"Discipline in American Colleges." *North American Review* 149: 10–15.

"Effects of Permanent Moisture on Certain Forest Trees." *Science* 13: 176–77.

"The Geology of Cape Ann, Massachusetts." *Ninth Annual Report of the United States Geological Survey, 1887–88*, pp. 529–611. Washington, D.C.: Government Printing Office.

The Geology of Nantucket. United States Geological Survey Bulletin no. 53. Washington, D.C.: Government Printing Office. (Abstracts, *American Geologist* 5 [1890]: 111–14; *Popular Science Monthly* 36: [1890]: 125–28.)

"The Geology of the Island of Mount Desert, Maine." In *Eighth Annual Report of the United States Geological Survey, 1886–'87*, pp. 987–1061. Washington, D.C.: Government Printing Office.

"The Last Gift of the Nineteenth Century." *North American Review* 149: 674–84.

"On the Occurrence of Fossils of the Cretaceous Age on the Island of Martha's Vineyard." *Bulletin of the Museum of Comparative Zoology* 16: 89–97.

"Physiography of North America." In *Narrative and Critical History of America*, edited by Justin Winsor, vol. 4, pp. i–xxx. 8 vols. Boston: Houghton Mifflin, 1884–89.

"The Problem of Discipline in Higher Education." *Atlantic Monthly* 64: 24–37.

"Report: Division of Coast Line Geology." In *Eighth Annual Report of the United States Geological Survey, 1886–'87*, pp. 125–28. Washington, D.C.: Government Printing Office.

"School Vacations." *Atlantic Monthly* 64: 824–33.

"The Sense of Honor in Americans." *North American Review* 149: 203–14.

"The Work of Underground Water." *Chautauquan* 10: 276–80.

"The Work of Waves." *Chautauquan* 10: 546–50.

1890

"The African Element in America." *Arena* 2: 660–73.

"Annual Report: Atlantic Coast Division." In *Tenth Annual Report of the United States Geological Survey, 1888–89*, pp. 117–19. Washington, D.C.: Government Printing Office.

"Critical Points in the Continuity of Natural Phenomena." *Unitarian Review* 33: 1–18.

"The Economic Future of the New South." *Arena* 2: 257–69.

"General Account of the Fresh-Water Morasses of the United States with a Description of the Dismal Swamp District of Virginia and North Carolina." In *Tenth Annual Report of the United States Geological Survey, 1888–89*, pp. 255–339. Washington, D.C.: Government Printing Office.

"The Knees of the Bald Cypress." *Garden and Forest* 3 (29 January): 57.

"Nature and Man in America." *Scribner's Magazine* 8: 360–76, 473–84, 645–56.

"The Nature of the Negro." *Arena* 3: 23–35.

"Note on Glacial Climate." *Proceedings of the Boston Society of Natural History* 24: 460–65.

"Note on the Value of Saliferous Deposits as Evidence of Former Climatal Conditions." *Proceedings of the Boston Society of Natural History* 24: 580–85.

"The Peculiarities of the South." *North American Review* 151: 477–88.

"Remarks on Conditions Attending a Pleistocene Submergence on the Atlantic Coast." *Bulletin of the Geological Society of America* 1: 409.

"Rock Gases." *Arena* 2: 631–42.
"Science and the African Problem." *Atlantic Monthly* 66: 36–45.
Soils of Massachusetts. Boston. (Lecture delivered at the annual meeting of the Massachusetts State Board of Agriculture, 6 February 1890.)
"Tertiary and Cretaceous Deposits of Eastern Massachusetts." *Bulletin of the Geological Society of America* 1: 443–52.
"The Topography of Florida." *Bulletin of the Museum of Comparative Zoology* 16: 139–56. (Abstract, *American Naturalist* 24: 768.)
"The Use and Limits of Academic Culture." *Atlantic Monthly* 66: 160–70.

1891

"The Antiquity of the Last Glacial Period." *Proceedings of the Boston Society of Natural History* 25: 258–67.
"College Examinations." *Atlantic Monthly* 68: 95–102.
"Individualism in Education." *Atlantic Monthly* 67: 82–90.
Nature and Man in America. New York: Scribner's Sons.
Report of the Massachusetts Topographical Survey Commission. Boston: Wright & Potter.
With F. A. Walker and H. L. Whiting. *Atlas of Massachusetts*. Boston: Wright & Potter. (Atlas prepared from topographical surveys made in cooperation by the United States Geological Survey and the Commissioners of the Commonwealth, 1885–88.)

1892

"The Betterment of Our Highways." *Atlantic Monthly* 70: 505–14.
"The Border State Men of the Civil War." *Atlantic Monthly* 69: 245–57.
"The Depths of the Sea." *Scribner's Magazine* 12: 77–95.
"Icebergs." *Scribner's Magazine* 12: 181–200.
"The Origin and Nature of Soils." In *Twelfth Annual Report of the United States Geological Survey, 1890–91*, pp. 219–345. Washington, D.C.: Government Printing Office. (Abstracts, *American Journal of Science*, 3d ser., 45 [1893]: 163–64; *American Geologist* 14 [1894]: 114–15.)
"Our Costly Geological Survey." *Engineering Magazine* 4: 221–32.
"Remarks on the Life of Samuel Dexter." *Proceedings of the Boston Society of Natural History* 25: 365–66.
"Sea and Land." *Scribner's Magazine* 11: 611–27.
"Sea Beaches." *Scribner's Magazine* 11: 728–75.
"Shall the Professions Be Regulated?" *Engineering Magazine* 4: 9–15.
The Story of Our Continent: A Reader in the Geography and Geology of North America. Boston: Ginn.

1893

"The Conditions of Erosion beneath Deep Glaciers, Based upon a Study of the Bowlder Train from Iron Hill, Cumberland, Rhode Island." *Bulletin of the Museum of Comparative Zoology* 16: 185–225. (Abstracts, *American Geologist* 12: 191–92; *American Naturalist* 12: 662.)

"European Peasants as Immigrants." *Atlantic Monthly* 71: 646–55.
"The Geological History of Harbors." In *Thirteenth Annual Report of the United States Geological Survey, 1891–'92,* pt. 2, pp. 93–209. Washington, D.C.: Government Printing Office.
"High Buildings and Earthquakes." *North American Review* 156: 338–45.
The Interpretation of Nature. New York: Houghton Mifflin.
"Man and the Glacial Period." *American Geologist* 11: 180–84.
"Relations of Academic and Technical Instruction." *Atlantic Monthly* 72: 259–68.
"The Summer Schools." *Harvard Graduates' Magazine* 1: 455–59.
"Undiscovered Mineral Wealth of the World." *Donahoe's Magazine,* May, pp. 689–94.
"What Is Geology?" *Chautauquan* 18: 284–87.
With G. S. Perkins and W. E. McClintock. *Report of the Commission to Improve the Highways of the Commonwealth of Massachusetts.* Boston: Wright & Potter.

1894

"Beasts of Burden." *Scribner's Magazine* 16: 83–100.
"Discussion on Facetted Pebbles." *Bulletin of the Geological Society of America* 5: 608.
"The Dog." *Scribner's Magazine* 16: 692–711.
"The Horse." *Scribner's Magazine* 16: 566–86.
"On the Distribution of Earthquakes in the United States since the Close of the Glacial Period." *Proceedings of the Boston Society of Natural History* 26: 246–56. (Abstract, *American Geologist* 14: 396–97.)
"Phenomena of Beach and Dune Sands." *Bulletin of the Geological Society of America* 5: 207–12.
"Pleistocene Distortions of the Atlantic Seacoast." *Bulletin of the Geological Society of America* 5: 199–202.
"Relation of Mountain Growth to Formation of Continents." *Bulletin of the Geological Society of America* 5: 203–6.
Sea and Land: Features of Coasts and Oceans with Special Reference to the Life of Man. New York: Scribner's Sons.
"The Transmission of Learning through the University." *Atlantic Monthly* 73: 115–24.
"The Value of Geological Science to Man." *Chautauquan* 20: 170–74.
Editor. *The United States of America: A Study of the American Commonwealth, Its Natural Resources, People, Industries, Manufactures, Commerce, and Its Work in Literature, Science, Education, and Self-Government.* 3 vols. New York: Appleton.
With W. B. Dwight, editor of geological terms. *Standard Dictionary of the English Language,* vol. 1. New York: Funk & Wagnalls.
With H. L. Whiting and D. Fitzgerald. *Report of the Commission of the Topographical Survey of the Commonwealth of Massachusetts for the Year 1893.* Boston: Wright & Potter.

1895

"Certain Features of Jointing and Veining of the Lower Silurian Limestones near Cumberland Gap, Tennessee." (Title of paper read, *Bulletin of the Geological Society of America* 6: 443.)

"Conditions and Effects of the Expulsion of Gases from the Earth." *Proceedings of the Boston Society of Natural History* 27: 87–106. (Title of paper read, *Bulletin of the Geological Society of America* 7: 11.)

"The Conquest of the Under Earth." *Chautauquan* 22: 280–84.

"The Direction of Education." *Atlantic Monthly* 75: 389–97.

"Dislocations of the Cretaceous and Tertiary Rocks of Martha's Vineyard." *Bulletin of the Geological Society of America* 6: 7.

Domesticated Animals: Their Relation to Man and to His Advancement in Civilization. New York: Scribner's Sons.

"Evidences as to Change of Sealevel." *Bulletin of the Geological Society of America* 6: 141–66.

"The Geology of the Road-Building Stones of Massachusetts, with Some Consideration of Similar Materials from Other Parts of the United States." In *Sixteenth Annual Report of the United States Geological Survey, 1894–'95,* pt. 2, pp. 277–341. Washington, D.C.: Government Printing Office.

"The March of Invention." *Chautauquan* 22: 151–54.

"Natural Science Training for Engineers." *Engineering Magazine* 7: 1021–26.

"Origin, Distribution, and Commercial Values of Peat Deposits." In *Sixteenth Annual Report of the United States Geological Survey, 1894–'95,* pt. 4, pp. 305–14. Washington, D.C.: Government Printing Office.

"Preliminary Report on the Geology of the Common Roads of the United States." In *Fifteenth Annual Report of the United States Geological Survey, 1893–'94,* pp. 255–306. Washington, D.C.: Government Printing Office.

"The Relation of Science to Industry." *Chautauquan* 22: 24–28.

"Scientific Aspects of the Negro Question." *Public Opinion* 18: 147.

"The Share of Volcanic Dust and Pumice in Marine Deposits." *Bulletin of the Geological Society of America* 7: 490–92.

"Some Causes of the Imperfection of the Geological Record." *Science,* n.s., 2: 858–59.

With W. B. Dwight and J. B. Woodworth, editor of geological terms. *Standard Dictionary of the English Language,* vol. 2. New York: Funk & Wagnalls.

With G. H. Perkins and W. C. McClintock. *Second Annual Report of the Massachusetts Highway Commission.* Boston: Wright & Potter.

1896

American Highways: A Popular Account of Their Conditions and of the Means by Which They May Be Bettered. New York: Century.

"Beaches and Tidal Marshes of the Atlantic Coast." *National Geographic Monographs* 1: 137–68.

"Discussion on the Carriage of Bowlders by the Indians." *American Geologist* 17: 104. (Abstract.)

"Discussion regarding Low Temperature Gradients in Mines." *American Geologist* 17: 100. (Abstract.)

"The Economic Aspects of Soil Erosion." *National Geographic Magazine* 7: 328–38, 368–77.

"Environment and Man in New England." *North American Review* 162: 726–37.

"Relations of Geologic Science to Education." *Bulletin of the Geological Society of America* 7: 315–26.

"The Scotch Element in the American People." *Atlantic Monthly* 77: 508–16.
With G. A. Perkins and W. E. McClintock. *Third Annual Report of the Massachusetts Highway Commission.* Boston: Wright & Potter.
With J. B. Woodworth and C. F. Marbut. "Glacial Brick Clays of Rhode Island and Southeastern Massachusetts." In *Seventeenth Annual Report of the United States Geological Survey, 1895–'96,* pt. 1, pp. 957–1004. Washington, D.C.: Government Printing Office.

1897

"Nansen's Heroic Journey." *Atlantic Monthly* 79: 610–17.
"The Physical Changes of Autumn." *Chautauquan* 26: 144–48.
"Water Supply of Eastern Massachusetts." *Science,* n.s., 5: 703. (Abstract.)
With T. C. Mendenhall and W. E. McClintock. *Fourth Annual Report of the Massachusetts Highway Commission.* Boston: Wright & Potter.

1898

"The Changes of the Seasons." *Chautauquan* 27: 16–20.
"Election of Studies in Secondary Schools." *Educational Review* 16: 417–23.
"Geology of the Cape Cod District." In *Eighteenth Annual Report of the United States Geological Survey, 1896–97,* pt. 2, pp. 503–93. Washington, D.C.: Government Printing Office.
"The Landscape as a Means of Culture." *Atlantic Monthly* 82: 777–85.
Outlines of the Earth's History: A Popular Study in Physiography. New York: Appleton.
"The True Measure of Valor." *Harvard Graduates' Magazine* 7: 192–203. (Address delivered in Sanders Theatre on Memorial Day, 30 May 1898.)

1899

"Formation of Dikes and Veins." *Bulletin of the Geological Society of America* 10: 253–62.
"Loess Deposits of Montana." *Bulletin of the Geological Society of America* 10: 245–52.
"Spacing of Rivers with Reference to Hypothesis of Baseleveling." *Bulletin of the Geological Society of America* 10: 262–76.
With J. B. Woodworth. "Geology of the Richmond Basin, Virginia." In *Nineteenth Annual Report of the United States Geological Survey, 1897–'98,* pt. 2, pp. 385–520. Washington, D.C.: Government Printing Office.
With J. B. Woodworth and A. F. Foerste. *Geology of Narragansett Basin.* United States Geological Survey Monograph, vol. 33. Washington, D.C.: Government Printing Office.

1900

"The Future of the Negro in the Southern States." *Popular Science Monthly* 57: 147–56.
The Individual: A Study of Life and Death. New York: Appleton.
"Influence of the Sun upon the Formation of the Earth's Surface." *International Monthly* 1: 41–82.

"The Negro since the Civil War." *Popular Science Monthly* 57: 29–39.
"Our Negro Types." *Current Literature* 29: 44–45.
"The Transplantation of a Race." *Popular Science Monthly* 56: 513–24.

1901

"American Quality." *International Monthly* 4: 48–67.
"Broad Valleys of the Cordilleras." *Bulletin of the Geological Society of America* 12: 271–300.
"Future Supply of Gold." (This item is listed under 1901 in the bibliography compiled by Shaler's wife as published in the *International Quarterly*. I have been unable, however, to locate any such article.)
"Huxley's Life and Letters." *Critic* 38: 253–55.
"The Proposed Appalachian Park." *North American Review* 173: 774–81.
With W. D. Howells and Mark Twain. *The Niagara Book.* New York: Doubleday, Page.

1902

"Faith in Nature." *International Quarterly* 6: 281–304.
"The Nature of Volcanoes." *North American Review* 175: 730–41.
"The Relations of Animals and Plants." *Harper's Monthly Magazine* 104: 721–24.
Reports of Professor N. S. Shaler on the Marshes and Swamps of Northern Long Island, between Port Washington and Cold Spring Harbor. New York: Press of Styles & Cash. (North Shore Improvement Association reports on plans for the extermination of mosquitoes on the north shore of Long Island between Hempstead Harbor and Cold Spring Harbor.)
"Valor." *Harvard Graduates' Magazine* 10 (26 June): 1–8. (Phi Beta Kappa Poem.)

1903

"A Comparison of the Features of the Earth and the Moon." *Smithsonian Contributions to Knowledge* 34, no. 1438: 1–130.
Elizabeth of England: A Dramatic Romance in Five Parts. 5 vols. Boston: Houghton Mifflin.
"General Description of the Moon." *Annual Report of the Smithsonian Institution*, pp. 103–12. Washington, D.C.: Smithsonian Institution.
"Human Personality and Its Survival." *New York Independent* 55: 1249–53.
"The Natural History of War." *International Quarterly* 8: 17–30.
"Plant and Animal Intelligence." *Harper's Monthly Magazine* 107: 183–87.

1904

The Citizen: A Study of the Individual and the Government. New York: Barnes.
The Neighbor: The Natural History of Human Contacts. Boston: Houghton Mifflin.

1905

"Earth and Man: An Economic Forecast." *International Quarterly* 10: 227–39.
"The Exhaustion of the World's Metals." *International Quarterly* 11: 230–47.
"The Future of Power." *International Quarterly* 11: 24–38.
"Gordon McKay." *Harvard Graduates' Magazine* 13: 596–75.
Man and the Earth. New York: Fox, Duffield.

1906

From Old Fields: Poems of the Civil War. Boston: Houghton Mifflin.

1908

"The Kentucky Campaign of 1862." *Papers of the Military Historical Society of Massachusetts* 7: 203–26. (Paper read 3 November 1891.)

1909

The Autobiography of Nathaniel Southgate Shaler with a Supplementary Memoir by His Wife. Boston: Houghton Mifflin.

Archival Material

Nathaniel Southgate Shaler Papers

Shaler, N. S. Papers. Harvard University Archives, Pusey Library, Harvard University, Cambridge, Massachusetts. These papers comprise the following materials:

N. S. Shaler. General File, HUG 1784. This file contains materials relating to the Shaler Memorial Fund, the four proposed scientific professorships at Berea College, and a brief biographical sketch by William Roscoe Thayer.

Shaler. Letters to A. C. Lane 1885–95, HUG 1784.2.

Shaler, N. S. Reports on Lunar Orography, HUG 1784.3.

Nathaniel Southgate Shaler. Box File, HUG 1784.10. This box file contains papers relating chiefly to work for the United States Geological Survey, 1884–98. It includes correspondence with, among others, T. C. Chamberlin, Henry Gannett, G. K. Gilbert, W J McGee, John Wesley Powell, R. S. Tarr, and Bailey Willis.

Shaler. Geology of the Genessee Valley, HUG 1784.15.

Woodworth, J. B. Life of N. S. Shaler, HUG 1784.96. Unpublished and incomplete manuscript.

Other Collections

Agassiz, Alexander. Papers. Archives of the Museum of Comparative Zoology, Harvard University, Cambridge.

Atkinson, Edward F. Papers. Massachusetts Historical Society, Boston.

Davis, William Morris. Papers. Houghton Library, Harvard University, Cambridge.

Eliot, Charles W. Papers. Harvard University Archives, Pusey Library, Harvard University, Cambridge.

Force, Manning. Papers. Houghton Library, Harvard University, Cambridge.

Gilman, Daniel Coit. Papers. Johns Hopkins University Library, Baltimore.

Gray, Asa. Papers. Gray Herbarium Letter File, Harvard University, Cambridge.

Higginson, Thomas Wentworth. Papers. Houghton Library, Harvard University, Cambridge.

Howells, William Dean. Papers. Houghton Library, Harvard University, Cambridge.

Immigration Restriction League, Boston. Papers. Houghton Library, Harvard University, Cambridge.

James, William. Papers. Harvard University Archives, Pusey Library, Harvard University, Cambridge.

Lane, Alfred C. Papers. Harvard University Archives, Pusey Library, Harvard University, Cambridge.

Lawrence Scientific School. Official Correspondence. Harvard University Archives, Pusey Library, Harvard University, Cambridge.

Leslie, Preston H. Papers. Kentucky Historical Society, Frankfort.

Loomis, Elias. Papers. Yale University Library, New Haven.

Marsh, Othniel C. Papers. Peabody Museum of Natural History, Yale University, New Haven.

Norton, Charles Eliot. Papers. Houghton Library, Harvard University, Cambridge.

Peabody, Endicott. Papers. Houghton Library, Harvard University, Cambridge.

Peirce, Charles S. Papers. Houghton Library, Harvard University, Cambridge.

Peter, Robert. Papers. University of Kentucky Library, Lexington.

Proctor, John R. Papers. Library of Congress, Washington, D.C.

Rotch, Abbott Lawrence. Papers. Houghton Library, Harvard University, Cambridge.

Scudder, Horace E. Papers. Houghton Library, Harvard University, Cambridge.

Shaler, Nathaniel S. File. Kentucky Historical Society, Frankfort.

Shaler, Nathaniel S. File. Library of the Museum of Comparative Zoology, Harvard University, Cambridge.

Stevenson, John W. Papers. Library of Congress, Washington, D.C.

United States Geological Survey Correspondence. National Archives, Washington, D.C.

Verrill, Addison E. Diary 1860–63. Harvard University Archives, Pusey Library, Harvard University, Cambridge.

Verrill, Addison E. Diary of Anticosti Expedition, 1861. Harvard University Archives, Pusey Library, Harvard University, Cambridge.

Wendell, Barrett. Papers. Houghton Library, Harvard University, Cambridge.

Secondary Sources

Aay, Henry. "Textbook Chronicles: Disciplinary History and the Growth of Geographic Knowledge." In *The Origins of Academic Geography in the United States*, edited by Brian W. Blouet, pp. 291–301. Hamden, Conn.: Archon Books, 1981.

Agassi, Joseph. "Towards an Historiography of Science." *History and Theory* 2 (1969): 1–117.

Agassiz, Alexander. "Abandonment of Penikese." *Popular Science Monthly* 42 (1892): 123.

Agassiz, Elizabeth Cary. *Louis Agassiz: His Life, Letters, and Correspondence*. 2 vols. London: Macmillan, 1885.

Agassiz, Louis. "On Brachiopods." *Proceedings of the Boston Society of Natural History* 9 (1865): 68–69.

Ahlstrom, Sydney E. *A Religious History of the American People*. New Haven: Yale University Press, 1972.

Aldrich, Michele L. "Geological Surveys, States." In *Dictionary of American History*, vol. 3, pp. 164–66. New York: Scribner's Sons, 1976.

Allen, Garland E. "Hugo De Vries and the Reception of the Mutation Theory." *Journal of the History of Biology* 2 (1969): 55–87.

Allen, Joel A. "The Influence of Physical Conditions in the Genesis of Species." *Radical Review* 1 (1877–78): 108–37.

Andrews, John T., ed. *Glacial Isostasy*. Stroudsburg, Pa.: Dowden, Hutchinson & Ross, 1974.

Argyll, Duke of [George John Douglas Campbell]. *The Reign of Law*. London: Strahan, 1867.

Bacon, H. Philip. "Fireworks in the Classroom: Nathaniel Southgate Shaler as a Teacher." *Journal of Geography* 54 (1955): 349–53.

Bagehot, Walter. *Physics and Politics; or, Thoughts on the Application of the Principles of "Natural Selection" and "Inheritance" to Political Society*. 11th ed. London: Kegan Paul, Trench, Trubner, 1896.

Bailey, Edward Battersby. *James Hutton—the Founder of Modern Geology*. Amsterdam: Elsevier, 1967.

Bailey, L. H. "Neo-Lamarckism and Neo-Darwinism." *American Naturalist* 28 (1894): 661–78.

Baker, J. N. L. "Nathanael Carpenter and English Geography." *Geographical Journal* 71 (1928): 261–71.

Ballard, Frank. *Haeckel's Monism False*. London: Kelly, 1905.

Bancroft, George. *History of the United States*. 14th ed. 10 vols. Boston: Little, Brown, 1854.

Bannister, Robert C. " 'The Survival of the Fittest Is Our Doctrine': History or Histrionics?" *Journal of the History of Ideas* 31 (1970): 377–98.

———. *Social Darwinism: Science and Myth in Anglo-American Social Thought*. Philadelphia: Temple University Press, 1979.

Barbour, Ian G. *Issues in Science and Religion*. New York: Harper & Row, 1971.

Barnes, Harry Elmer, ed. *An Introduction to the History of Sociology*. Chicago: University of Chicago Press, 1948.

Barrell, Joseph. "The Growth of Knowledge of Earth Structure." *American Journal of Science*, 4th ser., 46 (1918): 133–70.

Barton, G. H. "Remarks on Drumlins." *Proceedings of the Boston Society of Natural History* 26 (1893): 23–25.

———. "Original Origin of Channels on Drumlins." *Bulletin of the Geological Society of America* 6 (1895): 8–13.

Bascom, John. *Evolution and Religion; or, Faith as a Part of a Complete Cosmic System.* New York: Putnam's Sons, 1880.

———. *Natural Theology.* New York: Putnam's Sons, 1897.

Baym, Nina. "Thoreau's View of Science." *Journal of the History of Ideas* 26 (1965): 221–34.

Beaver, Stanley. "The Le Play Society and Field Work." *Geography* 47 (1962): 221–34.

Beck, Hanno. "Moritz Wagner als Geograph." *Erdkunde* 7 (1953): 125–28.

Beckinsale, Robert P. "Penck, Albrecht." In *Dictionary of Scientific Biography*, vol. 10, pp. 501–6. New York: Scribner's Sons, 1971.

Beecher, Henry Ward. *Evolution and Religion.* New York: Fords, Howard & Hulbert, 1885.

Bell, Ian F. A. "Divine Patterns: Louis Agassiz and American Men of Letters: Some Preliminary Explorations." *Journal of American Studies* 10 (1976): 349–81.

Ben-David, Joseph. "Introduction." *International Social Science Journal* 22 (1970): 7–27.

Benson, Lee A. "The Historical Background of Turner's Frontier Essay." *Agricultural History* 25 (1951): 59–82.

Berdoulay, Vincent. "Professionnalisation et institutionnalisation de la géographie." *Organon* 14 (1980): 149–56.

———. "The Contextual Approach." In *Geography, Ideology, and Social Concern*, edited by David R. Stoddart, pp. 8–16. Oxford: Blackwell, 1981.

Berg, Walter. "Nathaniel Southgate Shaler: A Critical Study of an Earth Scientist." Ph.D. diss. University of Washington, 1957.

———. "Shaler, Nathaniel Southgate." In *Dictionary of Scientific Biography*, vol. 12, pp. 343–44. New York: Scribner's Sons, 1971.

Bergson, Henri. *Creative Evolution.* Translated by Arthur Mitchell. London: Macmillan, 1911.

Berkhofer, Robert F. "Space, Time, Culture, and the New Frontier." *Agricultural History* 38 (1964): 21–30.

Berman, Milton. *John Fiske: The Evolution of a Popularizer.* Cambridge, Mass.: Harvard University Press, 1961.

Bernal, J. D. *Science in History.* London: Watts, 1957.

Bertocci, Peter A. "Creation in Religion." In *Dictionary of the History of Ideas*, vol. 1, pp. 571–77. New York: Scribner's Sons, 1973.

Billington, Ray Allen. *Genesis of the Frontier Thesis: A Study in Historical Creativity.* San Marino, Calif.: Huntington Library, 1971.

Bladen, Wilford A., and Karan, Pradyumna P., eds. *The Evolution of Geographic Thought in America: A Kentucky Root.* Dubuque, Iowa: Kendall/Hunt Publishing, 1983.

Bledstein, Burton J. *The Culture of Professionalism: The Middle Class and the Development of Higher Education in America.* New York: Norton, 1976.

Block, Robert. "Frederick Jackson Turner and American Geography." *Annals of the Association of American Geographers* 70 (1980): 31–42.

Böhm, August von. *Geschichte der Moränenkunde.* Vienna: Lechner, 1901.

Boller, Paul F., Jr. *American Thought in Transition: The Impact of Evolutionary Naturalism, 1865–1900.* Chicago: University of Chicago Press, 1969.

Bowen, Francis. "A Theory of Creation." *North American Review* 60 (1845): 426–78.

———. "Remarks on the Latest Form of the Development Theory." *Memoirs of the American Academy of Arts and Sciences,* n.s., 8 (1860): 98–107.

Bowler, Peter J. "The Changing Meaning of Evolution." *Journal of the History of Ideas* 36 (1975): 95–114.

———. *Fossils and Progress: Palaeontology and the Idea of Progressive Evolution in the Nineteenth Century.* New York: Science History Publications, 1976.

———. "Darwinism and the Argument from Design: Suggestions for a Re-evaluation." *Journal of the History of Biology* 10 (1977): 29–43.

———. "Edward Drinker Cope and the Changing Structure of Evolutionary Theory." *Isis* 68 (1977): 249–65.

———. "Hugo De Vries and Thomas Hunt Morgan: The Mutation Theory and the Spirit of Darwinism." *Annals of Science* 35 (1978): 55–73.

———. *The Eclipse of Darwinism: Anti-Darwinian Evolution Theories in the Decades around 1900.* Baltimore: Johns Hopkins University Press, 1983.

———. *Evolution: The History of an Idea.* Berkeley: University of California Press, 1984.

Bowyer, Carlton H. *Philosophical Perspectives for Education.* Glenview, Ill.: Scott, Foresman, 1970.

Brigham, Albert Perry. *Geographic Influences in American History.* Boston: Ginn, 1903.

———. "Geography and History in the United States." In *Report of the Eighth International Geographic Congress Held in the United States, 1904.* Washington, D.C.: Government Printing Office, 1905.

———. "William Morris Davis." *Geographen Kalender* 7 (1909): 1–73.

Broc, Numa. "Les classiques de Miss Semple: Essai dur les sources des *Influences of Geographic Environment,* 1911." *Annales de géographie* 497 (1981): 87–102.

Broca, Paul. *On the Phenomena of Hybridity in the Genus Homo.* Edited by C. Carter Blake. London: Anthropological Society of London, 1864.

Brooke, John Hedley. "The Natural Theology of the Geologists: Some Theological Strata." In *Images of the Earth: Essays in the History of the Environmental Sciences,* edited by L. J. Jordanova and Roy S. Porter, pp. 39–64. Chalfont St. Giles: British Society for the History of Science, 1979.

———. "Nebular Contraction and the Expansion of Naturalism." *British Journal for the History of Science* 12 (1979): 200–211.

Brown, Rollo Walter. *Harvard Yard in the Golden Age.* New York: Current Books, 1948.

Brubacher, John S. *Modern Philosophies of Education.* 3d ed. New York: McGraw-Hill, 1962.

Brush, Stephen G. "Nineteenth-Century Debates about the Inside of the Earth: Solid, Liquid, or Gas?" *Annals of Science* 36 (1979): 225–54.

Buck, Paul, ed. *Social Sciences at Harvard, 1860–1920: From Inculcation to the Open Mind.* Cambridge, Mass.: Harvard University Press, 1965.

Buckle, Thomas Henry. *History of Civilization in England.* 3 vols. London: Longmans, Green, 1894. First published in 1867–70.

Burkhardt, Richard W., Jr. *The Spirit of System: Lamarck and Evolutionary Biology.* Cambridge, Mass.: Harvard University Press, 1977.

Burrow, John W. "Evolution and Anthropology in the 1860's: The Anthropological Society of London, 1863–1871." *Victorian Studies* 7 (1963): 137–54.

———. *Evolution and Society: A Study in Victorian Social Theory.* Cambridge: Cambridge University Press, 1966.

Buttimer, Anne. "On People, Paradigms, and 'Progress' in Geography." In *Geography, Ideology, and Social Concern*, edited by David R. Stoddart, pp. 81–98. Oxford: Blackwell, 1981.

Büttner, Manfred. "Zum Gegenüber von Naturwissenschaft (insbesondere Geographie) und Theologie im 18 Jahrhundert." *Philosophia Naturalis* 14 (1973): 95–123.

———. "Die Neuausrichtung der Geographie im 17 Jahrhundert durch Bartholomäus Keckermann, ein Betrag zur Geographie in ihren Beziehungen zu Theologie und Philosophie." *Geographische Zeitschrift* 63 (1975): 1–12.

———. "Kant and the Physico-Theological Consideration of the Geographical Facts." *Organon* 11 (1975): 231–49.

———. "Kant und die Uberwindung der Physiko-Theologischen Betrachtung der Geographisch-Kosmologischen Fakten." *Erdkunde* 29 (1975): 53–60.

———. "Bartholomäus Keckermann, 1572–1609." In *Geographers: Biobibliographical Studies*, vol. 2, edited by T. W. Freeman and Philippe Pinchemel, pp. 73–79. London: Mansell, 1977.

Butts, R. F., and Cremin, L. A. *A History of Education in American Culture.* New York: Holt, 1959.

Cairnes, J. E. "The Negro Suffrage." *Macmillan's Magazine* 12 (1865): 334–43.

Campbell, John Angus. "Nature, Religion, and Emotional Response: A Reconsideration of Darwin's Affective Decline." *Victorian Studies* 18 (1974): 159–74.

Campbell, J. A., and Livingstone, D. N. "Neo-Lamarckism and the Development of Geography in the United States and Great Britain." *Transactions of the Institute of British Geographers*, n.s., 8 (1983): 267–94.

Cannon, Walter F. "The Problem of Miracles in the 1830's." *Victorian Studies* 4 (1960): 5–32.

———. "The Uniformitarian-Catastrophist Debate." *Isis* 51 (1960): 38–55.

———. "The Bases of Darwin's Achievement: A Revaluation." *Victorian Studies* 5 (1961): 109–34.

Capel, Horacio. "Institutionalization of Geography and Strategies for Change." In *Geography, Ideology, and Social Concern*, edited by David R. Stoddart, pp. 37–69. Oxford: Blackwell, 1981.

Cappon, Lester J. "The Historical Map in American Atlases." *Annals of the Association of American Geographers* 69 (1979): 622–34.

Carneiro, Robert L. "Herbert Spencer's *The Study of Sociology* and the Rise of Social Science in America." *Proceedings of the American Philosophical Society* 118 (1974): 540–54.

Carpenter, J. E. "Unitarianism." In *Encyclopaedia of Religion and Ethics*, edited by James Hastings, vol. 12, pp. 525–27. Edinburgh: Clark, 1921.

Carpenter, Nathanael. *Geography Delineated Forth in Two Bookes: Containing the Sphaericall and Topicall Parts Thereof.* Theatrum Orbis Terarum; the English Experience, no. 787. Norwood, N.J.: Johnson, 1976. First published in 1625.

Carroll, Peter N. *Puritanism and the Wilderness: The Intellectual Significance of the New England Frontier, 1629–1700.* New York: Columbia University Press, 1969.

Cashdollar, Charles D. "The Social Implications of the Doctrine of Divine Providence: A Nineteenth-Century Debate in American Theology." *Harvard Theological Review* 71 (1978): 265–84.

Caughey, John W. *Hubert Howe Bancroft, Historian of the West.* Berkeley: University of California Press, 1946.

Caverno, Charles. "Louis Agassiz and Charles Darwin: A Synthesis." *Bibliotheca Sacra* 73 (1916): 137–40.

Chadwick, W. Owen. "Religion and Science in the Nineteenth Century." In *Dictionary of the History of Ideas,* vol. 4, pp. 106–8. New York: Scribner's Sons, 1973.

Chamberlin, Thomas C. "Diastrophism as the Ultimate Basis of Correlation." *Journal of Geology* 17 (1909): 685–93.

———. "Soil Wastage." In *Proceedings of the Conference of Governors in the White House,* pp. 75–83. Washington, D.C.: Government Printing Office, 1909.

Chappell, John E., Jr. "Harlan Barrows and Environmentalism." *Annals of the Association of American Geographers* 61 (1971): 198–201.

———. "Environmental Causation." In *Themes in Geographic Thought,* edited by Milton E. Harvey and Brian P. Holly, pp. 163–86. London: Croom Helm, 1981.

Charlesworth, J. K. *The Quaternary Era with Special Reference to Its Glaciation.* 2 vols. London: Arnold, 1957.

Childs, John L. "Experimentalism and American Education." *Record* 44 (1943): 539–43.

Chorley, Richard J. "Diastrophic Background to Twentieth-Century Geomorphological Thought." *Bulletin of the American Geological Society* 74 (1963): 953–70.

Chorley, Richard J.; Dunn, A. J.; and Beckinsale, Robert P. *The History of the Study of Landforms; or, The Development of Geomorphology.* Vol. 1, *Geomorphology before Davis.* London: Methuen, 1964.

Chorley, Richard J.; Beckinsale, Robert P.; and Dunn, A. J. *The History of the Study of Landforms; or, The Development of Geomorphology.* Vol. 2, *The Life and Work of William Morris Davis.* London: Methuen, 1973.

Churchill, Frederick B. "The Weismann-Spencer Controversy over the Inheritance of Acquired Characteristics." *Proceedings of the International Congress of the History of Science* 15 (1978): 451–68.

Clapp, Frederick G. "Complexity of the Glacial Period in Northeastern New England." *Bulletin of the Geological Society of America* 18 (1907–8): 505–56.

Clark, Dan E. "Great Plains." In *Dictionary of American History,* vol. 3, pp. 220–21. New York: Scribner's Sons, 1976.

Clark, R. E. D. "Myers, Frederick William Henry (1843–1901)." In *The New International Dictionary of the Christian Church,* edited by J. D. Douglas, p. 690. Exeter: Paternoster Press, 1974.

Clark, Thomas D. *A History of Kentucky.* New York: Prentice-Hall, 1937.

Claval, Paul. *Essai sur l'évolution de la géographie humaine.* Paris: Belles Lettres, 1976.

———. "Epistemology and the History of Geographical Thought." In *Geography, Ideology, and Social Concern,* edited by David R. Stoddart, pp. 227–41. Oxford: Blackwell, 1981.

Clepper, Henry. "Forestry Education in America." *Journal of Forestry* 54 (1956): 455–57.

Cloud, Preston. "The Improbable Bureaucracy: The United States Geological Surveys, 1879–1979." *Proceedings of the American Philosophical Society* 124 (1980): 155–67.

Cohen, Maurice. *American Thought: A Critical Sketch.* Glencoe, Ill.: Free Press, 1954.

Colby, Charles C. "Ellen Churchill Semple." *Annals of the Association of American Geographers* 23 (1933): 229–40.

Coleman, William. "Science and Symbol in the Turner Frontier Hypothesis." *American Historical Review* 72 (1966): 22–49.

———. *Biology in the Nineteenth Century: Problems of Form, Function, and Transformation.* New York: Wiley, 1971.

Commager, Henry Steele. *The American Mind.* New Haven: Yale University Press, 1950.

Cooter, Roger. "The Power of the Body: The Early Nineteenth Century." In *Natural Order: Historical Studies of Scientific Culture,* edited by Barry Barnes and Steven Shapin, pp. 73–92. Beverly Hills: Sage Publications, 1979.

Cope, Edward Drinker. "Evolution and Its Consequences." In *Darwinism Comes to America,* edited by George Daniels, pp. 84–91. Waltham, Mass.: Blaisdell Publishing, 1968. First published in 1872.

Copeland, Charles T. "Biography of Prof. Shaler." *Boston Transcript,* 18 July 1906.

Cowham, J. H. *The School Journey: A Means of Teaching Geography, Physiography, and Elementary Science.* London: Westminster School Book Depot, 1900.

Coyle, Daniel Cushman. *Conservation: An American Story of Conflict and Accomplishment.* New Brunswick, N.J.: Rutgers University Press, 1957.

Crane, Diane. *Invisible Colleges: Diffusion of Knowledge in Scientific Communities.* Chicago: University of Chicago Press, 1972.

Cravens, Hamilton, and Burnham, John C. "Psychology and Evolutionary Naturalism in American Thought, 1890–1940." *American Quarterly* 23 (1971): 635–57.

Croll, James. *Climate and Time in their Geological Relations: A Theory of Secular Changes of the Earth's Climate.* New York: Appleton, 1886.

Cross, Whitney R. "W J McGee and the Idea of Conservation." *Historian* 15 (1953): 148–62.

Cruden, Robert. *The Negro in Reconstruction.* Englewood Cliffs, N.J.: Prentice-Hall, 1969.

Cruickshank, James G. *Soil Geography.* Newton Abbott: David & Charles, 1972.

Cunningham, Frank F. *The Revolution in Landscape Science.* British Columbia Geographical Series, no. 25. Vancouver: Tantalus Research, 1977.

Cupitt, Don. "Darwinism and English Religious Thought." *Theology* 78 (1975): 125–31.

Dall, William H. "On a Provisional Hypothesis of Saltatory Evolution." *American Naturalist* 11 (1877): 135–37.

Daly, R. A. "The Accordance of Summit Levels among the Alpine Mountains: The Fact and Its Significance." *Journal of Geology* 13 (1905): 105–25.

Dana, James D. "On Some Results of the Earth's Contraction from Cooling, Including a Discussion of the Origin of Mountains, and the Nature of the Earth's Interior. Part I." *American Journal of Science,* 3d ser., 5 (1873): 423–43.

———. "On Some Results of the Earth's Contraction from Cooling. Part II, The Condition of the Earth's Interior, and the Connection of the Facts with Mountain-Making. Part III, Metamorphism." *American Journal of Science,* 3d ser., 6 (1873): 6–14.

———. "On Some Results of the Earth's Contraction from Cooling. Part IV, Igneous Ejections, Volcanoes." *American Journal of Science,* 3d ser., 6 (1873): 104–15.

———. *Manual of Geology.* 4th ed. New York: Ivison, 1895.

Darling, Frank F. "Man's Responsibility for the Environment." In *Biology and Ethics,* edited by F. J. Ebling, pp. 117–22. Symposium of the Institute of Biology, no. 18. London: Academic Press, 1969.

Darrah, William Culp. *Powell of the Colorado.* Princeton: Princeton University Press, 1969.

Darwin, Charles. *On the Origin of Species by Means of Natural Selection; or, The Preservation of Favoured Races in the Struggle for Life.* London: Murray, 1859.

————. *The Descent of Man and Selection in Relation to Sex.* 2d ed. London: Murray, 1874.

————. *The Expression of the Emotions in Man and Animals.* London: Murray, 1889.

Darwin, Francis. *The Life and Letters of Charles Darwin.* 3 vols. London: Murray, 1887.

Darwin, Francis, and Seward, A. C. *More Letters of Charles Darwin: A Record of His Life and Work in a Series of Unpublished Letters.* 2 vols. New York: Appleton, 1903.

Dasmann, Raymond F. "Conservation of Natural Resources." In *Dictionary of the History of Ideas,* vol. 1, pp. 470–74. New York: Scribner's Sons, 1973.

Davies, Gordon L. *The Earth in Decay: A History of British Geomorphology, 1578–1878.* New York: Macdonald, 1969.

Davis, John F. "Constructing the British View of the Great Plains." In *Images of the Plains: The Role of Human Nature in Settlement,* edited by Brian W. Blouet and Merlin P. Lawson, pp. 181–85. Lincoln: University of Nebraska Press, 1975.

Davis, William Morris. "The Rivers and Valleys of Pennsylvania." *National Geographic Magazine* 1 (1889): 183–253.

————. "Bearing of Physiography on Uniformitarianism." *Bulletin of the Geological Society of America* 7 (1896): 8–11.

————. "Plains of Marine and Subaërial Denudation." *Bulletin of the Geological Society of America* 7 (1896): 377–98.

————. "Peneplains of Central France and Brittany." *Bulletin of the Geological Society of America* 12 (1901): 480–83.

————. "Nathaniel Southgate Shaler: Estimate of the Teacher and the Man." *Harvard Bulletin,* 16 May 1906.

————. "Professor Nathaniel S. Shaler." *American Journal of Science,* 4th ser., 21 (1906): 480–81.

————. "Professor Shaler and the Lawrence Scientific School." *Harvard Engineering Journal* 5 (1906): 129–38.

————. "The Progress of Geography in the United States." *Annals of the Association of American Geographers* 14 (1924): 158–215.

————. "The Faith of Reverent Science." In *The History of the Study of Landforms; or, The Development of Geomorphology.* Vol. 2, *The Life and Work of William Morris Davis,* edited by Richard J. Chorley, Robert P. Beckinsale, and A. J. Dunn, pp. 759–91. London: Methuen, 1973. First published in 1933.

Davis, W. M., and Daly, R. A. "Geology and Geography, 1858–1928." In *The Development of Harvard University since the Inauguration of President Eliot, 1869–1929,* edited by Samuel E. Morrison, pp. 307–28. Cambridge, Mass.: Harvard University Press, 1930.

Dawson, J. William. "Some Recent Discussions in Geology." *Bulletin of the Geological Society of America* 5 (1894): 101–16.

De Geer, Gerard. "On Pleistocene Changes of Level in Eastern North America." *Proceedings of the Boston Society of Natural History* 25 (1892): 454–77.

De Marrais, Robert. "The Double-Edged Effect of Sir Francis Galton: A Search for the Motives in the Biometrician-Mendelian Debate." *Journal of the History of Biology* 7 (1974): 141–74.

Dewey, John. *Democracy and Education.* New York: Macmillan, 1916.

————. *The Quest for Certainty*. New York: Minton, Balch, 1929.

Dexter, Ralph W. "Three Young Naturalists Afield: The First Expedition of Hyatt, Shaler, and Verrill." *Scientific Monthly* 79 (1954): 45–51.

————. "Historical Aspects of Studies on the Brachiopoda by E. S. Morse." *Systematic Zoology* 15 (1966): 241–44.

————. "From Penikese to the Marine Biological Laboratory at Woods Hole—the Role of Agassiz's Students." *Essex Institute Historical Collections* 110 (1974): 151–61.

Dickinson, Robert E. *The Makers of Modern Geography*. London: Routledge & Kegan Paul, 1969.

————. *Regional Concept: The Anglo-American Leaders*. London: Routledge & Kegan Paul, 1976.

Dillenberger, John. *Protestant Thought and Natural Science*. London: Collins, 1961.

Dobzhansky, T. "The Genetic Nature of Differences among Men." In *Evolutionary Thought in America,* edited by Stow Persons, pp. 86–154. New York: Braziller, 1956.

Dodds, Gordon B. "The Historiography of American Conservation: Past and Prospects." *Pacific Northwest Quarterly* 54 (1965): 75–81.

Dodge, Richard Elwood. "Albert Perry Brigham." *Annals of the Association of American Geographers* 20 (1930): 55–62.

Doughty, Robin W. "Environmental Theology: Trends and Prospects in Christian Thought." *Progress in Human Geography* 5 (1981): 234–48.

Draper, John William. *History of the Conflict between Religion and Science*. London: King, 1875. First published in 1874.

Drummond, Henry. *Natural Law in the Spiritual World*. London: Hodder & Stoughton, 1883.

————. *The Ascent of Man*. London: Hodder & Stoughton, 1894.

Dugdale, Richard. *"The Jukes": A Study in Crime, Pauperism, Disease, and Heredity*. New York: Putnam's Sons, 1877.

Dunbar, Gary S. "Credentialism and Careerism in American Geography, 1890–1915." In *The Origins of Academic Geography in the United States,* edited by Brian W. Blouet, pp. 71–88. Hamden, Conn.: Archon Books, 1981.

Dunbar, Robert G. "Irrigation." In *Dictionary of American History,* vol. 3, pp. 477–78. New York: Scribner's Sons, 1976.

Dupree, A. Hunter. *Science in the Federal Government: A History of Policies and Activities to 1940*. Cambridge, Mass.: Harvard University Press, 1957.

————. *Asa Gray*. Cambridge, Mass.: Harvard University Press, 1959.

Dutton, Clarence E. "On Some of the Greater Problems of Physical Geology." *Bulletin of the Philosophical Society of Washington* 11 (1889): 51–64.

Egerton, Frank N. "Ecological Studies and Observations before 1900." In *Issues and Ideas in America,* edited by Benjamin J. Taylor and Thurman J. White, pp. 311–51. Norman: University of Oklahoma Press, 1976.

Eiseley, Loren. *Darwin's Century: Evolution and the Men Who Discovered It*. New York: Anchor Books, 1961.

Ekirch, Arthur A. *Man and Nature in America*. New York: Columbia University Press, 1963.

Eliot, Charles W. "The New Education." *Atlantic Monthly* 23 (1869): 203–20, 358–67.

Ellegård, Alvar. "The Darwinian Theory and Nineteenth-Century Philosophies of Science." *Journal of the History of Ideas* 18 (1957): 362–93.

————. *Darwin and the General Reader: The Reception of Darwin's Theory of Evolution in the British Periodical Press, 1859–1872*. Göteborg: Göteborg Universitets Arsskrift, 1958.

Embleton, Clifford, and King, Cuchlaine A. M. *Glacial Geomorphology*. 2d ed. London: Arnold, 1975.

Emerson, Ralph Waldo. *On Nature*. Leipzig: Insel-Verlag, n.d. First published in 1836.

Erisman, Frank. "The Environmental Crisis and Present-Day Romanticism: The Persistence of an Idea." *Rocky Mountain Social Science Journal* 10 (1973): 7–14.

Evans, E. Estyn. "The Scotch-Irish in the New World: An Atlantic Heritage." *Journal of the Royal Society of Antiquaries of Ireland* 95 (1965): 39–49.

Fackre, Gabriel. "Ecology and Theology." *Religion in Life* 40 (1971): 210–24.

Fahl, Ronald J. *North American Forest and Conservation History: A Bibliography*. Santa Barbara: A.B.C.-Clio Press, 1977.

Fairbanks, Helen R., and Berkey, Charles P. *Life and Letters of R. A. F. Penrose, Jr.* New York: Geological Society of America, 1952.

Fairbridge, Rhodes W. "Gipfelflur." In *Encyclopedia of Geomorphology*, pp. 426–27. New York: Rinehold, 1968.

Farrall, Lyndsay A. "The History of Eugenics: A Bibliographical Review." *Annals of Science* 36 (1979): 111–23.

Feldman, Douglas A. "The History of the Relationship between Environment and Culture in Ethnological Thought: An Overview." *Journal of the History of the Behavioral Sciences* 11 (1975): 67–81.

Fisher, Osmond. "On the Effect upon the Ocean-Tides of a Liquid Substratum beneath the Earth's Crust." *Philosophical Magazine* 14 (1882): 213–15.

Fiske, John. *Outlines of Cosmic Philosophy*. 2 vols. London: Macmillan, 1874.

———. *The Discovery of America*. 2 vols. Boston: Houghton Mifflin, 1896.

———. *The Idea of God as Affected by Modern Knowledge*. London: Macmillan, 1901.

———. *Life Everlasting*. Boston: Houghton Mifflin, 1901.

Fitch, Robert E. "Charles Darwin: Science and the Saintly Sentiments." *Columbia University Forum* 2 (Spring 1959): 7–12.

Flemal, Ronald C. "The Attack on the Davisian System of Geomorphology: A Synopsis." *Journal of Geological Education* 19 (1971): 3–13.

Fleming, Donald. "Roots of the New Conservation Movement." *Perspectives in American History* 6 (1972): 7–91.

Flint, Richard Foster. *Glacial Geology and the Pleistocene Epoch*. New York: Wiley, 1947.

———. "Introduction." In *The Quaternary of the United States*, edited by H. E. Wright, Jr., and David G. Frey, pp. 3–11. Princeton: Princeton University Press, 1965.

Forrest, D. W. *Francis Galton: The Life and Work of a Victorian Genius*. New York: Elek, 1974.

Foster, Michael B. "The Christian Doctrine of Creation and the Rise of Modern Natural Science." *Mind* 43 (1934): 446–68.

Frederickson, George M. *The Black Image in the White Mind: Debates on Afro American Character and Destiny, 1817–1914*. New York: Irvington, 1971.

Freeden, Michael. "Eugenics and Progressive Thought: A Study in Ideological Affinity." *Historical Journal* 22 (1979): 645–71.

Freeman, T. W. *The Geographer's Craft*. Manchester: Manchester University Press, 1967.

Freund, Rudolf. "Turner's Theory of Social Evolution." *Agricultural History* 19 (1945): 78–87.

Froggatt, P., and Nevin, N. C. "The 'Law of Ancestral Heredity' and the Mendelian-Ancestrian Controversy in England, 1889–1906." *Journal of Medical Genetics* 8 (1971): 1–36.

Gallie, W. B. *Peirce and Pragmatism*. Harmondsworth: Penguin Books, 1952.

Galton, Francis. "Heredity, Talent, and Character." *Macmillan's Magazine* 12 (1865): 157–66, 318–27.

———. *Hereditary Genius: An Inquiry into Its Laws and Consequences*. London: Macmillan, 1869.

———. *English Men of Science: Their Nature and Nurture*. London: Macmillan, 1874.

———. "A Theory of Heredity." *Journal of the Anthropological Institute of Great Britain and Ireland* 5 (1875): 329–48.

———. "Eugenics: Its Definition, Scope, and Aims." *American Journal of Sociology* 10 (1904): 1–25.

———. *Essays in Eugenics*. London: Eugenics Education Society, 1909.

Gannett, Henry. "The Settled Area and the Density of Our Population." *International Review* 12 (1882): 70–77.

Garbarino, Marwyn S. *Sociocultural Theory in Anthropology: A Short History*. New York: Holt, Rinehart & Winston, 1977.

Garratt, George A. "Gifford Pinchot and Forestry Education." *Journal of Forestry* 63 (1965): 597–600.

Geddes, Patrick. "Nature Study and Geographical Education." *Scottish Geographical Magazine* 18 (1902): 525–36.

———. "Biology." In *Chambers's Encyclopaedia*, vol. 2, pp. 157–64. London: Chambers, 1925. First published in 1882.

Geikie, James. *Mountains: Their Origin, Growth, and Decay*. Edinburgh: Oliver & Boyd, 1913.

Gelfand, L. "Ellen Churchill Semple: Her Geographical Approach to American History." *Journal of Geography* 53 (1954): 30–41.

Ghiselin, Michael T. *The Triumph of the Darwinian Method*. Berkeley: University of California Press, 1969.

Gilbert, E. W. "Geographie Is Better than Divinity." *Geographical Journal* 128 (1962): 494–97.

Gillespie, Neal C. *Charles Darwin and the Problem of Creation*. Chicago: University of Chicago Press, 1979.

Gilley, Sheridan, and Loades, Ann. "Thomas Henry Huxley: The War between Science and Religion." *Journal of Religion* 61 (1981): 285–308.

Gillispie, Charles C. *Genesis and Geology: A Study in the Relations of Scientific Thought, Natural Theology, and Social Opinion in Great Britain, 1790–1850*. New York: Harper & Row, 1959.

Glacken, Clarence, J. "The Origins of the Conservation Philosophy." *Journal of Soil and Water Conservation* 11 (1956): 63–66.

———. "Changing Ideas of the Habitable World." In *Man's Role in Changing the Face of the Earth*, edited by William L. Thomas, Jr., pp. 70–92. Chicago: University of Chicago Press, 1960.

———. *Traces on the Rhodian Shore: Nature and Culture in Western Thought from Ancient Times to the End of the Eighteenth Century*. Berkeley: University of California Press, 1967.

———. "Man against Nature: An Outmoded Concept." In *Environmental Crisis: Man's Struggle to Live with Himself*, edited by Harold W. Helfrich, Jr., pp. 127–42. New Haven: Yale University Press, 1970.

———. "Environment and Culture." In *Dictionary of the History of Ideas*, vol. 2, pp. 127–34. New York: Scribner's Sons, 1973.

Glick, Thomas F. "History and Philosophy of Geography." *Progress in Human Geography* 8 (1984): 273–83.

Goetzmann, William H. *Exploration and Empire: The Explorer and the Scientist in the Winning of the American West.* New York: Knopf, 1971.

Goldman, Irving. "Evolution and Anthropology." *Victorian Studies* 3 (1959): 55–75.

Good, Harry G. *A History of Western Education.* New York: Macmillan, 1950.

Good, Harry G., and Teller, James D. *A History of American Education.* 3d ed. New York: Macmillan, 1973.

Gordon, Milton M. "Assimilation in America: Theory and Reality." *Daedalus* 90 (1961): 263–85.

———. *Assimilation in American Life: The Role of Race, Religion and National Origins.* New York: Oxford University Press, 1964.

Gossett, Thomas F. *Race: The History of an Idea.* Dallas: Southern Methodist University Press, 1963.

Gould, Stephen Jay. "Is Uniformitarianism Necessary?" *American Journal of Science* 263 (1965): 223–28.

———. *Ontogeny and Phylogeny.* Cambridge, Mass.: Harvard University Press, 1977.

———. "Flaws in a Victorian Veil." *New Scientist,* 31 August 1978, pp. 632–33.

———. *The Mismeasure of Man.* New York: Norton, 1981.

Graber, Linda H. *Wilderness as Sacred Space.* Washington, D.C.: Association of American Geographers, 1976.

Gräno, Olavi. "External Influence and Internal Change in the Development of Geography." In *Geography, Ideology, and Social Concern,* edited by David R. Stoddart, pp. 17–36. Oxford: Blackwell, 1981.

Gray, Asa. *Natural Science and Religion.* New York: Scribner's Sons, 1880.

———. "Louis Agassiz." *Andover Review* 9 (1886): 38.

Greene, John C. "Objectives and Methods in Intellectual History." *Mississippi Valley Historical Review* 44 (1957–58): 58–74.

———. *The Death of Adam: Evolution and Its Impact on Western Thought.* Ames. Iowa State University Press, 1959.

———. *Science, Ideology, and World View: Essays in the History of Evolutionary Ideas.* Berkeley: University of California Press, 1981.

Greene, Victor. "Immigration Restriction." In *Dictionary of American History,* vol 3, pp 342–43. New York: Scribner's Sons, 1976.

Gregory, Derek. *Ideology, Science, and Human Geography.* London: Hutchinson, 1978.

———. "Environmental Determinism." In *The Dictionary of Human Geography,* edited by R. J Johnston, p 103. Oxford: Blackwell, 1981.

Gregory, H. E. "A Century of Geology—Steps of Progress in the Interpretation of Land Forms." *American Journal of Science,* 4th ser., 46 (1918): 104–32.

Grene, Marjorie. "Darwin and Philosophy." In *Connaissance scientifique et philosophie colloque,* pp. 133–45. Publications du deuxième centenaire de l'Académie Royale des Sciences, des Lettres et des Beaux Arts de Belgique, no. 4, Brussels, 1973.

Gressley, Gene M. "The Turner Thesis—a Problem in Historiography." *Agricultural History* 32 (1958): 227–49.

Gulley, J. L. M. "The Turnerian Frontier—a Study in the Migration of Ideas." *Tijdschrift voor economische en sociale geografie* 50 (1959): 65–72, 81–91.

Guralnick, Stanley M. "Geology and Religion before Darwin: The Case of Edward Hitchcock: Theologian and Geologist (1793–1864)." *Isis* 63 (1972): 529–43.

Guyot, Arnold. *The Earth and Man*. Rev. ed. Boston: Scribner's Sons, 1897. First published in 1849.

Haddon, Alfred C. *History of Anthropology*. London: Watts, 1910.

Haller, John S., Jr. "Concepts of Race Inferiority in Nineteenth-Century Anthropology." *Journal of the History of Medicine and Allied Sciences* 25 (1970): 40–51.

———. "The Physician versus the Negro: Medical and Anthropological Concepts of Race in the Late Nineteenth Century." *Bulletin of the History of Medicine* 44 (1970): 154–67.

———. "Nathaniel Southgate Shaler: A Portrait of Nineteenth-Century Academic Thinking on Race." *Essex Institute Historical Collections* 107 (1971): 173–93.

———. *Outcasts from Evolution: Scientific Attitudes of Racial Inferiority, 1859–1900*. Chicago: University of Illinois Press, 1971.

Haller, Mark. *Eugenics: Hereditarian Attitudes in American Thought*. New Brunswick, N.J.: Rutgers University Press, 1963.

Halliday, R. J. "Social Darwinism: A Definition." *Victorian Studies* 14 (1971): 389–405.

Hallowell, A. Irving. "The Beginnings of Anthropology in America." In *Selected Papers from the American Anthropologist, 1888–1920*, edited by F. de Laguna, pp. 1–104. Evanston, Ill.: American Anthropologist, 1960.

Halpern, Ben. "America Is Different." In *The Jews: Social Patterns of an American Group*, edited by M. Sklare, pp. 23–39. Glencoe, Ill.: Free Press, 1958.

Handlin, Oscar, ed. *Immigration as a Factor in American History*. Englewood Cliffs, N.J.: Prentice-Hall, 1959.

Handlin, Oscar; Schlesinger, Arthur Meier; Morrison, Samuel Eliot; Merk, Frederick; Schlesinger, Arthur Meier, Jr.; and Buck, Paul Herman. *Harvard Guide to American History*. Cambridge, Mass.: Harvard University Press, 1954.

Hansen, B. "The Early History of Glacial Theory in British Geology." *Journal of Glaciology* 9 (1970): 135–41.

Harmon, Frank J. "Forestry." In *Dictionary of American History*, vol. 3, pp. 72–76. New York: Scribner's Sons, 1976.

Harris, George. *Moral Evolution*. New York: Houghton Mifflin, 1896.

Harris, Marvin. *The Rise of Anthropological Theory: A History of Theories of Culture*. New York: Crowell, 1968.

Harvey, Milton E., and Holly, Brian P. "Paradigm, Philosophy, and Geographic Thought." In *Themes in Geographic Thought*, edited by Milton E. Harvey and Brian P. Holly, pp. 11–37. London: Croom Helm, 1981.

Hawkins, Hugh. *Between Harvard and America: The Educational Leadership of Charles W. Eliot*. New York: Oxford University Press, 1972.

Hays, Samuel P. *Conservation and the Gospel of Efficiency: The Progressive Conservation Movement, 1890–1920*. Cambridge, Mass.: Harvard University Press, 1959.

Helm, June. "Ecological Approach in Anthropology." *American Journal of Sociology* 67 (1962): 630–39.

Herbst, Jurgen. "Social Darwinism and the History of American Geography." *Proceedings of the American Philosophical Society* 105 (1961): 538–44.

———. *The German Historical School in American Scholarship: A Study in the Transfer of Culture*. Ithaca: Cornell University Press, 1965.

Heyck, T. W. *The Transformation of Intellectual Life in Victorian England*. London: Croom Helm, 1982.

Higgins, Charles G. "Theories of Landscape Development: A Perspective." In *Theories of Landform Development*, edited by W. N. Melhorn and R. C. Flemal, pp. 1–29. Binghamton: State University of New York, 1975.

Higham, John. "Intellectual History and Its Neighbors." *Journal of the History of Ideas* 15 (1954): 339–47.

———. *Strangers in the Land: Patterns of American Nativism, 1865–1925*. New York: Atheneum, 1973.

———. *Send These to Me: Jews and Other Immigrants in Urban America*. New York: Atheneum, 1975.

Hill, Christopher. *Intellectual Origins of the English Revolution*. Oxford: Oxford University Press, 1965.

Himmelfarb, Gertrude. *Darwin and the Darwinian Revolution*. London: Chatto & Windus, 1959.

Hobbs, William H. "Nathaniel Southgate Shaler." *Transactions of the Wisconsin Academy of Sciences, Arts, and Letters* 15 (1907): 924–27.

———. *Earth Features and Their Meaning: An Introduction to Geology*. New York: Macmillan, 1926.

Hofstadter, Richard. *Social Darwinism in American Thought, 1860–1915*. Boston: Beacon Press, 1955. First published in 1944.

———. *The Progressive Historians: Turner, Beard, Parrington*. New York: Vintage Books, 1970.

Holt, W. Stull. *The Bureau of the Census: Its History, Activities, and Organization*. Washington, D.C.: Brookings Institution, 1929.

Hooykaas, R. "The Principle of Uniformity in Geology, Biology, and Theology." *Journal of the Transactions of the Victoria Institute* 88 (1956): 101–16.

———. "Science and Reformation." *Journal of World History* 3 (1956): 109–39.

———. "The Parallel between the History of the Earth and the History of the Animal World." *Archives internationales d'histoire des sciences* 10 (1957): 1–18.

———. *Philosophia Libera: Christian Faith and the Freedom of Science*. London: Tyndale Press, 1957.

———. "Geological Uniformitarianism and Evolution." *Archives internationales d'histoire des sciences* 19 (1966): 3–19.

———. *Religion and the Rise of Modern Science*. Edinburgh: Scottish Academic Press, 1972.

———. "Catastrophism in Geology, Its Scientific Character in Relation to Actualism and Uniformitarianism." In *Philosophy of Geohistory*, edited by Claude C. Albritton, pp. 310–56. Stroudsburg, Pa.: Dowden, Hutchinson & Ross, 1975.

Hopkins, William. "On the Thickness and Constitution of the Earth's Crust." *Philosophical Transactions of the Royal Society of London* 132 (1842): 43–55.

Horsman, Reginald. "Scientific Racism and the American Indian in the Mid-Nineteenth Century." *American Quarterly* 27 (1975): 152–68.

Hosmer, Ralph Sheldon. "The Progress of Education in Forestry in the United States." *Empire Forestry Journal* 2 (1923): 1–24.

Hoyme, L. E. "Physical Anthropology and Its Instruments." *Southwestern Journal of Anthropology* 9 (1953): 408–30.

Hull, David. *Darwin and His Critics: The Reception of Darwin's Theory by the Scientific Community*. Cambridge, Mass.: Harvard University Press, 1973.

———. "In Defense of Presentism." *History and Theory* 18 (1979): 1–15.

Humes, Walter. "Evolution and Educational Theory in the Nineteenth Century." In *The Wider Domain of Evolutionary Thought*, edited by David Oldroyd and Ian Langham, pp. 27–56. Dordrecht: Reidel Publishing, 1983.

Hunt, James. "On the Application of the Principle of Natural Selection in Anthropology." *Anthropological Review* 4 (1866): 320–40.

Hunt, Stanford B. "The Negro as a Soldier." *Anthropological Review* 7 (1869): 40–54.

Hunter, Michael. *Science and Society in Restoration England*. Cambridge: Cambridge University Press, 1981.

Huntington, Ellsworth. *Civilization and Climate*. New Haven: Yale University Press, 1915.

———. *Mainsprings of Civilization*. New York: Wiley, 1945.

Huth, Hans. *Nature and the American: Three Centuries of Changing Attitudes*. Berkeley: University of California Press, 1957.

Huxley, Thomas Henry. *Physiography: An Introduction to the Study of Nature*. London: Macmillan, 1877.

———. *Science and Education*. London: Macmillan, 1893.

———. *Man's Place in Nature and Other Anthropological Essays*. London: Macmillan, 1894.

———. *Evolution and Ethics and Other Essays*. London: Macmillan, 1898.

Hyatt, Alpheus. "Remarks on the Beatriceae, a New Division of Mollusca." *American Journal of Science and Arts* 39 (1865): 261–66.

———. "On the Parallelism between the Different Stages of Life in the Individual and Those in the Entire Group of the Molluscous Order Tetrabranchiata." *Memoirs of the Boston Society of Natural History* 1 (1866–67): 193–209.

Imbrie, John, and Imbrie, Katherine Palmer. *Ice Ages: Solving the Mystery*. London: Macmillan, 1979.

Irvine, William. *Apes, Angels, and Victorians: A Joint Biography of Darwin and Huxley*. London: Weidenfeld & Nicholson, 1956.

Jaki, Stanley L. *The Road of Science and the Ways of God*. Edinburgh: Scottish Academic Press, 1978.

———. "Science and Christian Theism: A Mutual Witness." *Scottish Journal of Theology* 32 (1979): 563–70.

Jakle, John A. "Salt on the Ohio Valley Frontier, 1770–1820." *Annals of the Association of American Geographers* 59 (1969): 687–709.

James, Henry. *Charles W. Eliot, President of Harvard University, 1869–1909*. 2 vols. Cambridge, Mass.: Constable, 1930.

James, Preston. *All Possible Worlds: A History of Geographic Ideas*. Indianapolis: Bobbs-Merrill, 1972.

James, Preston, and Martin, Geoffrey J., eds. *The Association of American Geographers: The First Seventy-Five Years, 1904–1979*. Washington, D.C.: Association of American Geographers, 1979.

James, Preston, and Mather, Cotton. "The Role of Periodic Field Conferences in the Development of Geographical Ideas in the United States." *Geographical Review* 67 (1977): 446–61.

James, William. *The Varieties of Religious Experience*. New York: Random House, 1929.

Jamieson, T. F. "On the History of the Last Geological Changes in Scotland." *Quarterly Journal of the Geological Society of London* 21 (1865): 161–203.

———. "On the Causes of the Depression and Re-elevation of the Land during the Glacial Period." *Geological Magazine* 9 (1882): 400–407.

Johnson, Douglas W., ed. *The Geographical Essays of William Morris Davis*. Boston: Ginn, 1909.

——. *Shore Processes and Shoreline Development*. New York: Wiley, 1919.

——. *The New England-Acadian Shoreline*. New York: Wiley, 1925.

Johnston, R. J. "Paradigms and Revolutions or Evolution? Observations on Human Geography since the Second World War." *Progress in Human Geography* 2 (1978): 189–206.

——. *Geography and Geographers: Anglo-American Human Geography since 1945*. London: Arnold, 1979.

Jones, Greta. *Social Darwinism and English Thought: The Interaction between Biological and Social Theory*. Sussex: Harvester Press, 1980.

Jones, L. Rodwell, and Bryan, P. W. *North America: An Historical, Economic, and Regional Geography*. London: Methuen, 1924.

Jordan, David Starr. *The Days of a Man; Being Memories of a Naturalist, Teacher, and Minor Prophet of Democracy*. 2 vols. New York: World Book, 1922.

Judson, Sheldon. "Davis, William Morris." In *Dictionary of Scientific Biography*, vol. 3, pp. 592–96. New York: Scribner's Sons, 1971.

Juricek, John T. "American Usage of the Word 'Frontier' from Colonial Times to Frederick Jackson Turner." *Proceedings of the American Philosophical Society* 110 (1966): 10–34.

Kellogg, C. E. "We Seek; We Learn." In *The Yearbook of Agriculture*, pp. 2–11. Washington, D.C.: United States Department of Agriculture, 1957.

Kellogg, Vernon L. *Darwinism Today: A Discussion of Present-Day Scientific Criticism of the Darwinian Selection Theories, together with a Brief Account of the Principal Other Proposed Auxiliary and Alternative Theories of Species Forming*. New York: Holt, 1908.

Keltie, J. Scott. "Report to the Council of the Royal Geographical Society." In *Report of the Proceedings of the Society in Reference to the Improvement of Geographical Education*. London: Royal Geographical Society, 1886.

Kemsley, Douglas S. "Religious Influence in the Rise of Modern Science: A Review and Criticism, Particularly of the 'Protestant-Puritan Ethic' Theory." *Annals of Science* 24 (1968): 199–226.

Kett, Joseph F. "Adolescence and Youth in Nineteenth-Century America." *Journal of Interdisciplinary History* 2 (1971): 283–98.

King, Clarence. "Catastrophism and Evolution." *American Naturalist* 11 (1877): 449–70.

Kittredge, Joseph. *Forest Influences*. New York: McGraw-Hill, 1948.

Klausner, Samuel Z. "Thinking Social-Scientifically about Environmental Quality." *Annals of the American Academy of Political and Social Science* 389 (1970): 1–10.

Knight, Edgar W. *Education in the United States*. Boston: Ginn, 1941.

Knox, H. M. *Introduction to Educational Method*. London: Oldbourne, 1961.

Koelsch, William A., ed. *Lectures on the Historical Geography of the United States As Given in 1933 by Harlan H. Barrows*. Department of Geography, Research Paper no. 77. Chicago: University of Chicago, 1962.

——. "The Historical Geography of Harlan H. Barrows." *Annals of the Association of American Geographers* 59 (1969): 632–51.

——. "Nathaniel Southgate Shaler." In *Geographers: Biobibliographical Studies*, edited by T. W. Freeman and Philippe Pinchemel, vol. 3, pp. 133–39. London: Mansell, 1979.

——. "'A Profound Though Special Erudition': Justin Winsor as Historian of Discovery." *Proceedings of the American Antiquarian Society* 93 (1983): 55–94.

Kohlstedt, Sally Gregory. "From Learned Society to Public Museum: The Boston Society of Natural History." In *The Organization of Knowledge in Modern America, 1860–1920*, edited by Alexandra Oleson and John Voss, pp. 386–406. Baltimore: Johns Hopkins University Press, 1979.

Kolata, Gina Bari. "Theoretical Ecology: Beginnings of a Predictive Science." *Science* 183 (1974): 400–401, 450.

Kollmorgen, Walter M. "The Woodsman's Assaults on the Domain of the Cattleman." *Annals of the Association of American Geographers* 59 (1969): 215–39.

Kropotkin, Peter. "On the Teaching of Physiography." *Geographical Journal* 2 (1893): 350–59.

———. *Mutual Aid*. London: Heinemann, 1902.

Kuhn, Thomas S. *The Structure of Scientific Revolutions*. Chicago: University of Chicago Press, 1962.

Kuklick, Bruce. *The Rise of American Philosophy: Cambridge, Massachusetts, 1860–1930*. New Haven: Yale University Press, 1977.

Laing, S. *Human Origins*. London: Chapman & Hall, 1902.

Lamb, H. H. *Climate: Present, Past, and Future*. Vol. 1, *Fundamentals and Climate Now*. London: Methuen, 1972.

———. *Climate: Present, Past, and Future*. Vol. 2, *Climatic History and the Future*. London: Methuen, 1977.

Lankester, E. Ray. "An American Sea-Side Laboratory." *Nature* 21 (1880): 498.

Lawrence, Philip. "Edward Hitchcock: The Christian Geologist." *Proceedings of the American Philosophical Society* 116 (1972): 21–34.

Le Conte, Joseph. *Elements of Geology*. New York: Appleton, 1888.

———. *Evolution: Its Nature, Its Evidences, and Its Relation to Religious Thought*. 2d ed. New York: Appleton, 1897.

Leighly, John. "Carl Ortwin Sauer, 1889–1975." *Annals of the Association of American Geographers* 66 (1976): 337–48.

Lewis, C. S. *The Abolition of Man; or, Reflections on Education with Special Reference to the Teaching of English in the Upper Forms of Schools*. London: Blis, 1943.

Lewis, G. Malcolm. "Changing Emphases in the Description of the Natural Environment of the American Great Plains Area." *Transactions of the Institute of British Geographers* 30 (1962): 75–90.

———. "Early American Exploration and the Cis-Rocky Mountain Desert, 1803–1823." *Great Plains Journal* 5 (1965): 1–11.

———. "Three Centuries of Desert Concepts in the Cis-Rocky Mountain West." *Journal of the West* 4 (1965): 457–68.

———. "Regional Ideas and Reality in the Cis-Rocky Mountain West." *Transactions of the Institute of British Geographers* 38 (1966): 135–50.

———. "William Gilpin and the Concept of the Great Plains Region." *Annals of the Association of American Geographers* 56 (1966): 33–51.

———. "First Impressions of the Great Plains and Prairies." In *The American Environment: Perceptions and Policies*, edited by J. Wreford Watson and Timothy O'Riordan, pp. 37–45. London: Wiley, 1976.

Lewthwaite, Gordon R. "Environmentalism and Determinism: A Search for Clarification." *Annals of the Association of American Geographers* 56 (1966): 1–23.

Life and Work of C. F. Marbut. Columbia, Mo.: Soil Science Society of America, 1942.

Lillie, Frank R. *The Woods Hole Marine Biological Laboratory.* Chicago: University of Chicago Press, 1944.

Livingstone, David N. "Some Methodological Problems in the History of Geographical Thought." *Tijdschrift voor economische en sociale geografie* 70 (1979): 226–31.

———. "Nature and Man in America: Nathaniel Southgate Shaler and the Conservation of Natural Resources." *Transactions of the Institute of British Geographers,* n.s., 5 (1980): 369–82.

———. "Environment and Inheritance: Nathaniel Southgate Shaler and the American Frontier." In *The Origins of Academic Geography in the United States,* edited by Brian W. Blouet, pp. 123–38. Hamden, Conn.: Archon Books, 1981.

———. "The History of Science and the History of Geography: Interactions and Implications." *History of Science* 22 (1984): 271–302.

———. "The Idea of Design: The Vicissitudes of a Key Concept in the Princeton Response to Darwin." *Scottish Journal of Theology* 37 (1984): 329–57.

———. "Natural Theology and Neo-Lamarckism: The Changing Context of Nineteenth-Century Geography in the United States and Great Britain." *Annals of the Association of American Geographers* 74 (1984): 9–28.

———. "Science and Society: Nathaniel S. Shaler and Racial Ideology." *Transactions of the Institute of British Geographers,* n.s., 9 (1984): 181–210.

———. *Darwin's Forgotten Defenders: The Encounter between Evangelical Theology and Evolutionary Thought.* Grand Rapids and Edinburgh: Eerdmans and Scottish Academic Press, 1987.

Livingstone, David N., and Harrison, Richard T. "Meaning through Metaphor: Analogy as Epistemology." *Annals of the Association of American Geographers* 71 (1981): 95–107.

Lockhead, Elspeth. "Scotland as the Cradle of Modern Academic Geography in Britain." *Scottish Geographical Magazine* 97 (1981): 98–109.

Logan, Rayford Whittingham. *The Negro in American Life and Thought: The Nadir, 1877–1901.* New York: Dial Press, 1954.

London, Herbert. "American Romantics: Old and New." *Colorado Quarterly* 18 (1969): 5–20.

Love, James Lee. "Summer Courses of Instruction." *Annual Report of the President and Treasurer of Harvard College, 1906–1907.* Cambridge, Mass.: Harvard University, 1908.

———. *The Lawrence Scientific School in Harvard University, 1847–1906.* Burlington, N.C.: n.p., 1944.

Lowenthal, David. "George Perkins Marsh and the American Geographic Tradition." *Geographical Review* 43 (1954): 207–13.

———. *George Perkins Marsh: Versatile Vermonter.* New York: Columbia University Press, 1958.

———. "Is Wilderness 'Paradise Enow'? Images of Nature in America." *Columbia University Forum* 7 (1964): 34–40.

Lucas, J. R. "Wilberforce and Huxley: A Legendary Encounter." *Historical Journal* 22 (1978): 313–30.

Ludmerer, Kenneth M. *Genetics and American Society: A Historical Appraisal.* Baltimore: Johns Hopkins University Press, 1972.

Lull, Howard W. "Forest Influences: Growth of a Concept." *Journal of Forestry* 47 (1949): 700–705.

Lurie, Edward. "Louis Agassiz and the Races of Mankind." *Isis* 45 (1954): 227–42.

———. "Louis Agassiz and the Idea of Evolution." *Victorian Studies* 3 (1959): 87–108.

———. *Louis Agassiz: A Life in Science.* Chicago: University of Chicago Press, 1960.

[Lyell, Katherine M.] *Life, Letters, and Journals of Sir Charles Lyell, Bart.* 2 vols. London: Murray, 1881.

McCaughey, Robert A. "The Transformation of American Academic Life: Harvard University, 1821–1892." *Perspectives in American History* 8 (1974): 239–332.

McGeary, Martin Nelson. *Gifford Pinchot: Forester/Politician.* Princeton: Princeton University Press, 1960.

McGee, W J. "The Classification of Geographic Forms by Genesis." *National Geographic Magazine* 1 (1888): 27–36.

———. "The Trend of Human Progress." *American Anthropologist* 1 (1888): 401–47.

McIntosh, R. P. "Ecology since 1900." In *Issues and Ideas in America,* edited by Benjamin J. Taylor and Thurman J. White, pp. 353–72. Norman: University of Oklahoma Press, 1976.

Mackay, C. "The Negro and the Negrophilists." *Blackwood's Edinburgh Magazine* 99 (1866): 581–97.

MacKenzie, Donald. "Eugenics in Britain." *Social Studies of Science* 6 (1976): 449–532.

Macleod, Roy M. "Evolutionism and Richard Owen, 1830–1868: An Episode in Darwin's Century." *Isis* 56 (1965): 259–80.

McMaster, John B. *A History of the People of the United States from the Revolution to the Civil War.* 8 vols. New York: Appleton, 1883–1913.

McPherson, James M. et al. *Blacks in America: Bibliographical Essays.* New York: Doubleday, 1971.

Macquarrie, John. *Twentieth Century Religious Thought: The Frontiers of Philosophy and Theology, 1900–1970.* London: SCM Press, 1963.

Maine, Henry Sumner. *Ancient Law: Its Connection with the Early History of Society and Its Relation to Modern Ideas.* London: Murray, 1861.

Malin, James C. "Space and History: Reflections on the Closed-Space Doctrines of Turner and Mackinder and the Challenge of Those Ideas by the Air Age." *Agricultural History* 17 (1943): 65–74, 107–26.

———. *Essays on Historiography.* Lawrence, Kans.: James C. Malin, 1948.

———. *The Grassland of North America: Prolegomena to Its History, with Addenda and Postscript.* Gloucester, Mass.: Smith, 1967. First published in 1947.

Mandelbaum, Maurice. "Concerning Recent Trends in the Theory of Historiography." *Journal of the History of Ideas* 16 (1955): 506–17.

Manning, Thomas G. "Geological Survey, United States." In *Dictionary of American History,* vol. 3, pp. 163–64. New York: Scribner's Sons, 1976.

Marcou, Jules. *A Little More Light on the United States Geological Survey.* Cambridge, Mass.: Privately printed, 1892.

Marsh, George Perkins. *Address Delivered before the Agricultural Society of Rutland County, 30th September, 1847.* Rutland, Vt.: Agricultural Society of Rutland County, 1848.

———. "The Study of Nature." *Christian Examiner* 68 (1860): 33–62.

Mayr, Ernst. "Agassiz, Darwin, and Evolution." *Harvard Library Bulletin* 13 (1959): 165–94.

————. "Isolation as an Evolutionary Factor." *Proceedings of the American Philosophical Society* 103 (1959): 221–30.

————. "Lamarck Revisited." *Journal of the History of Biology* 5 (1972): 55–94.

————. "The Nature of the Darwinian Revolution." *Science* 176 (1972): 981–89.

Merrens, H. Roy. "Historical Geography and Early American History." *William and Mary Quarterly*, 3d ser., 22 (1965): 529–48.

Merrill, George P. *The First One Hundred Years of American Geology.* New Haven: Yale University Press, 1924.

Merrill, G. P., and Dobson, E. R. *Dictionary of American Biography*, s.v. "Shaler, Nathaniel Southgate."

Merrill, Kenneth R. "From Edwards to Quine: Two Hundred Years of American Philosophy." In *Issues and Ideas in America*, edited by Benjamin J. Taylor and Thurman J. White, pp. 219–63. Norman: University of Oklahoma Press, 1976.

Merton, Robert K. "Science, Technology, and Society in Seventeenth-Century England." *Osiris* 4 (1938): 360–632.

Meyer, D. H. "American Intellectuals and the Victorian Crisis of Faith." *American Quarterly* 27 (1975): 585–603.

Mikesell, Marvin. "Ratzel, Friedrich." In *International Encyclopedia of the Social Sciences*, vol. 13, pp. 327–29. New York: Macmillan and Free Press, 1968.

Mill, Hugh Robert. *Encyclopaedia Britannica*, 14th ed., s.v. "Geography."

Miller, Perry. *Errand into the Wilderness.* New York: Harper Books, 1956.

Mivart, St. George. "The Origin of Intellect." *Edinburgh Review* 170 (1889): 359–88.

Moncrief, Lewis W. "The Cultural Basis for Our Environmental Crisis." *Science* 170 (1970): 508–12.

Mood, Fulmer. "The Development of Frederick Jackson Turner as a Historical Thinker." *Publications of the Colonial Society of Massachusetts* 34 (1943): 283–352.

————. "The Historiographic Setting of Turner's Frontier Essay." *Agricultural History* 17 (1943): 153–55.

————. "The Concept of the Frontier, 1871–1898." *Agricultural History* 19 (1945): 24–30.

————. "Notes on the History of the Word *Frontier.*" *Agricultural History* 22 (1948): 78–83.

Moor, Dean. "The Paxton Boys: Parkman's Use of the Frontier Hypothesis." *Mid-America* 36 (1954): 211–19.

Moore, James R. *The Post-Darwinian Controversies: A Study of the Protestant Struggle to Come to Terms with Darwin in Great Britain and America, 1870–1900.* Cambridge: Cambridge University Press, 1979.

————. "1859 and All That: Remaking the Story of Evolution-and-Religion." In *Charles Darwin, 1809–1882: A Centennial Commemorative*, edited by Roger G. Chapman and Cleveland T. Duval, pp. 167–94. Wellington, New Zealand: Nova Pacifica, 1982.

————. "On Revolutionising the Darwin Industry: A Centennial Retrospect." *Radical Philosophy* 37 (1984): 13–22.

Morgan, H. Wayne. "American History as Experiment." In *Issues and Ideas in America*, edited by Benjamin J. Taylor and Thurman J. White, pp. 5–18. Norman: University of Oklahoma Press, 1976.

Morison, Samuel Eliot. *Three Centuries of Harvard.* Cambridge, Mass.: Harvard University Press, 1937.

Morrill, R. L., and Donaldson, O. F. "Geographical Perspectives on the History of Black America." *Economic Geography* 48 (1972): 1–23.

Morris, R. C. "The Notion of a Great American Desert." *Mississippi Valley Historical Review* 8 (1926–27): 190–200.

Mumford, Lewis. *The Golden Day: A Study in American Literature and Culture*. New York: Dover Publications, 1968. First published in 1926.

———. *The Brown Decades: A Study of the Arts in America, 1865–1895*. New York: Dover Publications, 1971. First published in 1931.

Murphree, Idus. "The Evolutionary Anthropologists: The Progress of Mankind: The Concepts of Progress and Culture in the Thought of John Lubbock, Edward B. Tylor, and Lewis H. Morgan." *Proceedings of the American Philosophical Society* 105 (1961): 265–300.

Nash, Roderick. *Wilderness and the American Mind*. New Haven: Yale University Press, 1967.

"Nathaniel Southgate Shaler." *Geographical Journal* 28 (1906): 188

Needham, Joseph. *Time: The Refreshing River*. New York: Macmillan, 1943.

Nelson, C. M.; Rabbitt, M. C.; and Fryxell, F. M. "Ferdinand Vandeveer Hayden: The U.S. Geological Survey Years, 1879–1886." *Proceedings of the American Philosophical Society* 125 (1981): 238–43.

Nikiforoff, C. C. "Reappraisal of the Soil." *Science* 129 (1959): 186–96.

Numbers, Ronald L. *Creation by Natural Law: Laplace's Nebular Hypothesis in American Thought*. Seattle: University of Washington Press, 1977.

"Obituary: Nathaniel S. Shaler." *Bulletin of the American Geographical Society* 38 (1906): 366.

Olby, R. C. "Human Genetics, Eugenics, and Race: Essay Review." *British Journal for the History of Science* 10 (1977): 156–59.

Oleson, Alexandra, and Voss, John. "Introduction." In *The Organization of Knowledge in Modern America, 1860–1920*, edited by Alexandra Oleson and John Voss. Baltimore: Johns Hopkins University Press, 1979.

Olwig, Kenneth Robert. "Historical Geography and the Society/Nature 'Problematic': The Perspective of J. F. Schouw, G. P. Marsh, and E. Reclus." *Journal of Historical Geography* 6 (1980): 29–45.

Osborn, Henry Fairfield. "Paleontological Evidence for the Transmission of Acquired Characteristics." *American Naturalist* 23 (1889): 561–66.

Ospovat, Dov. "God and Natural Selection: The Darwinian Idea of Design." *Journal of the History of Biology* 13 (1980): 169–94.

———. *The Development of Darwin's Theory: Natural History, Natural Theology, and Natural Selection, 1838–1859*. Cambridge: Cambridge University Press, 1981.

———. "Perfect Adaptation and Teleological Explanation: Approaches to the Problem of the History of Life in the Mid-Nineteenth Century." *Studies in the History of Biology* 14 (1981): 193–230.

Ostrander, Gilman M. "Turner and the Germ Theory." *Agricultural History* 32 (1958): 258–61.

Otto, Rudolph. *The Idea of the Holy: An Inquiry into the Idea of the Divine and Its Relation to the Rational*. Oxford: Oxford University Press, 1925.

Passmore, John. "Darwin's Impact on British Metaphysics." *Victorian Studies* 3 (1959): 41–54.

———. *Man's Responsibility for Nature*. London: Duckworth, 1974.

Paterson, John H. *North America: A Geography of Canada and the United States*. 4th ed. London: Oxford University Press, 1970.

Pattee, Fred L., ed. *The Poems of Philip Freneau, Poet of the American Revolution*. 3 vols. Princeton: Princeton University Press, 1902.

Paul, Diane. "Eugenics and the Left." *Journal of the History of Ideas* 45 (1984): 567–90.

Persons, Stow. *American Minds: A History of Ideas.* New York: Holt, 1958.

———. *Free Religion: An American Faith.* Boston: Beacon Press, 1963. First published in 1946.

Pfeifer, Edward J. "The Genesis of American Neo-Lamarckism." *Isis* 56 (1965): 156–67.

———. "United States." In *The Comparative Reception of Darwinism*, edited by Thomas F. Glick, pp. 168–206. Austin: University of Texas Press, 1972.

Phillips, D. C. "Organicism in the Late Nineteenth and Early Twentieth Centuries." *Journal of the History of Ideas* 31 (1970): 413–32.

Pike, Richard J., and Rozema, Wesley. "Spectral Analysis of Landforms." *Annals of the Association of American Geographers* 65 (1975): 499–516.

Pinchot, Gifford. "Relation of Forests to Stream Control." *Annals of the American Academy of Political and Social Science* 31 (1908): 219–28.

———. *The Fight for Conservation.* Seattle: University of Washington Press, 1967. First published in 1910.

Platt, Robert S. "Determinism in Geography." *Annals of the Association of American Geographers* 38 (1948): 126–28.

Plewe, E. "Ritter, Carl." In *International Encyclopedia of the Social Sciences*, vol. 13, pp. 517–20. New York: Macmillan and Free Press, 1968.

Pomeroy, H. S. *The Ethics of Marriage.* New York: Funk & Wagnalls, 1888.

Powell, Arthur G. "The Education of Educators at Harvard, 1891–1912." In *Social Sciences at Harvard, 1860–1920: From Inculcation to the Open Mind*, edited by Paul Buck, pp. 223–74. Cambridge, Mass.: Harvard University Press, 1965.

Powell, John Wesley. *Report on the Lands of the Arid Region of the United States.* Washington, D.C.: Government Printing Office, 1878.

———. "On the Evolution of Language, As Exhibited in the Specialization of the Grammatic Processes, the Differentiation of the Parts of Speech, and the Integration of the Sentence; from a Study of the Indian Languages." *Bureau of Ethnology, Annual Report* 1 (1881): 1–16.

———. "Competition as a Factor in Human Evolution." *American Anthropologist* 1 (1888): 297–323.

———. "The North American Indians." In *The United States of America: A Study of the American Commonwealth, Its Natural Resources, People, Industries, Manufactures, Commerce, and Its Work in Literature, Science, Education, and Self-Government*, edited by Nathaniel S. Shaler, vol. 1, pp. 190–272. 3 vols. New York: Appleton, 1894.

"Professor Shaler Dead." *Harvard Bulletin*, 11 April 1906.

Pulliman, John D. "Shifting Patterns of Educational Thought." In *Issues and Ideas in America*, edited by Benjamin J. Taylor and Thurman J. White, pp. 161–87. Norman: University of Oklahoma Press, 1976.

Pumpelly, Raphael. *My Reminiscences.* New York: Holt, 1918.

Quick, R. H. *Some Thoughts concerning Education by John Locke.* Cambridge: Cambridge University Press, 1895.

The Race Question in Modern Science. New York: UNESCO, 1956.

Rainger, Ronald. "Race, Politics, and Science: The Anthropological Society of London in the 1860's." *Victorian Studies* 22 (1978): 51–70.

Raitz, Karl B. "Themes in the Cultural Geography of European Ethnic Groups in the United States." *Geographical Review* 69 (1979): 79–94.

Ratner, Sidney. "Evolution and the Rise of the Scientific Spirit in America." *Philosophy of Science* 3 (1936): 104–22.

Redway, J. W. "What Is Physiography?" *Educational Review* 10 (1895): 352–63.

Reinach, Theodore. *Histoires des Israelites depuis l'époque de leur dispersion jusqu'à nos jours.* Paris: n.p., 1885.

———. *Textes d'auteurs grecs et romains relatifs au Judaism.* Paris: Société des Etudes Juives, 1895.

Reist, Irwin. "Augustus Hopkins Strong and William Newton Clarke: A Study in Nineteenth Century Evolutionary and Eschatological Thought." *Foundations* 13 (1970): 26–43.

Rice, William North. "Darwinian Theory of the Origin of Species." *New Englander* 25 (1867): 607–33.

Richards, Robert J. "Lloyd Morgan's Theory of Instinct: From Darwinism to Neo-Darwinism." *Journal of the History of the Behavioral Sciences* 13 (1977): 12–32.

Ripley, William Z. "Geography as a Sociological Study." *Political Science Quarterly* 10 (1896): 636–55.

———. *The Races of Europe: A Sociological Study.* London: Kegan Paul, Trench, Trubner, 1899.

Rode, A. A. *Soil Science.* Translated by A. Gourevitch. Jerusalem: Israel Program for Scientific Translations, 1962.

Roe, Frank Gilbert. *The North American Buffalo: A Critical Study of the Species in Its Wild State.* Toronto: University of Toronto Press, 1951.

Romanes, George John. *Mental Evolution in Animals.* London: Kegan Paul, Trench, 1883.

Rose, Harold M. "Conservation in the United States." In *Conservation of Natural Resources,* edited by Guy-Harold Smith, pp. 3–14. 3d ed. New York: Wiley, 1965.

Rosenberg, Charles E. "The Bitter Fruit: Heredity, Disease, and Social Thought in Nineteenth-Century America." *Perspectives in American History* 8 (1974): 187–235.

Ross, James S. *Groundwork of Educational Theory.* London: Harrap, 1942.

Ross, John R. "Man over Nature: Origins of the Conservation Movement." *American Studies* 16 (1975): 49–62.

Rossiter, M. W. "Lawrence Scientific School." In *Dictionary of American History,* vol. 4, pp. 119–20. New York: Scribner's Sons, 1976.

Rostlund, Erhard. "The Myth of a Natural Prairie Belt in Alabama: An Interpretation of Historical Records." *Annals of the Association of American Geographers* 47 (1957): 392–411.

———. "The Geographical Range of the Historic Bison in the Southeast." *Annals of the Association of American Geographers* 50 (1960): 395–407.

Roszak, Theodore. *The Making of a Counter Culture: Reflections on the Technocratic Society and Its Youthful Opposition.* London: Faber, 1970.

———. "Ecology and Mysticism." *Humanist* 86 (1971): 134–36.

Rousseau, Jean Jacques. *Emile; or, On Education.* London: Everyman's Library, 1966. First published in 1762.

Rowley, Virginia. *J. Russell Smith: Geographer, Educator, and Conservationist.* Philadelphia: University of Pennsylvania Press, 1964.

Ruse, Michael. *The Darwinian Revolution: Science Red in Tooth and Claw.* Chicago: University of Chicago Press, 1979.

———. "Charles Darwin and Group Selection." *Annals of Science* 37 (1980): 615–30.

Rusk, Robert R. *Doctrines of the Great Educators.* 4th ed. London: Macmillan, 1969.

Russell, Israel C. *Volcanoes of North America.* New York: Macmillan, 1904.

———. *River Development.* New York: Putnam's Sons, 1907.

Russett, Cynthia Eagle. *Darwin in America: The Intellectual Response, 1865–1912.* San Francisco: Freeman, 1976.

Ryder, John A. "Proofs of the Effects of Habitual Use on the Modification of Animal Organisms." *Proceedings of the American Philosophical Society* 26 (1889): 541–49.

Sauer, Carl O. *Geography of the Pennyroyal: A Study of the Influence of Geology and Physiography upon the Industry, Commerce, and Life of the People.* Kentucky Geological Survey, ser. 6, vol. 25. Frankfort, 1927.

————. "The Settlement of the Humid East." In *Yearbook of Agriculture*, pp. 157–66. Washington, D.C.: Government Printing Office, 1941.

————. "The Agency of Man on the Earth." In *Man's Role in Changing the Face of the Earth*, edited by William L. Thomas, Jr., pp. 350–66. Chicago: University of Chicago Press, 1956.

————. "Conditions of Pioneer Life in the Upper Illinois Valley." In *Land and Life: A Selection from the Writings of Carl Ortwin Sauer*, edited by John Leighly, pp. 11–22. Berkeley: University of California Press, 1963.

————. "Theme of Plant and Animal Destruction in Economic Theory." In *Land and Life: A Selection from the Writings of Carl Ortwin Sauer*, edited by John Leighly, pp. 145–54. Berkeley: University of California Press, 1963.

————. "On the Background of Geography in the United States." In *Heidelberger Studien zur Kulturgeographie: Festgabe für Gottfried Pfeifer*, pp. 59–71. Heidelberger Geographische Arbeiten, no. 15. Wiesbaden, 1966.

————. "The Formative Years of Ratzel in the United States." *Annals of the Association of American Geographers* 61 (1971): 245–54.

Savage, Minot Judson. *The Religion of Evolution.* Boston: Lockwood, Brooks, 1876.

————. *Evolution and Religion from the Standpoint of One Who Believes Both: A Lecture Delivered in the Philadelphia Academy of Music, Seventh December, 1885.* Philadelphia: Buchanan, 1886.

Saveth, E. N. *American Historians and European Immigrants, 1875–1925.* New York: Columbia Press, 1948.

Schiller, Francis. *Paul Broca: Founder of French Anthropology, Explorer of the Brain.* Berkeley: University of California Press, 1979.

Schlesinger, Arthur M. *New Viewpoints in American History.* New York: Macmillan, 1922.

Schmitt, Peter J. *Back to Nature: The Arcadian Myth in Urban America.* New York: Oxford University Press, 1969.

Schneider, Herbert W. "The Influence of Darwin and Spencer on American Philosophical Theology." *Journal of the History of Ideas* 6 (1945): 3–18.

————. "Religious Enlightenment in American Thought." In *Dictionary of the History of Ideas*, vol. 4, pp. 109–12. New York: Scribner's Sons, 1973.

Schuchert, C. "A Century of Geology—the Progress of Historical Geology in North America." *American Journal of Science*, 4th ser., 46 (1918): 45–103.

Schultz, Arthur R. "Immigration." In *Dictionary of American History*, vol. 3, pp. 332–41. New York: Scribner's Sons, 1976.

Searle, G. R. *The Quest for National Efficiency, 1899–1914.* Oxford: Oxford University Press, 1971.

————. *Eugenics and Politics in Britain, 1900–1914.* Leiden: Noordhoff International Publishing, 1976.

Seltzer, Carl C. "The Jew—His Racial Status." *Harvard Medical Alumni Bulletin*, April 1939, pp. 10–11.

Semple, Ellen Churchill. "The Anglo Saxons of the Kentucky Mountains: A Study in Anthropogeography." *Geographical Journal* 17 (1901): 588–623.

———. *American History and Its Geographic Conditions.* Boston: Houghton Mifflin, 1903.

———. *Influences of Geographic Environment: On the Basis of Ratzel's Anthropo-geography.* New York: Holt, 1911.

Shelley, Bruce L. "Andover Controversy." In *The New International Dictionary of the Christian Church,* edited by J. D. Douglas, p. 40. Exeter: Paternoster Press, 1974.

Shepard, Paul. *Man in the Landscape: A Historic View of the Esthetics of Nature.* New York: Knopf, 1967.

Shin, Suk Han. "American Conservation Viewpoints." *Environmental Conservation* 4 (1977): 273–77.

Simpson, George G. "Uniformitarianism: An Inquiry into Principle, Theory, and Method in Geohistory and Biohistory." In *Philosophy of Geohistory,* edited by Claude C. Albritton, pp. 256–309. Stroudsburg, Pa.: Dowden, Hutchinson & Ross, 1975.

Sinnhuber, Karl A. "Carl Ritter, 1779–1859." *Scottish Geographical Magazine* 75 (1959): 153–63.

Skinner, Quentin. "Meaning and Understanding in the History of Ideas." *History and Theory* 8 (1969): 3–53.

Smith, Henry Nash. "Clarence King, John Wesley Powell, and the Establishment of the United States Geological Survey." *Mississippi Valley Historical Review* 34 (1947): 37–58.

———. *Virgin Land: The American West as Symbol and Myth.* New York: Vintage Books, 1950.

Smith, J. Russell. *Industrial and Commercial Geography.* New York: Holt, 1913.

———. *North America: Its People and the Resources, Development, and Prospects of the Continent as an Agricultural, Industrial, and Commercial Area.* London: Bell, 1924.

Smith, Roger. "Alfred Russel Wallace: Philosophy of Nature and Man." *British Journal for the History of Science* 6 (1972): 177–99.

Snyder, Louis L. *The Idea of Racialism: Its Meaning and History.* New York: Van Nostrand Reinhold, 1962.

Solomon, Barbara M. "The Intellectual Background of the Immigration Restriction Movement in New England." *New England Quarterly* 25 (1952): 47–59.

———. *Ancestors and Immigrants: A Changing New England Tradition.* Chicago: University of Chicago Press, 1956.

Spate, O. H. K. "Toynbee and Huntington: A Study in Determinism." *Geographical Journal* 108 (1952): 406–28.

———. "Environmentalism." In *International Encyclopedia of the Social Sciences,* vol. 5, pp. 93–97. New York: Macmillan and Free Press, 1968.

Spencer, Herbert. *Principles of Psychology.* 2 vols. London: Williams & Norgate, 1870.

———. *Education: Intellectual, Moral, and Physical.* New York: Allison [c. 1880].

———. *Principles of Ethics.* 2 vols. London: Williams & Norgate, 1904.

Speth, William W. "Carl Ortwin Sauer on Destructive Exploitation." *Biological Conservation* 11 (1977): 145–60.

Sprout, Harold H. "Political Geography." In *International Encyclopedia of the Social Sciences,* vol. 6, pp. 116–23. New York: Macmillan and Free Press, 1968.

Stanton, William. *The Leopard's Spots: Scientific Attitudes toward Races in America, 1815–59.* Chicago: University of Chicago Press, 1960.

Stegner, Wallace. *Beyond the Hundredth Meridian: John Wesley Powell and the Second Opening of the West.* Boston: Houghton Mifflin, 1954.

Stephens, Lester D. "Joseph Le Conte on Evolution, Education, and the Structure of Knowledge." *Journal of the History of the Behavioral Sciences* 12 (1976): 103–19.

———. "Joseph Le Conte's Evolutional Idealism: A Lamarckian View of Cultural History." *Journal of the History of Ideas* 39 (1978): 465–80.

———. *Joseph Le Conte: Gentle Prophet of Evolution*. Baton Rouge: Louisiana State University Press, 1982.

Stewart, T. D. "The Effect of Darwin's Theory of Evolution on Physical Anthropology." In *Evolution and Anthropology: A Centennial Appraisal*, edited by Betty J. Meggars, pp. 11–25. Washington, D.C.: Anthropological Society of Washington, 1959.

Stocking, George W., Jr. "Lamarckianism in American Social Science: 1890–1915." *Journal of the History of Ideas* 23 (1962): 239–56.

———. *Race, Culture, and Evolution: Essays in the History of Anthropology*. London: Free Press, 1968.

———. "What's in a Name? The Origins of the Royal Anthropological Institute, 1837–1871." *Man: The Journal of the Royal Anthropological Institute* 6 (1971): 369–90.

Stoddart, David R. "Darwin's Impact on Geography." *Annals of the Association of American Geographers* 56 (1966): 683–98.

———. "Organism and Ecosystem as Geographical Models." In *Models in Geography*, edited by Richard J. Chorley and Peter Haggett, pp. 511–48. London: Methuen, 1967.

———. "'That Victorian Science': Huxley's *Physiography* and Its Impact on Geography." *Transactions of the Institute of British Geographers* 66 (1975): 17–40.

———. "Darwin's Influence on the Development of Geography in the United States, 1859–1914." In *The Origins of Academic Geography in the United States*, edited by Brian W. Blouet, pp. 265–78. Hamden, Conn.: Archon Books, 1981.

———. "Ideas and Interpretation in the History of Geography." In *Geography, Ideology, and Social Concern*, edited by David R. Stoddart, pp. 1–7. Oxford: Blackwell, 1981.

———. "The Paradigm Concept and the History of Geography." In *Geography, Ideology, and Social Concern*, edited by David R. Stoddart, pp. 70–80. Oxford: Blackwell, 1981.

Suess, Eduard. *The Face of the Earth*. Translated by Hertha B. C. Sollas under the direction of W. J. Sollas. 5 vols. Oxford: Clarendon Press, 1904–24.

Sugden, David E., and John, Brian S. *Glaciers and Landscape: A Geomorphological Approach*. London: Arnold, 1976.

Sulloway, Frank J. "Geographic Isolation in Darwin's Thinking: The Vicissitudes of a Crucial Idea." *Studies in the History of Biology* 3 (1979): 23–65.

Sutton, S. B. *Charles Sprague Sargent and the Arnold Arboretum*. Cambridge, Mass.: Harvard University Press, 1970.

Swain, Donald C. *Wilderness Defender: Horace M. Albright and Conservation*. Chicago: University of Chicago Press, 1970.

Tarr, Ralph Stockman. "The Origins of Drumlins." *American Geologist* 8 (1894): 393–407.

———. "The Peneplain." *American Geologist* 21 (1898): 351–70.

———. *College Physiography*. New York: Macmillan, 1931.

Tatham, George. "Geography in the Nineteenth Century." In *Geography in the Twentieth Century*, edited by Griffith Taylor, pp. 28–69. New York: Philosophical Library, 1951.

Taylor, E. G. R. *Late Tudor and Early Stuart Geography, 1583–1650*. London: Methuen, 1934.

————. *Mathematical Practitioners of Tudor and Stuart England.* Cambridge: Cambridge University Press, 1954.

Thayer, William Roscoe. "Nathaniel Southgate Shaler." *Harvard Graduates' Magazine* 15 (1906): 1–9.

————. "Nathaniel Southgate Shaler." *Nation* 82 (1906): 318–19.

Thomas, Franklin. *The Environmental Basis of Society: A Study of the History of Sociological Theory.* New York: Century, 1925.

Thomas, William L., ed. *Man's Role in Changing the Face of the Earth.* Chicago: University of Chicago Press, 1956.

Thomson, J. "Theoretical Considerations on the Effect of Pressure in Lowering the Freezing Point of Water." *Transactions of the Royal Society of Edinburgh* 16, pt. 5 (1849): 575–80.

Thomson, W. "On the Rigidity of the Earth." *Proceedings of the Glasgow Philosophical Society* 5 (1862): 169–70.

Thwing, Charles F. *A History of Education in the United States since the Civil War.* Boston: Houghton Mifflin, 1910.

Tobin, Gregory M. *The Making of a History: Walter Prescott Webb and the Great Plains.* Austin: University of Texas Press, 1976.

Torrance, Thomas F. *Theological Science.* London: Oxford University Press, 1969.

Tuan, Yi-Fu. "Discrepancies between Environmental Attitude and Behaviour: Examples from Europe and China." *Canadian Geographer* 12 (1968): 176–91.

————. *The Hydrologic Cycle and the Wisdom of God: A Theme in Geoteleology.* Department of Geography Research Paper no. 1. Toronto: University of Toronto, 1968.

Turner, Frank Miller. *Between Science and Religion: The Reaction to Scientific Naturalism in Late Victorian England.* New Haven: Yale University Press, 1974.

————. "Rainfall, Plagues, and the Prince of Wales: A Chapter in the Conflict of Religion and Science." *Journal of British Studies* 13 (1974): 46–65.

————. "The Victorian Conflict between Science and Religion: A Professional Dimension." *Isis* 69 (1978): 356–76.

Turner, Frederick Jackson. "The Significance of the Frontier in American History." *Proceedings of the State Historical Society of Wisconsin* 41 (1894): 79–112.

————. "Geographical Interpretations of American History." *Journal of Geography* 3 (1905): 34–37.

————. *Frontier and Section: Selected Essays of Frederick Jackson Turner.* Edited and with an introduction and notes by Ray A. Billington. Englewood Cliffs, N.J.: Prentice-Hall, 1961.

Upham, Warren. "Drumlins and Marginal Moraines of Ice Sheets." *Bulletin of the Geological Society of America* 7 (1896): 17–30.

Van Hise, Charles R. *The Conservation of Natural Resources in the United States.* New York: Macmillan, 1910.

Verrill, Addison E. "Notes on the Natural History of Anticosti." *Proceedings of the Boston Society of Natural History* 9 (1862–63): 132–35.

Veysey, Lawrence R. *The Emergence of the American University.* Chicago: University of Chicago Press, 1965.

Voget, Fred W. "Man and Culture: An Essay in Changing Anthropological Interpretation." *American Anthropologist* 62 (1960): 943–65.

————. *A History of Ethnology.* New York: Holt, Rinehart & Winston, 1975.

Vorzimmer, Peter J. "Charles Darwin and Blending Inheritance." *Isis* 54 (1963): 371–90.

————. *Charles Darwin: The Years of Controversy: The Origin of Species and Its Critics, 1859–1882*. Philadelphia: Temple University Press, 1975.

Walker, Francis A. "The Indian Question." *North American Review* 116 (1873): 329–88.

————. "The Technical School and the University." *Atlantic Monthly* 72 (1893): 390–94.

————. *Restriction of Immigration*. Immigration Restriction League Publications, no. 33. New York, 1899.

Wallace, Alfred Russel. "The Origin of Human Races and the Antiquity of Man Deduced from the Theory of 'Natural Selection.'" *Journal of the Anthropological Society of London* 2 (1864): clviii–clxx.

————. *Contributions to the Theory of Natural Selection*. London: Macmillan, 1870.

————. *Man's Place in the Universe: A Study of the Results of Scientific Research in Relation to the Unity or Plurality of Worlds*. 4th ed. London: Chapman & Hall, 1904.

Wanklyn, Harriet. *Friedrich Ratzel: A Biographical Memoir and Bibliography*. Cambridge: Cambridge University Press, 1961.

Ward, Lester Frank. *Psychic Factors of Civilization*. Boston: Houghton Mifflin, 1893.

————. *Glimpses of the Cosmos*. 6 vols. New York: Putnam's Sons, 1913–18.

Warmington, Eric H., ed. *Greek Geography*. London: Dent, 1934.

Watson, J. Wreford. "The Image of Nature in America." In *The American Environment*, edited by W. R. Mead, pp. 1–16. London: Athlone Press, 1974.

Webster, Charles, ed. *The Intellectual Revolution of the Seventeenth Century*. London: Routledge & Kegan Paul, 1974.

Weinryb, Bernard D. "Jewish Immigration and Accommodation to America." In *The Jews: Social Patterns of an American Group*, edited by M. Sklare, pp. 4–22. Glencoe, Ill.: Free Press, 1958.

Westfall, Richard S. *Science and Religion in Seventeenth Century England*. New Haven: Yale University Press, 1958.

Whicher, George F., ed. *The Transcendentalist Revolt against Materialism*. Boston: Heath, 1949.

Whitaker, J. Russell. "World View of Destruction and Conservation of Natural Resources." *Annals of the Association of American Geographers* 30 (1940): 143–62.

————. "The Way Lies Open." *Annals of the Association of American Geographers* 44 (1954): 231–44.

White, Edward A. *Science and Religion in American Thought*. Stanford: Stanford University Press, 1952.

White, Leslie A. "The Concept of Evolution in Cultural Anthropology." In *Evolution and Anthropology: A Centennial Appraisal*, edited by Betty J. Meggers, pp. 106–25. Washington, D.C.: Anthropological Society of Washington, 1959.

White, Morton. *Science and Sentiment in America: Philosophical Thought from Jonathan Edwards to John Dewey*. New York: Oxford University Press, 1972.

White, Morton, and White, Lucia. "The American Intellectual versus the American City." *Daedalus* 90 (1961): 166–79.

————. *The Intellectual versus the City*. Cambridge, Mass.: Harvard University Press, 1962.

Whitehead, Alfred North. *Science and the Modern World*. Cambridge: Cambridge University Press, 1926.

Wiener, Philip P. *Evolution and the Founders of Pragmatism*. Cambridge, Mass.: Harvard University Press, 1949.

————. "Some Problems and Methods in the History of Ideas." *Journal of the History of Ideas* 22 (1961): 531–48.

Williams, Daniel Day. *The Andover Liberals: A Study in American Theology.* New York: King's Crown Press, 1941.

Willoughby, W. W. "The History of Summer Schools in the United States." In *Annual Report of the Commissioner of Education for the Year 1891–1892.* Washington, D.C.: Government Printing Office, 1894.

Wilm, H. G. "The Status of Watershed Management Concepts." *Journal of Forestry* 44 (1946): 968–71.

Wilson, John B. "Darwin and the Transcendentalists." *Journal of the History of Ideas* 26 (1965): 286–90.

Wolff, John E. "Memoir of Nathaniel Southgate Shaler." *Bulletin of the Geological Society of America* 18 (1907): 592–609.

Worster, Donald, ed. *American Environmentalism: The Formative Period, 1860–1915.* New York: Wiley, 1973.

Wright, John Kirtland. "A Plea for the History of Geography." *Isis* 8 (1925): 477–91.

Wright, W. B. *The Quaternary Ice Age.* London: Macmillan, 1937.

Wright, W. L., Jr. *Dictionary of American Biography,* s.v. "Shaler, William."

Wyllie, Irvine. "Social Darwinism and the Businessman." *Proceedings of the American Philosophical Society* 103 (1959): 629–35.

Yeo, Richard. "William Whewell, Natural Theology, and the Philosophy of Science in Mid Nineteenth Century Britain." *Annals of Science* 36 (1979): 493–516.

Young, Robert M. "The Impact of Darwin on Conventional Thought." In *The Victorian Crisis of Faith,* edited by Anthony Symondson, pp. 13–35. London: S.P.C.K., 1970.

———. *Mind, Brain, and Adaptation in the Nineteenth Century: Cerebral Localization and Its Biological Context from Gall to Ferrier.* Oxford: Oxford University Press, 1970.

———. "Darwin's Metaphor: Does Nature Select?" *Monist* 55 (1971): 442–503.

———. "The Historiographic and Ideological Contexts of the Nineteenth-Century Debate on Man's Place in Nature." In *Changing Perspectives in the History of Science,* edited by M. Teich and R. M. Young, pp. 344–438. London: Heinemann, 1973.

———. "Natural Theology, Victorian Periodicals, and the Fragmentation of a Common Context." In *Darwin to Einstein: Historical Studies on Science and Belief,* edited by Colin Chant and John Fauvel, pp. 69–107. Harlow: Longman, 1980.

Zabilka, Ivan L. "Nathaniel Southgate Shaler and the Kentucky Geological Survey." *Register of the Kentucky Historical Society* 80 (1982): 408–31.

Zon, Raphael. "Climate and the Nation's Forests." In *Yearbook of Agriculture,* pp. 477–98. Washington, D.C.: United States Department of Agriculture, 1941.

Index

ABOUT THE AUTHOR

David N. Livingstone is a Research Officer
and Curator of Maps, Department of Geography, The Queen's University of Belfast. He
received his doctorate from The Queen's University of Belfast.